Progress and Prospects in Evolutionary Biology

Progress and Prospects in Evolutionary Biology

Edited by Noah Rodriguez

SYRAWOOD
PUBLISHING HOUSE

New York

Published by Syrawood Publishing House,
750 Third Avenue, 9th Floor,
New York, NY 10017, USA
www.syrawoodpublishinghouse.com

Progress and Prospects in Evolutionary Biology
Edited by Noah Rodriguez

International Standard Book Number: 978-1-64740-100-9 (Hardback)

Cataloging-in-Publication Data

Progress and prospects in evolutionary biology / edited by Noah Rodriguez.
    p. cm.
Includes bibliographical references and index.
ISBN 978-1-64740-100-9
1. Evolution (Biology). 2. Biology. 3. Evolution. I. Rodriguez, Noah.
QH366.2 .P76 2022
575--dc23

TABLE OF CONTENTS

PREFACE

Evolution is at the core of all biological studies. Biology can be divided at the level of biological organization, taxonomic group and manner of approach. Many of these fields can be integrated with evolutionary biology to create the subfields of evolutionary developmental biology and evolutionary ecology. The merger between applied sciences and biological sciences has resulted in specialized domains such as evolutionary robotics, evolutionary engineering, architecture, algorithms and economics. It is now possible to understand the effects of each gene and their contribution to complex evolutionary phenomena of speciation and adaptation because of the better understanding of their molecular basis. The interdependence of individual genes and the changes that can occur in each of these, such as gene mutation, gene duplication or genome duplication can also be better understood now. This book covers in detail some existing theories and innovative concepts revolving around evolution. Different approaches, evaluations, methodologies and advanced studies have been included herein. Students, researchers, experts and all associated with evolutionary biology will benefit alike from this book.

This book is a result of research of several months to collate the most relevant data in the field.

When I was approached with the idea of this book and the proposal to edit it, I was overwhelmed. It gave me an opportunity to reach out to all those who share a common interest with me in this field. I had 3 main parameters for editing this text:

1. Accuracy – The data and information provided in this book should be up-to-date and valuable to the readers.

2. Structure – The data must be presented in a structured format for easy understanding and better grasping of the readers.

3. Universal Approach – This book not only targets students but also experts and innovators in the field, thus my aim was to present topics which are of use to all.

Thus, it took me a couple of months to finish the editing of this book.

I would like to make a special mention of my publisher who considered me worthy of this opportunity and also supported me throughout the editing process. I would also like to thank the editing team at the back-end who extended their help whenever required.

Editor

Dynamics of growth factor production in monolayers of cancer cells and evolution of resistance to anticancer therapies

Marco Archetti

School of Biological Sciences, University of East Anglia, Norwich, UK

Keywords

anticancer therapy, cancer, cooperation, evolutionary game theory, growth factors, heterogeneity, polymorphism, public goods, somatic evolution

Correspondence

Marco Archetti, School of Biological Sciences, University of East Anglia, Norwich Research Park, Norwich NR4 7TJ, UK.

e-mail: m.archetti@uea.ac.uk

Abstract

Tumor heterogeneity is well documented for many characters, including the production of growth factors, which improve tumor proliferation and promote resistance against apoptosis and against immune reaction. What maintains heterogeneity remains an open question that has implications for diagnosis and treatment. While it has been suggested that therapies targeting growth factors are robust against evolved resistance, current therapies against growth factors, like antiangiogenic drugs, are not effective in the long term, as resistant mutants can evolve and lead to relapse. We use evolutionary game theory to study the dynamics of the production of growth factors by monolayers of cancer cells and to understand the effect of therapies that target growth factors. The dynamics depend on the production cost of the growth factor, on its diffusion range and on the type of benefit it confers to the cells. Stable heterogeneity is a typical outcome of the dynamics, while a pure equilibrium of nonproducer cells is possible under certain conditions. Such pure equilibrium can be the goal of new anticancer therapies. We show that current therapies, instead, can be effective only if growth factors are almost completely eliminated and if the reduction is almost immediate.

Introduction

Tumor heterogeneity

Heterogeneity of cells within a tumor is well documented for many types of cancers and many distinguishable phenotypes (Marusyk and Polyak 2010) and has important implications for disease progression (Maley et al. 2006), diagnosis, and therapeutic responses (Dexter and Leith 1986). As diagnostic biopsies sample only a small region of the tumor, treatments based upon such samples might not be effective against all tumor cells. Understanding the origin, extent, and dynamics of tumor heterogeneity therefore is essential for the development of successful anticancer therapies.

A basic question about heterogeneity is still unsolved (Merlo et al. 2006): how can more than one clone stably coexist in a neoplasm? Given that the development of cancer is a process of clonal selection (Cairns 1975; Nowell 1976; Crespi and Summers 2005; Merlo et al. 2006; Greaves and Maley 2012) in which cells compete for resources, space, and nutrients, one would predict that a mutant clone with a fitness advantage should drive other clones extinct and go to fixation. Current explanations for the maintenance of heterogeneity include the possibility that different clones are evolutionarily neutral (Iwasa and Michor 2011), specialize on different niches (Nagy 2004; Gatenby and Gillies 2008) or are not in equilibrium (Gonzalez-Garcia et al. 2002), or that mutations have small effect (Durrett et al. 2011); which, if any, of these mechanisms are at work in neoplasms remains an open question (Merlo et al. 2006).

Here, we show that stable heterogeneity for the production of growth factors arises as a direct consequence of the fact that growth factors are nonlinear public goods. We develop a model of public goods production in the framework of evolutionary game theory and extend it to take into account specific features of the production of growth factors by cancer cells growing on a monolayer. We show how the evolutionary dynamics of the system can explain the maintenance of stable heterogeneity, how this affects the development of resistance to anticancer therapies that target growth factors, and its implications for the development of stable therapies.

Game theory of cancer

Mathematical models of cancer were first developed to explain the relationship between the time of exposure to carcinogens and the number of tumors (Charles and Luce-Clausen 1942) to understand the number of mutations necessary to cause cancer (Nordling 1953) and the observed age-incidence patterns (Armitage and Doll 1954; Fisher 1958). The statistical study of age-incidence of hereditary versus sporadic cancers (Ashley 1969; Knudson 1971) was instrumental for the introduction of the idea of tumor suppressor genes. Following this line of research, most current models of cancer dynamics developed by ecologists and evolutionary biologists (Frank 2007; Byrne 2010) study the effect of mutations, selection, population size, and tissue architecture on the dynamics of cancer.

While game theory has often been mentioned (e.g., Gatenby and Maini 2003; Axelrod et al. 2006; Merlo et al. 2006; Basanta and Deutsch 2008; Lambert et al. 2011) as a promising avenue for cancer research, only a few studies actually develop game theoretical models of cancer. Tomlinson (1997) and Tomlinson and Bodmer (1997) used the game of chicken (Rapoport and Chammah 1966) [or hawk-dove game (Maynard Smith and Price 1973) or snowdrift game (Sugden 1986)] to describe interactions between cancer cells. The model has been extended to up to four types of cells, using different types of cancer as examples, including multiple myeloma, prostate cancer, glioma, and glioblastoma (Basanta et al. 2008a,b, 2011, 2012; Dingli et al. 2009; Gerstung et al. 2011). Interactions among cancer cells for the production of diffusible growth factors, however, are not pairwise, but multiplayer, collective interactions for the production of a public good (Archetti 2013a). It is known that results from the theory of two-player games cannot be extended to games with collective interactions, and that this can actually lead to fundamental misunderstandings (Archetti and Scheuring 2012).

Growth factors as public goods

Consider a population of cells in which a fraction of the cells (producers: +/+) secrete a growth factor. If the benefit of this factor is not restricted to the producers, we can consider it a public good that can be exploited by all other individuals (or cells) within the diffusion range of the factor, including nonproducers (−/−). Public goods are studied in economics, where rational, self-interested behavior may lead to the overexploitation of common pool resources [the 'tragedy of the commons' (Hardin 1968)], and in evolutionary biology in cases like the production of

diffusible molecules in microbes (Crespi 2001). Diffusible public goods raise a collective action problem: an individual can free ride on the goods produced by his neighbors. Why, then, do noncontributors not increase in frequency and go to fixation? What factors influence the production of these public goods?

Similar collective action problems arise during cancer development, where growth factors support tumor growth by protecting cells from apoptosis (for example, IGF-II), by stimulating the growth of new blood vessels (for example, VEGF), by impairing immune system reaction (for example, TGFβ), or by promoting the epithelial–mesenchimal transition. While cooperation for the production of growth factors has been shown directly only in one case (FGF) (Jouanneau et al. 1994), it stands to reason that many diffusible factors produced by cancer cells benefit producers and nonproducers (Axelrod et al. 2006). Self-sufficiency of growth factor production is one of the hallmarks of cancer (Hanahan and Weinberg 2000) and, like for other characters, there is evidence of heterogeneity in the ability to produce diffusible factors (Achilles et al. 2001; Marusyk and Polyak 2010). What maintains this heterogeneity? And what are the implications for anticancer therapies?

It has been suggested that treatments that attack growth factors may be less susceptible than traditional drugs to the evolution of resistance (Pepper 2012; Aktipis and Nesse 2013). Current drugs that target growth factors, however, like the anti-angiogenic drug Avastin, have a limited effect, with only a few months of overall survival extension (Amit et al. 2013). Limited theoretical analysis has been devoted to investigating the problem of the evolution of resistance to therapies (Aktipis et al. 2011). The rationale of this study is that analyzing the production of growth factors in cancer as a public goods game can explain both stable heterogeneity and the long-term failure of antigrowth factor therapies and reveal conditions that can lead to evolutionarily stable therapies.

Public goods games

Archetti and Scheuring (2012) review public goods games (PGGs) in well-mixed populations and Perc et al. (2013) review PGGs in structured populations. The current literature on PGGs often assumes that the benefit of the public good is a linear function of the number of contributors (the N-person prisoner's dilemma: NPD). The simplest cases of nonlinear benefits, synergistic, and discounting benefits (Motro 1991; Foster 2004; Hauert et al. 2006), as well as threshold PGGs (in which a benefit is produced if a number of contributors is above a fixed threshold) (Archetti 2009a,b; Pacheco et al. 2009; Boza and Szamado 2010; see also Palfrey and Rosenthal 1984 for a similar model in economics) have been studied extensively (Archetti and Scheuring 2012). The benefit produced by growth factors, however, is likely to be a sigmoid function of the number of producer cells because the effect of enzyme production is generally a saturating function of its concentration (e.g., Hemker and Hemker 1969), specifically, a sigmoid function (Ricard and Noat 1986); signaling pathways often follow a highly nonlinear on–off behavior, which is a steep sigmoid function of signal concentration (e.g., Mendes 1997; Eungdamrong and Iyengar 2004). Similar nonlinearities are known in microbes (Chuang et al. 2010). Sigmoid PGGs are somewhat intermediate between linear and threshold PGGs, while synergistic/discounting PGGs can be thought of as special, degenerate cases of sigmoid PGGs (Archetti and Scheuring 2011). Linear, threshold, and synergistic/discounting benefits can lead to dramatically different dynamics and equilibria in multiplayer games (Archetti and Scheuring 2012); the dynamics and equilibria of multiplayer sigmoid PGGs in well-mixed populations have been described analytically only recently (Archetti 2013a).

While this literature analyses PGGs in well-mixed populations, the study of PGGs in spatially structured populations generally assumes linear benefits (the NPD). The few exceptions (see Perc et al. 2013) using nonlinear benefits in spatially structured populations assume, as is standard in the current approach, that each individual belongs to n different groups, each group centered on one of that individual's one-step neighbors, and that an individual's fitness is the sum of all the payoffs accumulated in all the groups she belongs to (Perc et al. 2013). While this assumption is reasonable for interactions in human social networks, it is not appropriate for modeling interactions in cell populations, where the growth factors produced by one individual can diffuse beyond one-step neighbors, and the benefit an individual gets as a result of the diffusible factors is a function of the number of producers within the diffusion range of the factor, not of all individuals belonging to her neighbors' group. The only exceptions to the use of the standard framework are Ifti et al. (2004) and Ohtsuki et al. (2007): they study the prisoner's dilemma (that is, a two-person game with a linear benefit function) on lattices in which the interacting group is decoupled from the update neighborhood. Here, we need to analyze the more general case of sigmoid benefits (rather than linear) with collective interactions (rather than pairwise). We will not assume a particular type of cancer but describe the dynamics of growth factors like insulin-like growth factor II (IGF-II) that confer a direct beneficial effect to the cells, for example by protecting against apoptosis. Other growth factors confer a benefit to the tumor indirectly by stimulating the development of blood vessels or the release of other growth factors by stromal cells. The dynamics of these growth factors would be more complex.

Model

The game

A cell can be a producer $(+/+)$ or a nonproducer $(-/-)$ of a growth factor. Producers pay a cost c that nonproducers do not pay $(0 < c < 1)$. All cells $(+/+$ and $-/-)$ benefit from the public good produced by all the cells in their group (of size n; this depends on the diffusion range of the factor – see below). The benefit for an individual is given by the logistic function $V(j)=1/[1 + e^{-s(j-k)/n}]$ of the number j of producers among the other individuals (apart from self) in the group, normalized as $b(j)=[V(j)-V(0)]/[V(n)-V(0)]$. The parameter k controls the position of the inflection point ($k \rightarrow n$ gives strictly increasing returns and $k \rightarrow 0$ strictly diminishing returns) and the parameter s controls the steepness of the function at the inflection point ($s \rightarrow \infty$ models a threshold public goods game; $s \rightarrow 0$ models an N-person prisoner's dilemma) (Archetti and Scheuring 2011). It is useful to define $h=k/n$.

Evolution in spatially structured populations

We model a monolayer of cancer cells as a two-dimensional regular lattice obtained using a modification of the GridGraph implementation in Mathematica 8.0 (Wolfram Research Inc.) connecting opposing edges to form a toroidal network, to avoid edge effects. As in the standard approach, individuals occupy the nodes of the network (population size is fixed at 900) and social interactions proceed along the edges connecting the nodes. Differently from the standard approach, however, [in which an individual's group is limited to her one-step neighbors and an individual plays multiple games centered on each of her neighbors (Perc et al. 2013)], the interaction neighborhood and the update neighborhood are decoupled: a cell's group (of size n) is not limited to her one-step neighbors but is defined by the diffusion range (d) of the growth factor, that is, the number of edges between the focal cell and the most distant cell whose contribution affects the fitness of the focal cell. A cell's payoff is a function of the amount of factor produced by the group she belongs to. The process starts with a number of nonproducer cells placed on the graph; at each round, a cell x with a payoff P_x is selected (at random) for update (death) and a cell y (with a payoff P_y) is then chosen among x's neighbors. Two types of update are used: in the deterministic case, if $P_x > P_y$, no update occurs, while if $P_x < P_y$, x will adopt y's strategy (unconditional imitation); in the stochastic case, replacement occurs with a probability given by $(P_y - P_x)/M$, where M ensures the proper normalization and is given by the maximum possible difference between the payoffs of x and y (Perc et al. 2013). Results are obtained averaging the final 200 of 1000 generations per cell, averaged over 10 different runs.

Gradient of selection in well-mixed populations

In a finite population, the gradient of selection can be calculated following Traulsen et al. (2006). Sampling of individuals follows a hypergeometric distribution and the average fitness of +/+ and −/− can be written as, respectively

$$W_{+/+} = \binom{Z-1}{n-1}^{-1} \sum_{j=0}^{n-1} \binom{i-1}{j}\binom{Z-i}{n-j-1} \cdot b(j+1) - c$$

$$W_{-/-} = \binom{Z-1}{n-1}^{-1} \sum_{j=0}^{n-1} \binom{i}{j}\binom{Z-j-1}{n-j-1} \cdot b(j)$$

where i is the number of +/+ individuals in the population. Assuming a stochastic birth–death process combined with a pairwise comparison rule, two individuals from the population, A and B, are randomly selected for update. The strategy of A will replace that of B with a probability given by the Fermi function

$$p \equiv \frac{1}{1 + e^{-\beta(W_A - W_B)}}$$

and the reverse will happen with probability 1-p. The quantity β specifies the intensity of selection (for $\beta \ll 1$, selection is weak, and in the limit $Z \to \infty$ one recovers the replicator equation) (Traulsen et al. 2007). In finite populations, the quantity corresponding to the 'gradient of selection' in the replicator dynamics is given by

$$g(i) = \frac{i}{Z}\frac{Z-i}{Z} \tanh \frac{\beta(W_{+/+} - W_{-/-})}{2}$$

In an infinitely large population, the gradient of selection can be written as (Archetti and Scheuring 2012)

$$\sum_{j=0}^{n-1} \binom{n-1}{j} x^j (1-x)^{n-1-j} \cdot \Delta b_j - c$$

where $\Delta b_j = b(j+1) - b(j)$

Results

Evolutionary dynamics of growth factor production

Decoupling the interaction and update networks
A comparison between the standard framework and the one used here (decoupling interaction and replacement networks) is possible if we assume that d = 2 (Fig. 1). Group size with d = 2 on a lattice with connectivity 4 is the same as the total number of individuals participating in the five PGGs in the standard approach, counting each individual only once ($n = 25$). In the standard approach, however, the focal individual contributes to all PGGs she is involved in, her one-step neighbors contribute to two PGGs that

affect the focal individual, and her two-step neighbor contribute only to one PGG that affects the focal individual. In the case of diffusible factors instead, all individuals contribute equally to a single, larger PGG. Fig. 1 shows the differences between the two systems. Cooperation evolves for a wider parameter set in the standard approach than in the case of diffusible goods, and the fraction of producers is larger. This is not surprising, given the smaller group size implied by the standard approach. If d > 2 of course, the standard approach cannot be defined, and the two systems are not directly comparable. All the results are based on the new approach in which the interaction and replacement graphs are decoupled, and the diffusion range of the growth factor can be larger than 1.

Heterogeneity
When a nonproducer (−/−) is introduced in a population of producers (+/+), in most cases −/− cells increase in frequency and coexist with +/+ cells; this change in frequency of the two types is accompanied by a change in the relative position of the +/+ and −/− cells, as shown by the degree centrality (the number of neighbors) and the closeness centrality (the inverse of the sum of the distance to all other vertices) of the +/+ subgraphs (Fig. 1). In most cases after about 100 generations per cell, the frequencies remain relatively stable, even though the position of producers and nonproducers on the lattice continues to change (Fig. 2). In certain cases, the −/− type goes to fixation. The frequency of the two types, or the extinction of the +/+ type, depends on the diffusion range (d), the cost of growth factor production (c), the position of the inflection point of the benefit function (h), and the steepness of the benefit function (s), the update rule and the initial frequency of the two types, as described below.

Effect of the diffusion range
Both the frequency of producers and fitness at equilibrium decline with increasing d (the diffusion range of the public good), that is, increasing group size (n); −/− cells form clusters whose size increases with d (Fig. 3). A short diffusion range enables a mixed equilibrium (coexistence of +/+ and −/− cells) for a larger set of parameters (higher c and more extreme h values). That is, a short diffusion range favors cooperation.

Effect of the initial frequencies
The stable equilibrium described above does not depend on the initial frequency of the two types (Fig. 4), unless the initial frequency of +/+ cells is below a certain threshold; in this case, the +/+ type goes extinct. In other words, the system has an internal stable equilibrium, to which the population evolves if and only if +/+ cells are above a critical threshold. Inspection of the gradient of selection shows the

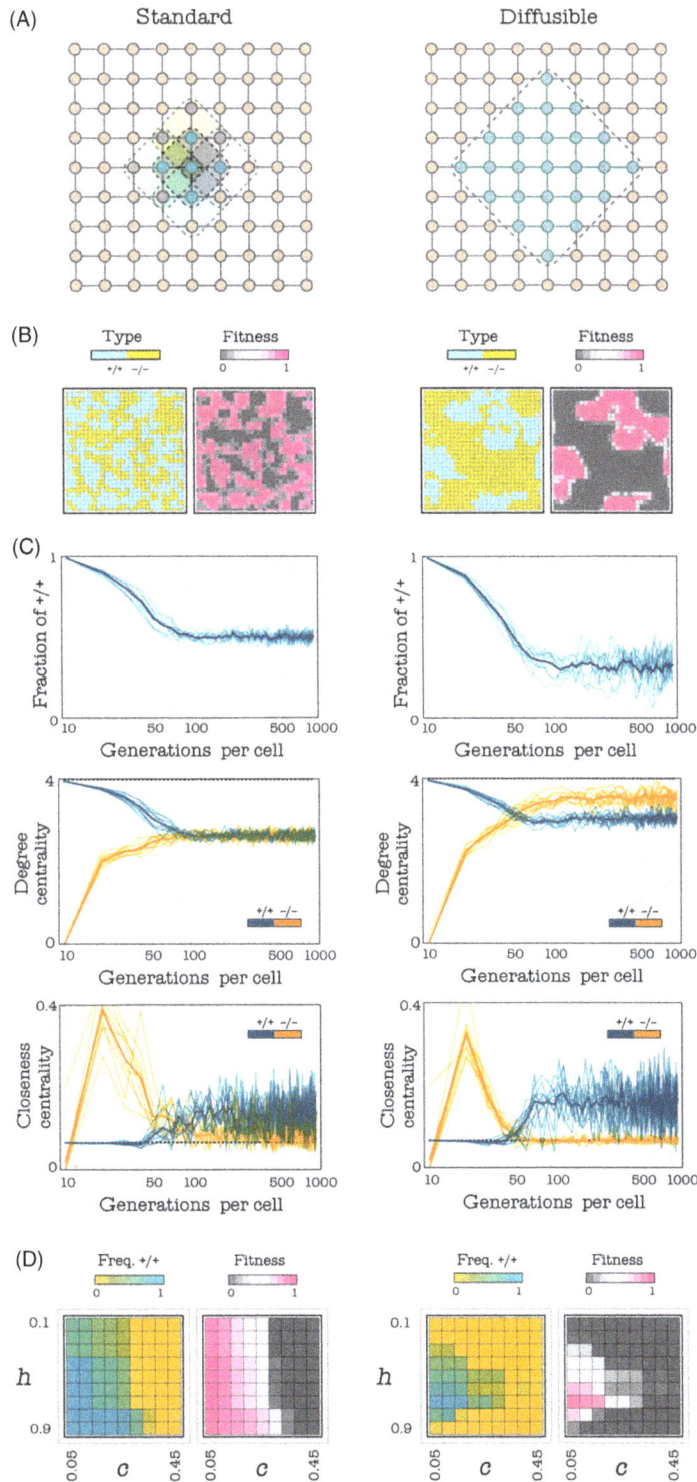

Figure 1 Growth factors as public goods. (A) In the standard approach, a cell's payoff is determined by the games played by the groups centered on that cell and on its one-step neighbors; in the case of diffusible factors, the group (the interaction neighborhood) is defined by the diffusion range (d) of the factor (here d = 3) and is larger than the update group (the one-step neighbors). (B) The structure of the population after 1000 generations per cell (c = 0.25, h = 0.5, d = 2, s = 20, deterministic update) (C) The change in frequency of +/+ cells, degree centrality, and closeness centrality of the +/+ and −/− subgraphs (c = 0.25, h = 0.5, d = 2, s = 20, deterministic update). (D) The equilibrium frequency of +/+ and average fitness as a function of h (the position of the threshold) and c (the cost of production) when 10 −/− cells are introduced in the population (d = 2, s = 20, deterministic update).

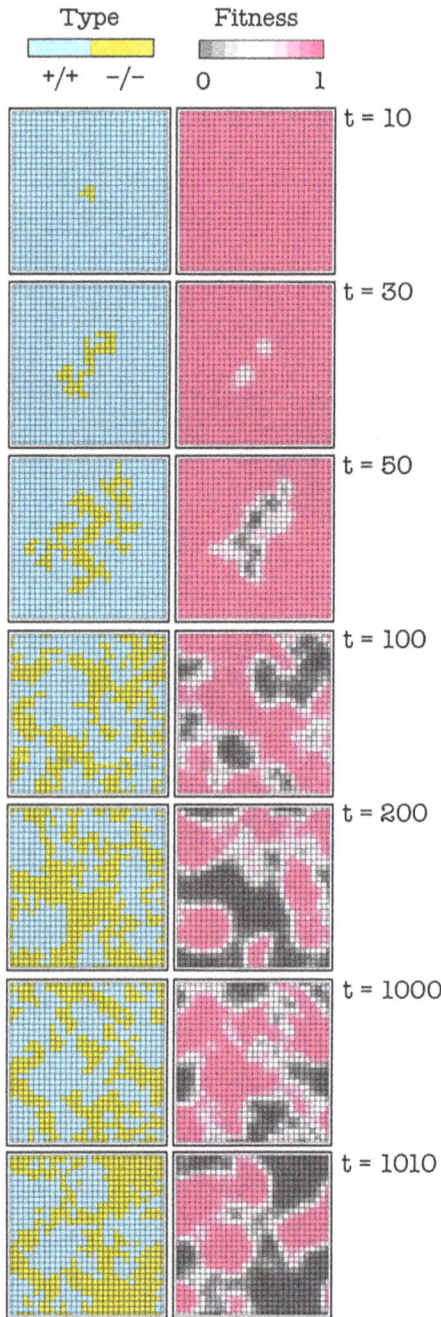

Figure 2 Dynamic heterogeneity. Snapshots of the population at different times (t is the number of generations per cell). The frequency of the two types remains relatively stable after about 100 generations per cell, but the position of +/+ and −/− cells changes. c = 0.02, h = 0.5, s = 20, d = 4; deterministic update.

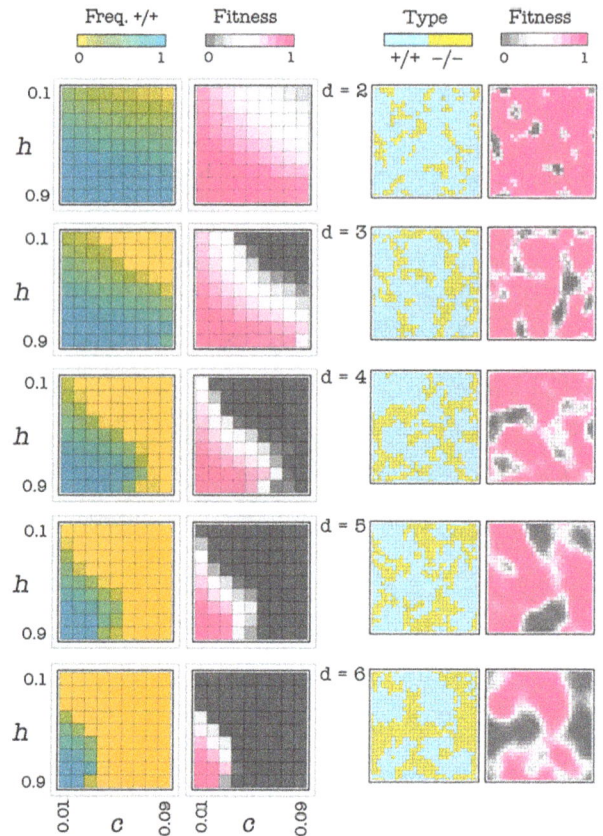

Figure 3 Effect of the diffusion range. Left: Each square in each plot shows the frequency of +/+ cells or the average fitness of the population (its growth rate) as a function of h (the position of the threshold) and c (the cost of production), for a given d (diffusion range) and for s = 20 (deterministic update). The frequency of +/+ cells and the average fitness are higher at intermediate levels of h and at low values of c. Both decline with increasing d, that is, increasing group size. Right: Snapshots of the population after 1000 generations per cell; c = 0.01, h = 0.5, s = 20, deterministic update.

increase in frequency up to a stable mixture of +/+ and −/−. Below the unstable internal equilibrium, +/+ cells go extinct.

Effect of the benefit function
The internal stable equilibrium disappears for high values of h; the internal stable equilibrium disappears for low values of h, especially for low values of s. In the extreme case s→0, the game approaches the N-person prisoner's dilemma (the benefit function is a linear function of the frequency of cooperators), and both internal equilibria disappear. Both the frequency of producers and fitness are higher at intermediate levels of h (the position of the threshold) (Fig. 5). A shallow benefit function (low s) can favor cooperation, especially in the deterministic update rule. The reason can be understood more easily if we consider a step function with threshold k as an approximation

reason for the existence of mixed equilibria and bistability (Fig. 5): +/+ cells decrease in frequency when the gradient of selection is positive, that is when there are too few or too many +/+ cells; at intermediate frequencies of +/+ cells, however, +/+ cells have a selective advantage and can

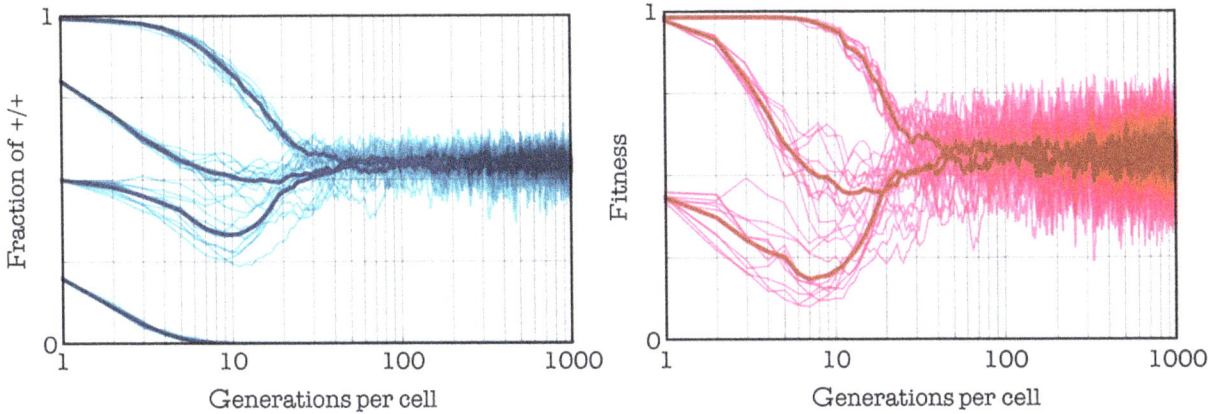

Figure 4 Stable heterogeneity. The change over time of the fraction of producers and of the average fitness of the tumor. At the stable mixed equilibrium, producers and nonproducers coexist, unless the initial fraction of producers is lower than an internal unstable equilibrium (here approximately 0.25). c = 0.02, h = 0.5, s = 20, d = 3.

of a very steep benefit function: in this case, it is convenient to be +/+ only when one is pivotal for the production of the public good, that is, only when there are exactly k-1 other +/+ cells. If the benefit function is a smooth sigmoid function, instead, it pays to be a +/+ even when not pivotal for reaching the threshold. In spatially structured populations, it easily happens that a mutant −/− arising in a group centered on one individual with few −/− also affect the number of +/+ in an adjacent group that was previously at equilibrium; in this other group, the frequency of +/+ will be now below the unstable equilibrium and therefore in the basin of attraction of the pure −/− equilibrium. This process is buffered in the stochastic update process but relatively fast in the deterministic update rule, which is therefore less permissive for the stability of cooperation. The deterministic update rule, therefore, is less conductive to cooperation than stochastic update, especially for very steep public good functions (Fig. 6).

Effect of the cost of production
As expected, increasing the cost of production (c) reduces both the frequency of producers and fitness at equilibrium (Fig. 5). A critical value of c exists above which no public goods production can be sustained and producers go extinct. This critical value is higher for intermediate values of the position of the threshold (that is, for h around 0.5) and for lower values of the diffusion range (d), and it also depends on the type of benefit function and update rule (Fig. 6)

Evolutionary dynamics of resistance to therapies that target growth factors
Effect of therapies that increase the threshold
An anticancer therapy that acts by impairing circulating growth factors will increase the amount of growth factors

that the cells must produce to achieve a certain benefit, that is, it will increase the threshold h. Two results are possible. In the first case, the population adapts to the new threshold, that is, +/+ cells increase in frequency and fitness increases - the opposite of the scope of the drug; only if the threshold increase is substantial does the +/+ type go extinct (Fig. 7).

Effect of the speed of change
The speed of the transition from the original to the new threshold is also essential for the success of a therapy that targets growth factors. While a fast transition to the new threshold can lead to a successful, stable therapy, a slower delivery can lead to relapse (Fig. 8). In summary, therapies are only effective when the initial threshold is low and the increase in threshold is substantial, and if the transition to the new threshold is fast enough (Fig. 9).

Dynamics of the evolution of resistance
The logic of these two effects (magnitude and speed of the shift in threshold) can be understood by looking at the gradient of selection (Fig. 10 shows the logic for a large, well-mixed population, but the logic is the same in finite populations). The therapy is successful (the +/+ cells go extinct) if and only if the new (posttherapy) unstable equilibrium is above the original (pretherapy) stable equilibrium; if this is not the case, the system will move to the new stable equilibrium. This can happen for two reasons: either the increase in h is not large enough; or the increase is slow enough that the current, transient stable equilibrium remains within the basin of attraction of the new, transient stable equilibrium until the change is completed (Fig. 10A). Note that the evolution of resistance is, therefore, more likely for low values of c (the production cost) and s (the steepness of the benefit function).

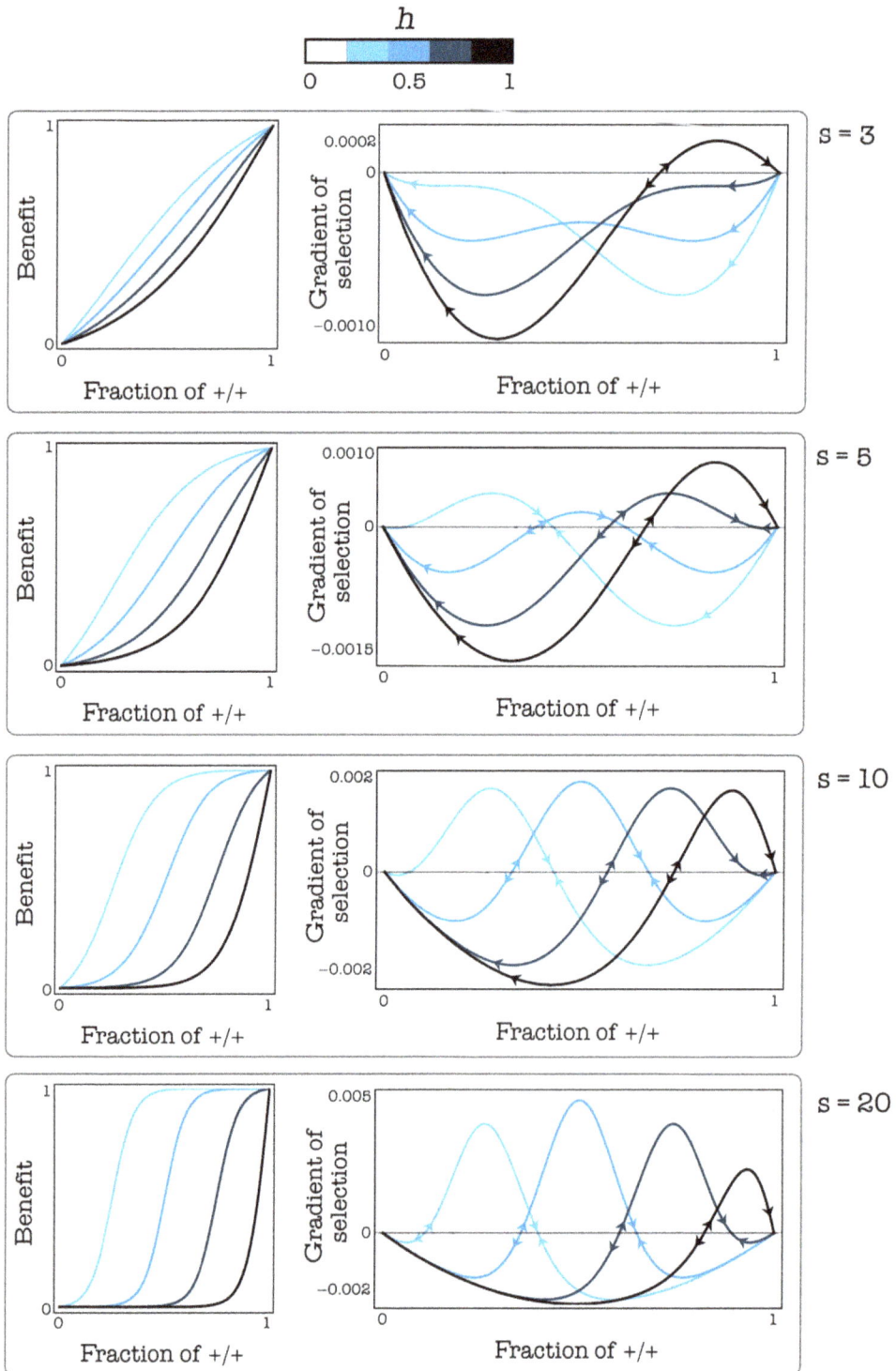

Figure 5 Evolutionary dynamics and equilibria. Left: The benefit functions and the gradients of selection for given values of s (the steepness of the benefit function) and h (the position of the threshold). The sign of the gradient of selection determines the dynamics (arrows show the direction of the change in frequency of +/+ individuals); equilibria occur where the gradient of selection is zero. $c = 0.02$; $\beta = 1$; $d = 3$

Instability of successful treatments

If the treatment is effective and fast enough that the population does reach the stable pure $-/-$ equilibrium, a mutant $+/+$ will not be able to invade an infinitely large, well-mixed population. In a finite, structured population, however, it is possible that random fluctuations change

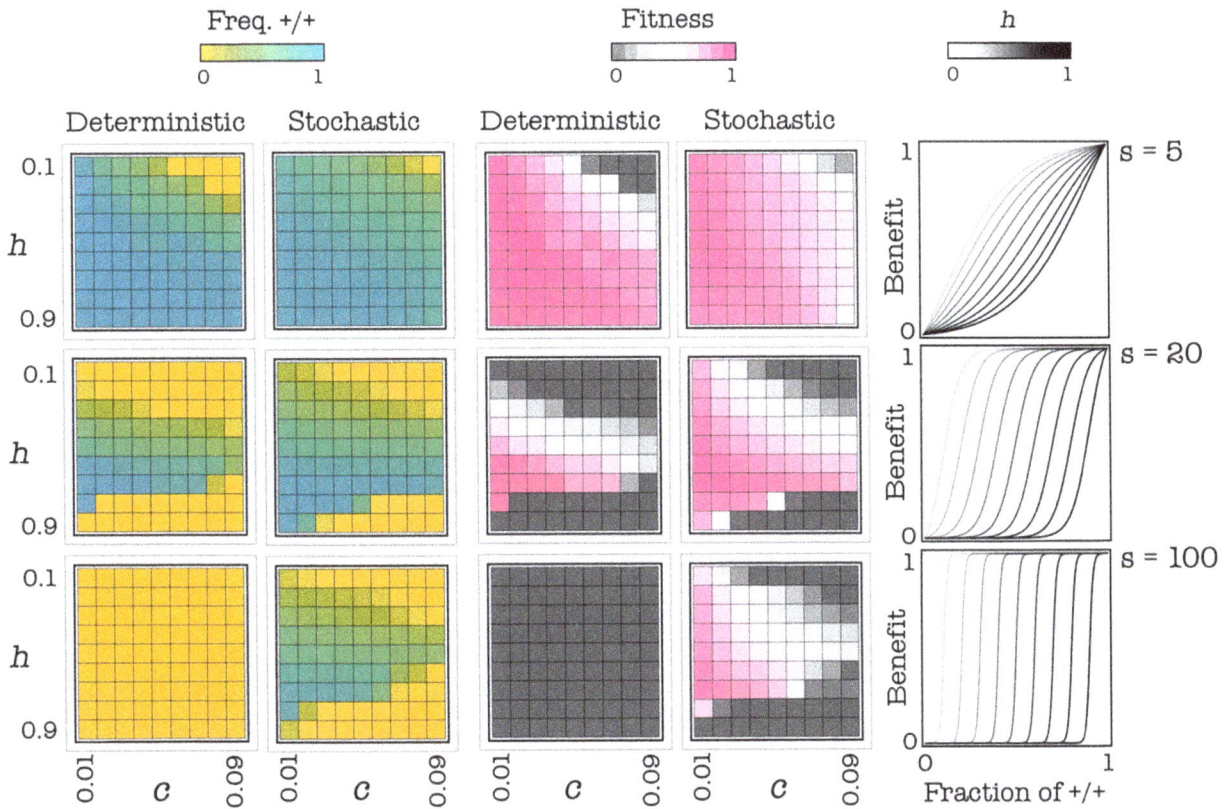

Figure 6 Effect of stochastic events. Each cell in each plot shows the frequency of +/+ cells or the average fitness of the population (its growth rate) as a function of h and c (the cost of production), for a given s at equilibrium. The deterministic update rule makes the internal stable equilibrium sensitive to stochastic fluctuations and therefore not robust when the benefit function is steep (high s); a more realistic stochastic update rule increases the robustness of the equilibrium. d = 3.

the fraction of +/+ cells within a cluster above the unstable equilibrium, which would lead that cluster to the mixed equilibrium in which +/+ and −/− cells coexist (Figs 10B and 11). The opposite effect is also possible, that is, random fluctuations can move the frequency of +/+ at a mixed equilibrium within a group below the unstable internal equilibrium and therefore to the fixation of −/− cells. The relative importance of these two effects depends on the shape of the benefit function, that is on the value of s (the steepness of the benefit function), and on the cost of production (c): low s and c favor the stability of the mixed equilibrium, whereas high s and c make the mixed equilibrium less robust to random fluctuations. An exception to this occurs in the case of very low values of s (that is, for almost linear benefits, similar to the NPD), because such system has only a stable equilibrium: pure +/+ if the cost is low enough, and pure −/− if the cost is low (Fig. 10B); in the latter case, the equilibrium would be immune to invasion by +/+ mutants and therefore arguably stable against the evolution of resistance: a +/+ mutant would only invade if the cost of production decreased.

Discussion

Analyzing the production of growth factors as a nonlinear public goods game reveals that tumor heterogeneity can be maintained by the frequency-dependent selection that arises as a natural consequence of the fact that growth factors are diffusible, and therefore public goods. Tumor heterogeneity has important implications for diagnosis and treatment. The results help us understand anticancer therapies that attack growth factors, either directly (using drugs like Avastin that target the growth factors) or indirectly (using RNA interference). While it has been suggested that attacking growth factors may be less susceptible to the evolution of resistance (Pepper 2012; Aktipis and Nesse 2013), the results shown here suggest that the issue is not so simple.

The rationale of the analysis is that when one reduces the amount of a growth factor, the immediate result is a sudden reduction in tumor growth, because the threshold necessary to achieve the original benefit is not reached; as a consequence, the growth rate of the tumor immediately declines. At the same time, however, the amount of growth

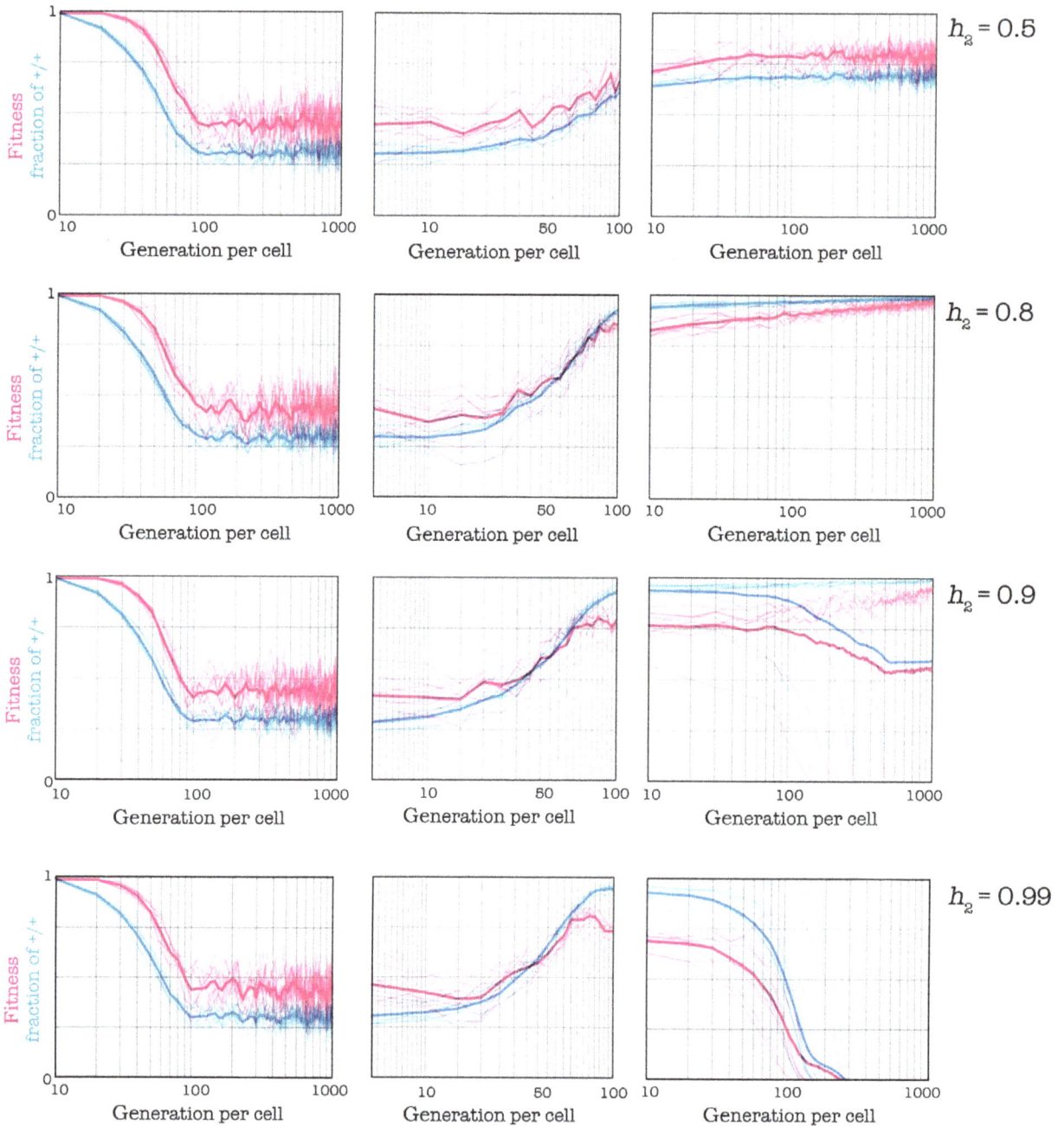

Figure 7 Effect of therapies that target growth factors. Reducing the amount of available growth factor increases the threshold from $h_1 = 0.3$ to h_2. $d = 3$, $h_1 = 0.3$, $c = 0.01$, $s = 20$. In all cases, the shift from h_1 to h_2 occurs gradually, after 1000 generations per cell, in 100 generations.

factors necessary for the population to grow increases (because part of them are disrupted by the drug), which changes the dynamics of the system; unfortunately, it changes into the wrong direction: by increasing the threshold, one increase the frequency of producers at equilibrium, which explains relapse simply as the new equilibrium reached by the system under the new conditions. While it is too early to evaluate the efficacy of RNAi treatments, it

seems reasonable that even silencing the gene for a growth factor should incur a similar problem and be susceptible to the evolution of resistance.

As pointed out by André and Godelle (2005)and Pepper (2012), therapies that target diffusible factors are a more evolutionarily robust approach than conventional drugs that target cells directly. The logic is that (i) drugs that target growth factors can disrupt cooperation between cells

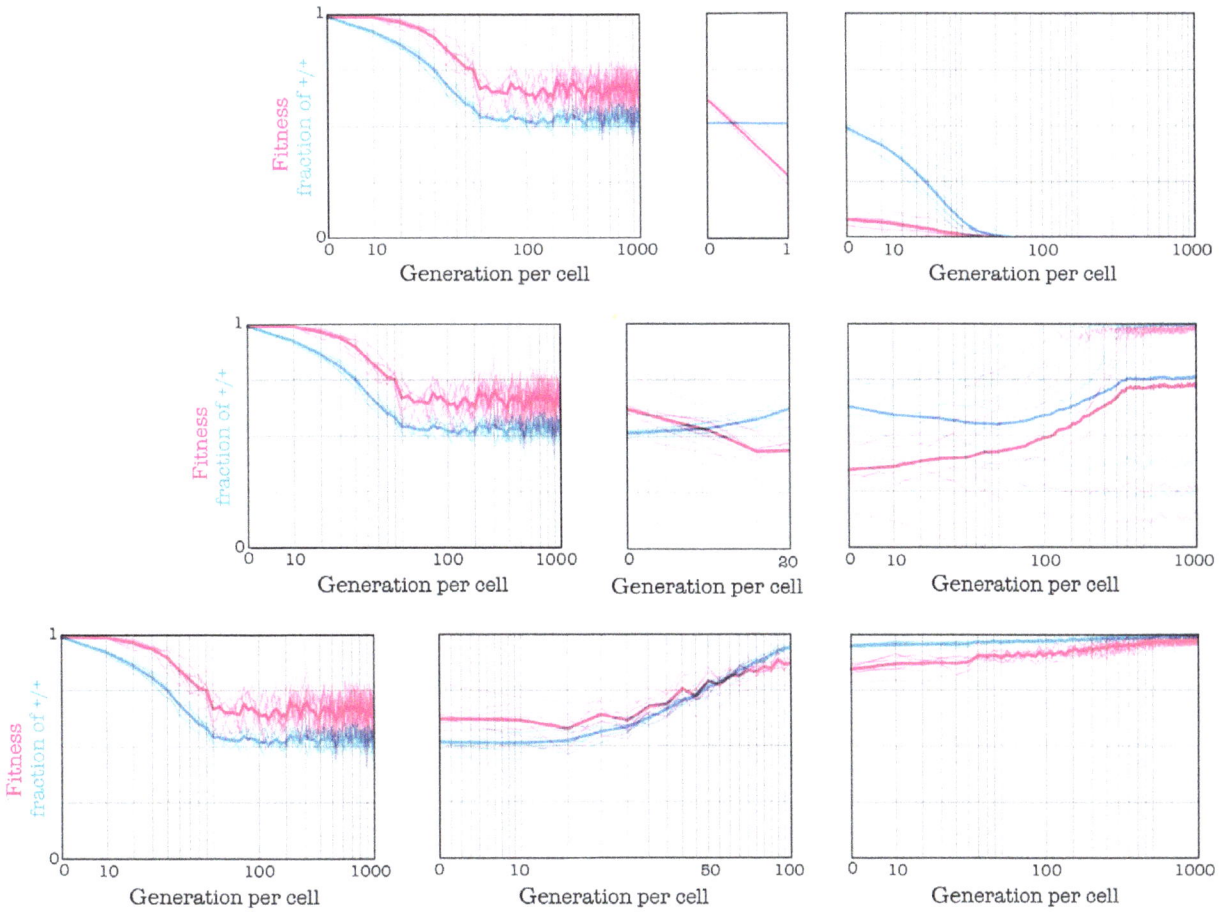

Figure 8 Importance of the speed of change. Reducing the amount of available growth factor increases the threshold from $h_1 = 0.4$ to $h_2 = 0.8$; $d = 3$, $c = 0.01$, $s = 20$. If the change occurs immediately (in the following generation), the +/+ type goes extinct; if the change takes 100 generation per cell to be completed, the population moves to a new equilibrium with a higher fraction of +/+ cell and higher fitness (the contrary of the desired effect); if the change takes 20 generations per cell, results are intermediate.

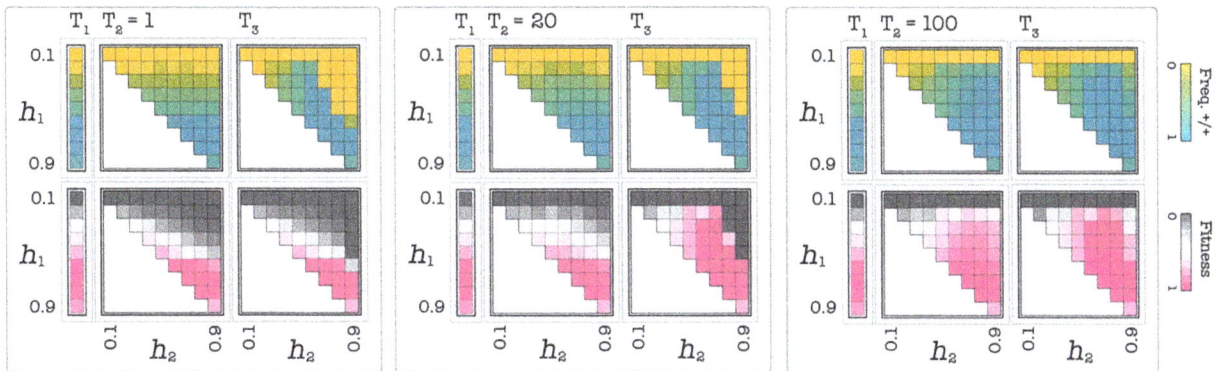

Figure 9 Combined effect of the amount and speed of change. Each cell in each plot shows the frequency of +/+ cells or the average fitness of the population (its growth rate) after T_i generations per cell, as a function of h_1 (the threshold before the treatment) and h_2 (the threshold after the treatment), for $s = 20$ and $c = 0.01$. $T_1 = T_3 = 1000$; before T_1 $h_2 = h_1$. A therapy that increases the threshold is effective only when the initial threshold is low, the new threshold is high enough and the shift to the new threshold is fast enough.

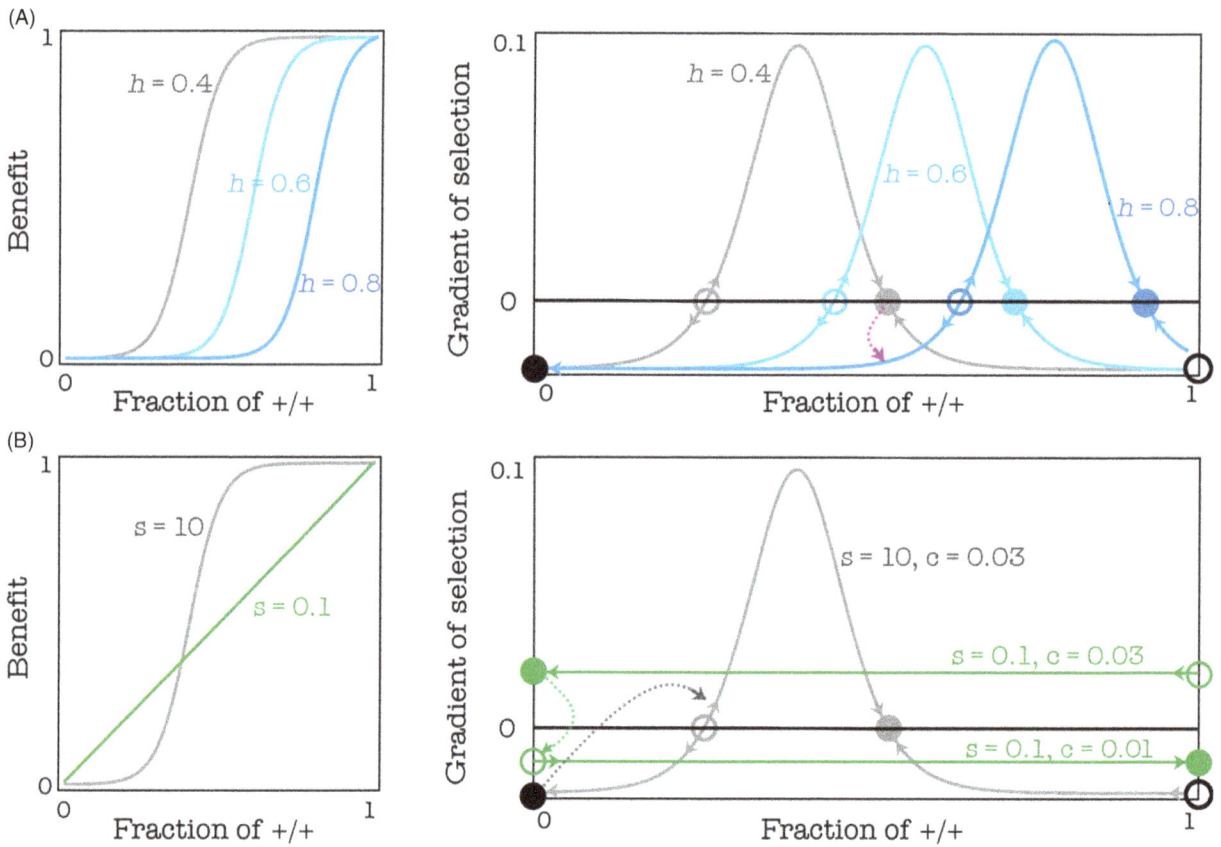

Figure 10 Dynamics of the evolution of resistance to therapies that target growth factors. In a well-mixed population, the gradient of selection determines the direction of the dynamics: where it is positive, the frequency of producers increases; where it is negative it decreases; equilibria (empty circle: unstable; full circle: stable) are found where the gradient of selection is zero (A) Targeting growth factors directly increases the threshold (h) of the public goods game. The therapy is successful (the +/+ cells go extinct) if the new unstable equilibrium is above the original stable equilibrium; if this is not the case, the system will move to the new internal equilibrium. ($c = 0.01$, $s = 10$, $n = 50$) (B) If the population is at the pure −/− equilibrium, a mutant +/+ can only invade if the cost declines below the gradient of selection, if the benefit function is linear; if the benefit is nonlinear, random fluctuations can allow small clusters of +/+ to invade and thus allow the population to reach the internal stable equilibrium. ($h = 0.5$, $n = 50$)

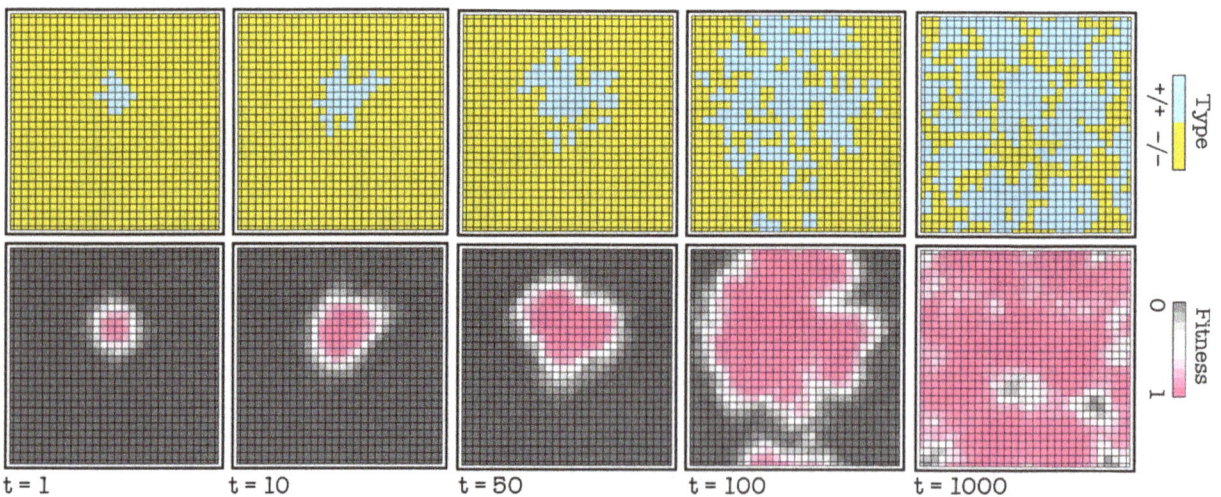

Figure 11 Instability of successful treatments due to random fluctuations. Snapshots (after t generations per cell) of a population initially fixed on −/− in which a mutant +/+ cell line arises and manages to expand. $d = 3$, $c = 0.01$, $h = 0.5$, $s = 10$, stochastic update.

and lead to a pure −/− equilibrium, which will make the population collapse and that (ii) such equilibrium is stable against the invasion of +/+ mutants because the cost paid by the mutant is a private cost, while the benefit it provides is a public benefit. As we have seen, however, (i) requires the therapy to be extremely efficient and fast, and (ii) is not necessarily the case, unless the benefit of the public good is a linear function of the amount of diffusible factors.

As we have shown, the details of the benefit function, diffusion range, cost of production, and update rule that drive the dynamics of growth factor production are critical to determine the type of dynamics and equilibria. More precise theoretical prediction therefore is necessary to understand under what conditions resistance will evolve, including the use of three-dimensional Voronoi graphs to model interactions within the tumor and gradients of diffusion to model the effect of growth factors. Furthermore, the results reported here only apply to growth factors that confer a direct advantage to the tumor, such as factors that protect against apoptosis and promote proliferation; other growth factors, however, act indirectly by inducing the production of other growth factors by stromal cells or by promoting the development of blood vessels. Finally, we have assumed competition due to constant population size, which may describe cancer cell populations that have reached a carrying capacity, but not early stages of tumor growth. Analyzing the dynamics of these cases requires more complex models.

Understanding the production of growth factors as a public goods game suggests that an evolutionarily stable treatment could be achieved through autologous cell therapy (Archetti 2013b): harvesting cancer cells from the patient, knocking out genes coding growth factors in these cells, and reinserting these modified cells inside the tumor. Such therapy, differently from current therapies that target growth factors, would not directly reduce the amount of growth factors produced by the tumor but would change the dynamics of the population. As we have seen, by introducing a critical amount of −/− cells within the tumor, the mixed equilibrium can be destabilised so that the +/+ cells will go extinct; this may lead to the collapse of the tumor due to lack of essential growth factors.

Acknowledgements

Thanks to David Haig for discussing the importance of costs and benefits, to Steve Frank for discussing the difference between early-stage and late-stage tumors, to Daniela Ferraro for discussing tumor–stroma interactions, to Gerhard Christofori for discussing the properties of growth factors. My work is funded in part by NERC grant NE/H015701/1.

Literature cited

Achilles, E. G., A. Fernandez, E. N. Allred, O. Kisker, T. Udagawa, W. D. Beecken, and E. Flynn et al. 2001. Heterogeneity of angiogenic activity in a human liposarcoma: a proposed mechanism for "no take" of human tumors in mice. Journal of the National Cancer Institute 93:1075–1081.

Aktipis, C. A. and R. M. Nesse 2013. Evolutionary foundations for cancer biology. Evolutionary Applications 6:144–159.

Aktipis, C. A., V. S. Y. Kwan, K. A. Johnson et al. 2011. Overlooking evolution: a systematic analysis of cancer relapse and therapeutic resistance research. PLoS ONE 6:e26100.

Amit, L., I. Ben-Aharon, L. Vidal, L. Leibovici, and S. Stemmer 2013. The Impact of Bevacizumab (Avastin) on Survival in Metastatic Solid Tumors - A Meta-Analysis and Systematic Review. PLoS ONE 8: e51780.

André, J. B., and B. Godelle 2005. Multicellular organization in bacteria as a target for drug therapy. Ecology Letters 8:800–810.

Archetti, M. 2009a. The volunteer's dilemma and the optimal size of a social group. Journal of Theoretical Biology 261:475–480.

Archetti, M. 2009b. Cooperation as a volunteer's dilemma and the strategy of conflict in public goods games. Journal of Evolutionary Biology 22:2192–2200.

Archetti, M., and I. Scheuring 2011. Coexistence of cooperation and defection in public goods games. Evolution 65:1140–1148.

Archetti, M., and I. Scheuring 2012. Review: game theory of public goods in one-shot social dilemmas without assortment. Journal of Theoretical Biology 299:9–20.

Archetti, M. 2013a. Evolutionary game theory of growth factor production: implications for tumor heterogeneity and resistance to therapies. British Journal of Cancer. doi:10.1038/bjc.2013.336 (in press).

Archetti, M. 2013b. Evolutionarily stable anti-cancer therapies by autologous cell defection. Evolution, Medicine and Public Health. doi: 10.1093/emph/eot014 (in press).

Armitage, P., and R. Doll 1954. The age distribution of cancer and a multi-stage theory of carcinogenesis. British Journal of Cancer 8:1–12.

Ashley, D. J. 1969. The two "hit" and multiple "hit" theories of carcinogenesis. British Journal of Cancer 23:313–328.

Axelrod, R., D. E. Axelrod, and K. J. Pienta 2006. Evolution of cooperation among tumor cells. Proceedings of the National Academy of Sciences of the United States of America 103:13474–13479.

Basanta, D., and A. Deutsch (2008) A game theoretical perspective on the somatic evolution of cancer. In N. Bellomo, M. Chaplain, and E. De Angelis, eds. Selected Topics on Cancer Modelling: Genesis Evolution Immune Competition Therapy, pp. 97–112. Birkhauser, Boston.

Basanta, D., H. Hatzikirou, and A. Deutsch 2008a. Studying the emergence of invasiveness in tumours using game theory. The European Physical Journal 63:393–397.

Basanta, D., M. Simon, H. Hatzikirou, and A. Deutsch 2008b. Evolutionary game theory elucidates the role of glycolysis in glioma progression and invasion. Cell Proliferation 41:980–987.

Basanta, D., J. G. Scott, R. Rockne, K. R. Swanson, and A. R. A. Anderson 2011. The role of IDH1 mutated tumour cells in secondary glioblastomas: an evolutionary game theoretical view. Physical Biology 8:015016.

Basanta, D., J. G. Scott, M. N. Fishman, G. Ayala, S. W. Hayward, and A. R. A. Anderson 2012. Investigating prostate cancer tumour-stroma interactions: clinical and biological insights from an evolutionary game. British Journal of Cancer 106:174–181.

Boza, G., and S. Szamado 2010. Beneficial laggards: multilevel selection, cooperative polymorphism and division of labour in threshold public good games. BMC Evolutionary Biology 10:336.

Byrne, H. 2010. Dissecting cancer through mathematics: from cell to the animal model. Nature Reviews Cancer 10:221–230.

Cairns, J. 1975. Mutation selection and the natural history of cancer. Nature 255:197–200.

Charles, D. R., and E. M. Luce-Clausen 1942. The kinetics of papilloma formation in benzpyrene-treated mice. Cancer Research 2:261–263.

Chuang, J. S., O. Rivoire, and S. Leibler 2010. Cooperation and Hamilton's rule in a simple synthetic microbial system. Molecular Systems Biology 6:398. doi: 10.1038/msb.2010.57.

Crespi, B. J. 2001. The evolution of social behavior in microorganisms. Trends in Ecology & Evolution 16:178–183.

Crespi, B. J., and K. Summers 2005. Evolutionary biology of cancer. Trends in Ecology & Evolution 20:545–552.

Dexter, D. L., and J. T. Leith 1986. Tumor heterogeneity and drug resistance. Journal of Clinical Oncology 4:244–257.

Dingli, D., F. A. Chalub, F. C. Santos, S. Van Segbroeck, and J. M. Pacheco. 2009. Cancer phenotype as the outcome of an evolutionary game between normal and malignant cells. British Journal of Cancer 101:1130–1136.

Durrett, R., J. Foo, K. Leder, J. Mayberry, and F. Michor 2011. Intratumor heterogeneity in evolutionary models of tumor progression. Genetics 188:461–477.

Eungdamrong, N. J., and R. Iyengar 2004. Modeling cell signalling networks. Biology of the Cell 96:355–362.

Fisher, J. C. 1958. Multiple-mutation theory of carcinogenesis. Nature 181:651–652.

Foster, K. R. 2004. Diminishing returns in social evolution: the not-so-tragic commons. Journal of Evolutionary Biology 17:1058–1072.

Frank, S. A. (2007) Dynamics of Cancer. Princeton University Press, Princeton, NJ.

Gatenby, R., and R. J. Gillies 2008. A microenvironmental model of carcinogenesis. Nature Reviews Cancer 8:56–61.

Gatenby, R. A., and P. Maini 2003. Cancer summed up. Nature 421:321.

Gerstung, M., H. Nakhoul, and N. Beerenwinkel 2011. Evolutionary games with affine fitness functions: applications to cancer. Dynamic Games and Applications 1:370–385.

Gonzalez-Garcia, I., R. V. Sole, and J. Costa 2002. Metapopulation dynamics and spatial heterogeneity in cancer. Proceedings of the National Academy of Sciences of the United States of America 99:13085–13089.

Greaves, M., and C. C. Maley 2012. Clonal evolution in cancer. Nature 481:306–313.

Hanahan, D., and R. A. Weinberg 2000. The hallmarks of cancer. Cell 100:57–70.

Hardin, J. 1968. The tragedy of the commons. Science 162:1243–1248.

Hauert, C., F. Michor, M. A. Nowak, and M. Doebeli 2006. Synergy and discounting in social dilemmas. Journal of Theoretical Biology 239:195–202.

Hemker, H. C., and P. W. Hemker 1969. General kinetics of enzyme cascades. Proceedings of the Royal Society of London. Series B 173:411–420.

Ifti, M., T. Killingback, and M. Doebeli 2004. Effects of neighbourhood size and connectivity on the spatial continuous prisoner's dilemma. Journal of Theoretical Biology 231:97–106.

Iwasa, Y., and F. Michor 2011. Evolutionary dynamics of intratumor heterogeneity. PLoS ONE 6:e17866.

Jouanneau, J., G. Moens, Y. Bourgeois, M. F. Poupon, and J. P. Thiery 1994. A minority of carcinoma cells producing acidic fibroblast growth factor induces a community effect for tumor progression. Proceedings of the National Academy of Sciences of the United States of America 91:286–290.

Knudson, A. G. 1971. Mutation and cancer: statistical study of retinoblastoma. Proceedings of the National Academy of Sciences of the United States of America 68:820–823.

Lambert, G., L. Estévez-Salmeron, S. Oh, D. Liao, B. M. Emerson, T. D. Tlsty, and R. H. Austin 2011. An analogy between the evolution of drug resistance in bacterial communities and malignant tissues. Nature Reviews Cancer 11:375–382.

Maley, C. C., P. C. Galipeau, J. C. Finley, V. J. Wongsurawat, X. Li, C. A. Sanchez, and T. G. Paulson 2006. Genetic clonal diversity predicts progression to esophageal adenocarcinoma. Nature Genetics 38:468–473.

Marusyk, A., and K. Polyak 2010. Tumor heterogeneity: causes and consequences. Biochimica et Biophysica Acta 1805:105–117.

Maynard Smith, J., and G. R. Price 1973. The logic of animal conflict. Nature 246:15–18.

Mendes, P. 1997. Biochemistry by numbers: simulation of biochemical pathways with Gepasi 3. Trends in Biochemical Sciences 22:361–363.

Merlo, L. M. F., J. W. Pepper, B. J. Reid, and C. C. Maley 2006. Cancer as an evolutionary and ecological process. Nature Reviews Cancer 6:924–935.

Motro, U. 1991. Co-operation and defection: playing the field and ESS. Journal of Theoretical Biology 151:145–154.

Nagy, J. D. 2004. Competition and natural selection in a mathematical model of cancer. Bulletin of Mathematical Biology 66:663–687.

Nordling, C. O. 1953. A new theory on the cancer-inducing mechanism. British Journal of Cancer 7:68–72.

Nowell, P. C. 1976. The clonal evolution of tumor cell populations. Science 194:23–28.

Ohtsuki, H., J. M. Pacheco, and M. A. Nowak 2007. Evolutionary graph theory: breaking the symmetry between interaction and replacement. Journal of Theoretical Biology 246:681–694.

Pacheco, J. M., F. C. Santos, M. O. Souza, and B. Skyrm 2009. Evolutionary dynamics of collective action in N-person stag hunt dilemmas. Proceedings of the Royal Society of London. Series B 276:315–321.

Palfrey, T. R., and H. Rosenthal 1984. Participation and the provision of public goods: a strategic analysis. Journal of Public Economics 24:171–193.

Pepper, J. W. 2012. Drugs that target pathogen public goods are robust against evolved drug resistance. Evolutionary Applications 5:757–761.

Perc, M., J. Gómez-Gardeñes, A. Szolnoki, L. M. Floría, and Y. Moreno 2013. Evolutionary dynamics of group interactions on structured populations: a review. Journal of the Royal Society, Interface 10:20120997.

Rapoport, A., and A. M. Chammah 1966. The game of chicken. American Behavioral Scientist 10:10–28.

Ricard, J., and G. Noat 1986. Catalytic efficiency, kinetic co-operativity of oligomeric enzymes and evolution. Journal of Theoretical Biology 123:431–451.

Sugden, R. 1986. The Economics of Rights. Co-operation and Welfare, Blackwell, Oxford, UK.

Tomlinson, I. P. 1997. Game-theory models of interactions between tumour cells. European Journal of Cancer 33:1495–1500.

Tomlinson, I. P., and W. F. Bodmer 1997. Modelling consequences of interactions between tumour cells. British Journal of Cancer 75:157–160.

Traulsen, A., J. C. Claussen, and C. Hauert 2006. Coevolutionary dynamics in large, but finite populations. Physical Review E 74:011901.

Traulsen, A., J. M. Pacheco, and M. A. Nowak 2007. Pairwise comparison and selection temperature in evolutionary game dynamics. Journal of Theoretical Biology 246:522–529.

Contemporary evolution and the dynamics of invasion in crop–wild hybrids with heritable variation for two weedy life–histories

Lesley G. Campbell,[1] Zachary Teitel[1,3] and Maria N. Miriti[2]

1 Department of Chemistry & Biology, Ryerson University, Toronto, ON, Canada
2 Department of Evolution, Ecology and Organismal Biology, The Ohio State University, Columbus, OH, USA
3 Present address: Department of Integrative Biology, University of Guelph, Guelph, ON Canada

Keywords
agriculture, artificial selection, evolutionary demography, hybridization, invasive species, life table response experiment, life-history evolution.

Correspondence
Lesley G. Campbell, Department of Chemistry & Biology, Ryerson University, 350 Victoria Street, Toronto, ON, Canada.

e-mail: lesley.g.campbell@ryerson.ca

Abstract

Gene flow in crop–wild complexes between phenotypically differentiated ancestors may transfer adaptive genetic variation that alters the fecundity and, potentially, the population growth (λ) of weeds. We created biotypes with potentially invasive traits, early flowering or long leaves, in wild radish (*Raphanus raphanistrum*) and F$_5$ crop–wild hybrid (*R. sativus* × *R. raphanistrum*) backgrounds and compared them to randomly mated populations, to provide the first experimental estimate of long-term fitness consequences of weedy life-history variation. Using a life table response experiment design, we modeled λ of experimental, field populations in Pellston, MI, and assessed the relative success of alternative weed strategies and the contributions of individual vital rates (germination, survival, seed production) to differences in λ among experimental populations. Growth rates (λ) were most influenced by seed production, a trait altered by hybridization and selection, compared to other vital rates. More seeds were produced by wild than hybrid populations and by long-leafed than early-flowering lineages. Although we did not detect a biotype by selection treatment effect on lambda, lineages also exhibited contrasting germination and survival strategies. Identifying life-history traits affecting population growth contributes to our understanding of which portions of the crop genome are most likely to introgress into weed populations.

Introduction

The ecological processes of population growth and persistence are shaped by the evolutionary characteristics of a population, that is, phenotypic frequencies and their relative fitness (Darwin and Wallace 1858; Simpson 1944; Gould 1989). In fact, rapid, adaptive evolution in response to environmental variation is expected to result in altered demography, which has implications for population growth rates. The Galapagos finches (*Geospiza fortis*) experienced catastrophic demographic decline during a drought (Grant and Grant 2002); the population crash was subsequently explained, in large part, by slowed evolution due to genetic load (Hairston et al. 2005). To complement such natural 'experiments', reciprocal translocation experiments show that local adaptation can

dramatically affect reproductive success (Kinnison et al. 2008; Hereford 2009), a correlation of population growth for many annual plants. Several experimental microcosms have manipulated genetic diversity (presumably neutral and adaptive) to determine that its very presence positively influences population persistence (e.g., in small versus large common toad [*Bufo bufo*] populations, Hitchings and Beebee 1998; predator–prey ecosystems consisting of algae and rotifers, Hitchings and Beebee 1998; Yoshida et al. 2003). Finally, introduction of predators into natural populations of *Poecilia reticulata* resulted in rapid evolution of key phenotypic (e.g., dulled male coloration) and demographic traits (i.e., delayed maturation and fewer, larger offspring (Reznick and Bryga 1987). Thus, trait evolution is likely a significant driver of population demography (Frankham 2005; Kinnison and

Nelson 2007). Yet, to our knowledge, there are no published descriptions of experimental manipulations of adaptive trait variation in populations that subsequently explore the consequences for population growth.

Gene flow is one evolutionary mechanism that alters the quality and quantity of adaptive trait variation (Arnold 1997). When gene flow is high, it has a homogenizing effect (Burgess et al. 2005), constrains adaptive evolution (e.g., Slatkin 1987; Kirkpatrick and Barton 1997; Lenormand 2002), and may cause population declines (e.g., Hanski 1999). In contrast, episodic events of gene flow may hasten adaptive evolution (e.g., Ehrlich and Raven 1969; Gomulkiewicz et al. 1999; Rieseberg et al. 2003; Hendry 2004; Whitney et al. 2006) and thus contribute to weedy population growth (Hovick and Whitney 2014). Therefore, gene flow between genetically distinct crops and sexually compatible weedy relatives may contribute to the evolution of more problematic weeds when gene flow alters phenotypic frequencies and the relative fitness of phenotypes within weed populations.

Crop-to-wild gene flow has served as a model system to evaluate the ecological and evolutionary consequences of gene flow and the potential for hybridization to lead to rapid evolution of life-history and fitness-related traits (e.g., Campbell et al. 2006; Hovick et al. 2012). The impact of crop–wild gene flow depends on rates of gene flow and the relative success of heritable, migrant phenotypes when compared with the recipient population's adaptive optima (Snow et al. 2003; Hooftman et al. 2005; Mercer et al. 2006). Studies on crop–wild hybrid populations show that the successful phenotypes are not always a random subset of genotypes from parental populations (Ellstrand and Schierenbeck 2000; Hovick and Whitney 2014), and hybrid populations may facilitate the transfer of novel, adaptive traits to recipient weed populations (e.g., Snow et al. 2003; Hooftman et al. 2011; Owart et al. 2014). Indeed, over short-time scales, crop-to-wild gene flow can be a more significant source of adaptive genetic variation than mutation (Gomulkiewicz et al. 1999; Holt et al. 2004).

Risk assessment of crop–wild hybridization often explores the evolution of increased fecundity of weeds (Pilson and Prendeville 2004; Snow et al. 2005; Ellstrand et al. 2014). Although it is difficult to predict an invader based on a suite of traits (Perrins et al. 1992), a quintessential, annual weed often reaches sexual maturity quickly or grows large quickly to compete for limited resources (or both, Roff 1992; Stearns 1992). From a demographic perspective, weedy populations may exhibit high population growth rates, a strong capacity to colonize new locations, and/or high population persistence (Campbell et al. 2014). Certainly, a short life cycle will reduce the likelihood of death before reproduction, but individuals may reproduce at a size smaller than is adaptive and therefore curtail

reproduction. In contrast, large size at reproduction may not only provide a competitive benefit, but allometric consequences of large size may also result in the production of more offspring (Weiner et al. 2009). Yet, there are few controlled experiments linking genetic variation in life-history traits to population demographic consequences. Life table response experiments (LTRE) offer a robust tool to measure the demographic significance of life-history variation in experimental populations (Hooftman et al. 2007; Campbell et al. 2014).

Here, we build on our studies of fitness components (Campbell et al. 2006; Hovick et al. 2012), evolutionary responses of two key life-history traits to directional selection (Campbell et al. 2009a,b), and demographic analyses of populations experiencing natural selection (Campbell et al. 2014) to determine the influence of heritable variation for early flowering or long leaves (as an indicator of plant size) on the relative population growth of advanced-generation hybrid and wild radish biotypes grown in a common garden in Michigan, USA.

Methods

Study species

We used the crop–wild complex of cultivated radish (*Raphanus sativus*), an open-pollinated vegetable selected for large, colorful roots and high seed production (Snow and Campbell 2005), and its weedy relative, wild radish (*Raphanus raphanistrum*, also known as jointed charlock), a cosmopolitan, agricultural weed that also colonizes disturbed sites and coastal beaches (Warwick and Francis 2005). These two radish species have emerged as model systems in plant evolutionary ecology and in the assessment of ecological consequences of crop-to-wild gene flow (Mazer et al. 1986; Klinger and Ellstrand 1994; Snow et al. 2001, 2010). Although *R. raphanistrum* and *R. sativus* share many phenotypic characters, they exhibit divergent life histories in several key traits associated with weediness. Many *R. sativus* cultivars germinate quickly and develop large rosettes before bolting and flowering late in the growing season, whereas *R. raphanistrum* plants germinate slowly and inconsistently, form narrow, branching taproots and develop smaller rosette sizes before flowering early in the growing season (Panetsos and Baker 1967; Campbell et al. 2009a,b).

Using genotypes from natural selection experiments, we have studied many aspects of crop-to-wild gene flow and hybrid fitness of radishes in Michigan, California, and Texas, where *R. raphanistrum* is non-native and, sometimes, weedy. Unlike in Michigan where wild and hybrid phenotypes were equally successful, hybrids grown in California exhibited ~22% greater survival and ~270% greater fecundity than wild plants. Furthermore, in Texas, hybrids were more successful at colonizing this novel location, due to earlier, increased germination, and increased survival, despite producing fewer seeds per plant (Hovick et al. 2012). Finally, hybrid populations had faster population growth than wild plants in Michigan, under low, but not high, competition conditions (Campbell et al. 2014). These results are consistent with the hypothesis that crop–wild hybrid biotypes have the potential to displace their wild parent in certain environments (Ellstrand and Schierenbeck 2000; Hovick and Whitney 2014).

Biotypes

Detailed descriptions of the wild and hybrid populations are available in Campbell et al. (2009a,b). Briefly, control and artificially selected populations were generated by hand-pollinating 100 wild *R. raphanistrum* plants with either wild pollen to create F_1 wild biotype populations, or pollen from 100 *R. sativus* var. 'Red Silk' plants (Harris-Moran Seed Co., Modesto, CA, USA) to create F_1 hybrid biotype populations. Based on hybridization in this first generation, we refer to radish biotypes as wild or hybrid. Physical separation and unpollinated control flowers were used to ensure that crosses between these self-incompatible plants were uncontaminated.

Artificial selection was imposed on glasshouse-grown plants for three generations (F_2–F_4). Randomly mated 'control' populations and artificial selection populations were initiated in the F_2 generation after 100 individuals from each F_1 biotype were cross-pollinated (Campbell et al. 2009a,b). During three generations of mating (F_3–F_5), populations were initiated with 130–200 F_2–F_4 individuals and were propagated with a subset (10%) of individuals from each replicate each generation (Table 1). For the purposes of imposing selection and following trait evolution, we recorded dates of germination and anthesis, and leaf length at anthesis of each plant. Age at flowering was calculated as the difference, in days, between germination and anthesis.

Table 1. Summary of wild and hybrid populations included in this experiment.

Biotype	Selection treatment	Number of generations of artificial selection or random mating	Number of populations
Wild	Early-flowering	3	3
	Control*	3	3
	Long-leaf	3	3
F_5 Crop–Wild hybrid	Early	3	3
	Control*	3	3
	Long-leaf	3	3

*Note that these populations did not experience selection but rather random mating for three generations.

As the length of the longest leaf in wild and crop–wild hybrid radish is correlated with several measures of plant size at the time of reproduction (e.g., number of flowers, stem diameter at harvest, Campbell et al. 2009a,b), the length of the longest leaf on the first day of flowering served as an early indicator of plant size at the time of reproduction. Applying truncation selection, we selected 10% of the plants from each lineage that represented the earliest flowering individuals for early lineages, 10% of the plants from each lineage that represented the longest leafed individuals for long lineages, and randomly selected 10% of the plants from the control lineages to produce the following generation.

Selected plants were cross-pollinated within a lineage in a complete diallel design. To account for drift as a possible evolutionary mechanism, we created three independent populations for each treatment combination (wild or hybrid; early, large or control) for a total of eighteen populations (Table 1). Populations represented the variation in evolutionary trajectories of randomly mated or artificially selected populations typically associated with genetic drift. We assumed that if control populations became adapted to experimental conditions, this had only minor effects on the phenotypic and demographic traits of interest. Thus, we used the control populations to determine the expected variation in traits without selection in advanced-generation hybrid and nonhybrid populations.

Demographic experimental design

We measured vital rate dynamics of populations from the wild and hybrid artificial selection populations in a common garden. As in previous studies (Campbell et al. 2009a, b), the common garden was located at the University of Michigan Biological Station in Pellston, Michigan, USA. The proximity of the common garden to our original experimental plots helped to assure that the phenotypic variation observed was typical for these plants (e.g., Campbell et al. 2006). In 2004, we collected F_5 seeds from F_4 artificial selection population plants (see Campbell et al. 2009a,b).

The common garden included F_5 wild and hybrid artificial selection populations. Whole fruits were planted on May 30, 2005 in 3.54 L of local sandy soil in an aluminum foil pan (22.9 cm × 30.34 cm × 5.1 cm, Walmart, Cheboygan, MI, USA) with holes puncturing the bottom surface, allowing plant roots to grow into local soil and excess water to drain easily. The number of seeds within a fruit was estimated based on the number of visible locules from the outside of the fruit. For the artificial selection lineages, we planted six locules per pan. Each artificial selection lineage (e.g., Hybrid Control Rep 1) was represented by five replicate pans (a total of 30 seeds). Pans were arranged in a complete randomized block design. Within a pan, fruits were spaced out as evenly as possible. Pans were separated by at least 30 cm from neighboring pans to minimize root and shoot competition. Pans were watered every other day until August 31st. Insecticide (0.0033% esfenvalerate, 20 g/ 9.5 L, Scotts Miracle-Gro Co., Marysville, OH, USA) was used to control insect herbivory three times during the first month after planting, when aphid herbivory was highest. Aphids were present at low densities later in the season but did not colonize any plant heavily. Pollinators were abundant throughout the experiment (as in Lee and Snow 1998). Plants were individually harvested as they senesced, until the first hard frost (September 16th–20th, 2005), when we harvested all remaining plants. Harvested radish plants were dried at 60°C.

Censuses and data collection

From May 10, 2005 to September 19, 2005, we censused plots weekly to record changes in demographic status of the experimental individuals. Each week, new individuals were flagged to identify them in future censuses and all flagged individuals were categorized as either dead or in one of the three stages mentioned below. Once plants were harvested, we recorded flower number, fruit number, and seeds per fruit as measures of lifetime fecundity. To estimate the number of seeds per plant, we multiplied the average number of locules per fruit (for 10 randomly chosen fruits per plant) by the number of fruits.

Matrix construction

We classified plants into three stages, chosen after several years of observing this species. The stages were seeds, germinating cotyledonous plants (plants with only cotyledons), and flowering plants (plants with open flowers). Four demographic transitions were included in our model for *Raphanus* populations using data from the 2005 field season: seed dormancy/mortality, germination, survival to flowering, and fecundity. Because our methods could not distinguish between seed dormancy and mortality, we maintain both terms in a single demographic parameter. Lumping these terms was justified by the results of a recent study exploring the seedbank dynamics of *Raphanus raphanistrum* and F$_3$ crop–wild hybrids; seed dormancy was ~58% lower in hybrid versus wild populations whereas seed mortality did not differ among biotypes (~8% of seed, Teitel 2014). If the artificial selection lineages used here differed in seed mortality and dormancy, comparing mortality and dormancy between lineages is inappropriate because the effect of seeds in this stage are not equivalent. Note in the results, however, that the relative contribution of

mortality and dormancy to lambda is the smallest among all life-history stages, and this is consistent with our findings in other studies (Campbell et al. 2014; Teitel 2014). Therefore, differences in seedbank dynamics tend to have a relatively small impact on population growth compared with juvenile survival or fecundity. Our analysis synthesizes the dynamic vital rates across the annual summer growing season, from planting on June 7 to harvesting on September 19.

Matrix algebra, LTRE, and sensitivity analyses

We used a fixed-effect LTRE (Caswell 2001) to model lambda (λ) of each experimental population (18 constructed matrices; e.g., wild early-flowering replicate 1, hybrid large replicate 2, Appendix 1) as a linear function of biotype (g), selection treatment (s), and their interaction (gs): $-\lambda^{gs} = \lambda^{(..)} + \alpha^g + \beta^s + \alpha\beta^{gs}$ where α^g is the effect of the *gth* level of the biotype, β^s is the effect of the *sth* level of the selection treatment, and $\alpha\beta^{gs}$ is the interaction of the *gth* biotype and *sth* selection treatment, measured relative to the projected growth rate of a reference matrix $^{(..)}$. We obtained our reference matrix by combining data from randomly mating wild or hybrid populations into a mean (calculated by averaging transition frequencies) matrix (Miriti et al. 2001). To obtain the treatment matrices, we first averaged all replicates of matrices belonging to a given treatment combination (e.g., the transition frequencies of wild early replicates 1, 2, and 3 were averaged). We then averaged common treatment groups of these matrices to give us mean representative matrices for a given treatment (mean wild type, mean early flowering). We estimated treatment effects as:

$$\alpha^g = \lambda^{g.} - \lambda^{..}$$
$$\approx \sum [a_{ij}^{g.} - a_{ij}^{..}] \cdot (\delta\lambda/\delta a_{ij})|_{[A^{g.} + A^{..}]/2}$$
$$\beta^s = \lambda^{.s} - \lambda^{..}$$
$$\approx \sum [a_{ij}^{.s} - a_{ij}^{..}] \cdot (\delta\lambda/\delta a_{ij})|_{[A^{.s} + A^{..}]/2}$$
$$\alpha\beta^{gs} = \lambda^{gs} - \lambda^{..} - \alpha^g - \beta^s$$
$$\approx \sum [a_{ij}^{gs} - a_{ij}^{..}] \cdot (\delta\lambda/\delta a_{ij})|_{\frac{[A^{gs} + A^{..}]}{2}} - \alpha^g - \beta^s$$

where we obtained sensitivities ($\delta\lambda/\delta a_{ij}$) from the relationship $\delta\lambda/\delta a_{ij} = v_i w_j/<w,v>$, and v and w are the right and left eigenvectors of the matrix. We then evaluated the sensitivities, halfway between the reference and treatment matrices (Caswell 2001). We obtained treatment matrices (e.g., A^g, A^s) by pooling data across all levels of the other treatments. Finally, the contributions were calculated by weighting the differences in vital rates by their sensitivities. In general, a vital rate will increase or decrease lambda relative to some standard model (i.e., the mean matrix). For instance, a positive contribution of fecundity suggests that fecundity in the 'experimental' population made lambda more positive relative to the mean. Therefore, we interpreted the above equations as how both observed variation in matrix elements, and the sensitivity of population growth to variation in those elements, influence the effect of the treatments on population growth. A particular matrix element a_{ij} may contribute little to variation in lambda in cases when a_{ij} was invariant among treatment classes or when lambda was insensitive to variation in a_{ij}. Additionally, a_{ij} may contribute little to variation in lambda even if lambda was highly sensitive to the element if the vital rate did not differ among treatments. In alternate scenarios, even small amounts of variation in a_{ij} may drive variation in lambda when there are consistent differences among treatments and when lambda is highly sensitive to that matrix element. One must note that contributions can differ in direction and magnitude even if there is no significant difference among lambdas of the experimental and mean matrices (Caswell, 2001). It is important to recognize that these contributions show the consequences of vital rates on population growth and are not a measure of the statistical significance of a vital rate. Matrix algebra and analyses were performed using MATLAB (v.2012a; The Mathworks, Inc., Natick, MA, USA).

Vital rate comparisons

To test whether the estimates of the three vital rates and lambda were similar among artificial selection wild and hybrid populations, we ran a Type III multivariate ANOVA in which biotype and selection treatment, and their interaction were fixed effects. As the proportion of seeds that remains in the seed bank is correlated with the proportion of seeds that germinate, we only tested the effects of biotype and selection on germination. Germination and survival to flowering were arcsine square root transformed to normalize data; fecundity and lambda were $\log_{10}$ transformed. When significant differences were detected, *post hoc* comparisons were performed using Tukey's correction for multiple hypothesis tests. All statistical analyses were performed using SPSS v. 21. (SPSS Inc., Chicago, IL, USA)

Results

Weed population demography responded to our artificial selection treatments (Multivariate ANOVA, $F_{6,22} = 2.864$, $P = 0.032$), hybrid ancestry ($F_{3,10} = 2.745$, $P = 0.099$), and their interaction ($F_{6,22} = 2.44$, $P = 0.058$). Although all populations exhibited positive population growth, lambda was marginally significantly higher in wild (λ mean $\pm$ SE = 5.34 $\pm$ 0.17) than hybrid populations (4.55 $\pm$ 0.43,

Fig. 1, Table 2). Whereas, the biotypes did not differ significantly in germination rate or survival to flowering, wild plants produced significantly more seeds than hybrid plants (Fig. 1, Table 2). Population growth rate did not differ between selection treatments, and there were no significant differences in vital rates among selection treatments. Finally, although we did not detect a significant biotype by selection treatment effect on lambda, lineages exhibited contrasting demographic strategies. Germination rates in wild lineages were highest when the population had experienced artificial selection for long leaves whereas the pattern was opposite in hybrid lineages (Fig. 1, Table 2). Survival to flowering was marginally significantly higher for wild control lineages than hybrid control lineages, but did not differ among biotypes across early or long-leaf treatments. Long-leafed populations produced marginally more seeds than the early-flowering lineages (Table 2, Fig. 1). Therefore, demographic growth and relative invasiveness may be significantly altered by genotypic frequencies and artificial selection for weedy traits within populations.

Population growth rates are a consequence of contributions from each vital rate, and evaluating these consequences allows us to understand the influence of each vital rate on the relative weediness of populations. Contributions of fecundity and flowering to differences in lambda between early flowering and control populations were greater in hybrid than wild populations, whereas contributions from seeds (either remaining in the seed bank or germinating) were roughly equivalent and minor between biotypes (Fig. 2A). Fecundity negatively contributed to change in population growth rate between early flowering and control lineages for both wild and hybrid biotypes, although flowering only negatively contributed to changes in population growth rate in wild biotype populations. Contributions of fecundity to differences in lambda between large-leafed and control populations were greater in hybrid than wild populations, whereas contributions of survival and germination were greater in wild than hybrid populations (Fig. 2B). In other words, in populations selected for early flowering, fecundity reduced population growth relative to controls (the mean matrix), but in populations selected for large leaves, fecundity increased population growth relative to control populations. Therefore, the demography of these populations differed substantially. The results from the long leaf length treatments reveal that in hybrid populations, relative to the wild populations, fecundity increased population growth, whereas germination rates reduced population growth in hybrid relative to wild lineages. Fecundity positively contributed to differences in lambda between large-leafed and control lineages for both wild and hybrid

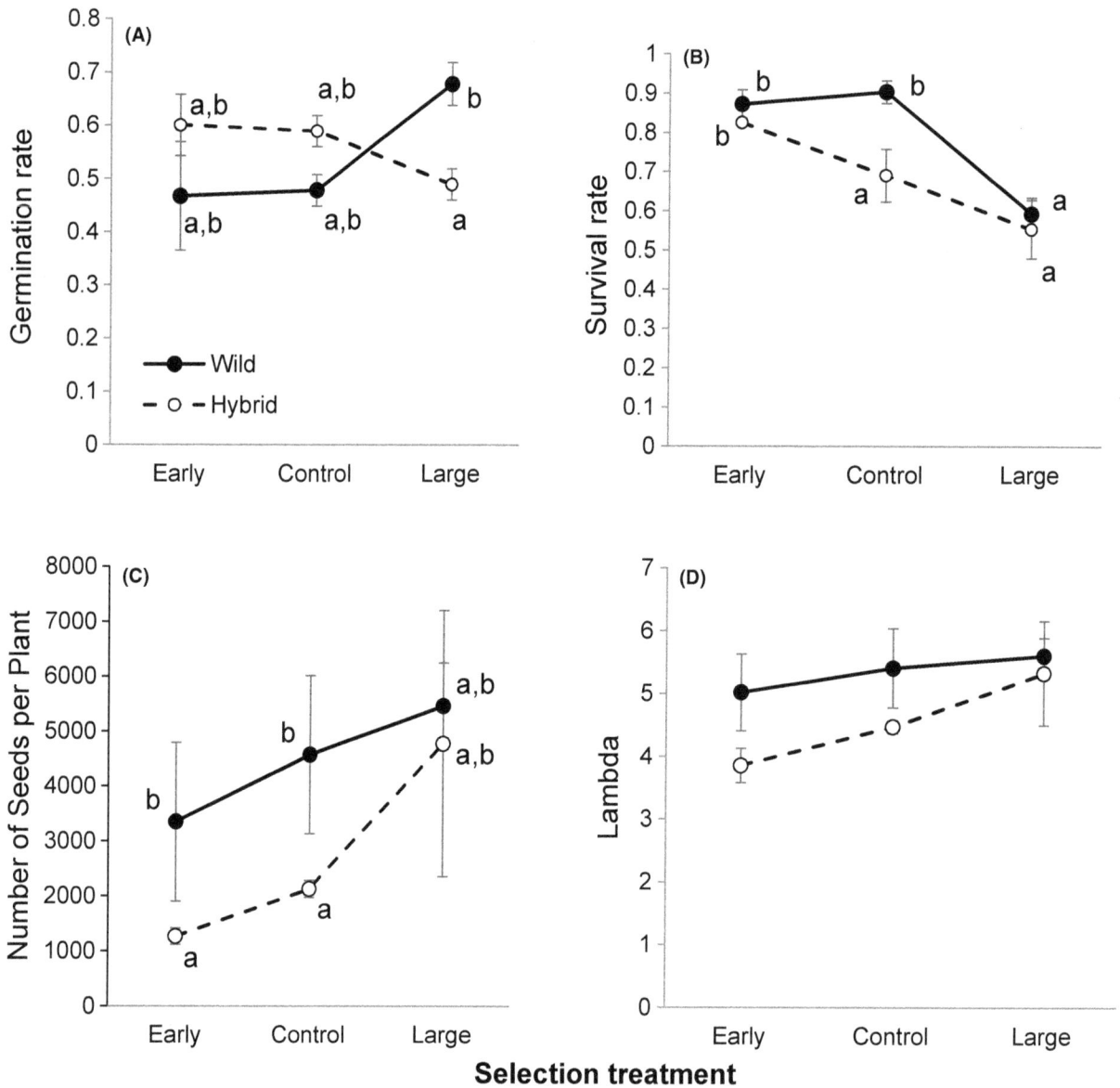

Figure 1 Comparison of least square mean vital rates and population growth rates of wild (solid line) and F$_5$ hybrid populations (dashed line) after selection for early flowering or long leaves, relative to random mating grown in a field experiment in Pellston, MI, USA. (A) Germination rate; (B) Proportion of the population that survived to flower; (C) Number of seeds per population; (D) Lambda. Error bars represent the SE of the mean; $n = 3$ replicate populations per biotype and selection treatment combination.

Table 2. Summary of F-statistics and P-values (indicated with superscript symbols: *$P < 0.05$; †$P < 0.10$) from ANOVAS to test for significant differences across biotypes and artificial selection treatments (and their interaction) in the rates of germination, survival to flowering, number of seeds, and population growth (lambda).

Factor	df (numerator, denominator)	Germination	Survival to flowering	Number of seeds	Lambda
Biotype (B)	1, 2	0.03	2.96	27.64*	10.58†
Selection (S)	2, 2	0.10	6.88	19.64*	5.67
B × S	2, 12	5.43*	2.77†	0.17	0.38

biotypes, whereas survival to flowering only negatively contributed to differences in lambda in wild biotype populations and the direction of contributions from germination to differences in lambda were affected by biotype. Across all comparisons, seed dormancy/mortality had little impact on lambda (Fig. 2).

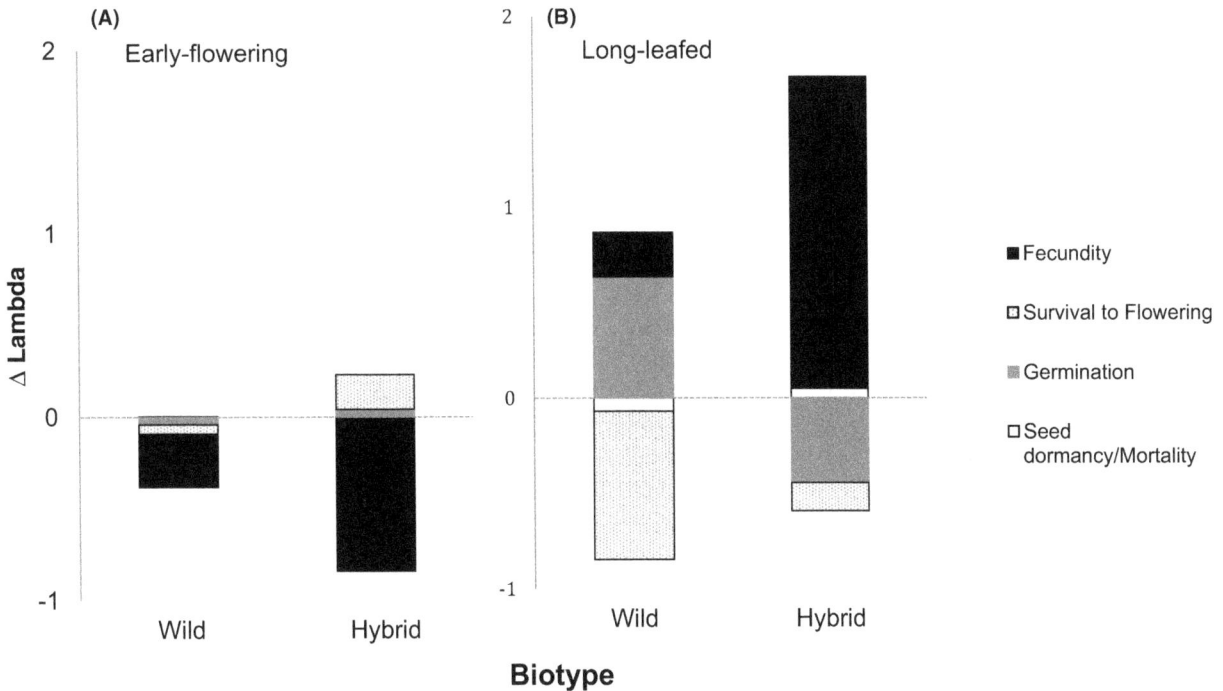

Figure 2 Contributions from seed dormancy/mortality, germination, flowering and fecundity vital rates to differences in lambda between control and F₅ lineages selected for (A) early flowering or (B) long leaves in two biotypes (hybrid and wild).

Discussion

Over the last two decades, questions have arisen over the evolutionary and ecological consequences of gene flow from transgenic crops to weeds, and thus weed ecologists have been asked to evaluate ongoing effects of gene flow from nontransgenic cultivars to weeds. These questions are increasingly important as crop breeding becomes more sophisticated and weedy biotypes disperse around the globe. It is likely that fitness-enhancing traits such as resistance to pests, pathogens, and herbicides will be engineered or bred into crops more widely in the future. This research fills a significant gap in our understanding of weeds that can hybridize with crops. For instance, strong directional selection for either early flowering or long leaves did not result in crop–wild hybrid progeny that surpassed wild progeny in their expression of weediness. Further, we have, for the first time, estimated the long-term fitness consequences of variation in life history in weeds. Identifying life-history traits that affect population growth contributes to our understanding of which portions of the crop genome are most likely to introgress into wild populations.

Both gene flow and adaptation may contribute to positive population growth of nascent, weedy populations. From an agronomic standpoint, annual weeds that exhibit high population growth rates and also have heritable variation in key life-history traits are likely to be most difficult to manage. Here, after three generations of random mating or artificial selection for early flowering or long leaves, we found crop–wild hybrid *Raphanus* populations tended to exhibit lower population growth rates, than nonhybrid weed populations in similar selection environments (Fig. 1D, Table 2). These differences may be due to the strength and direction of selection. Previously, we found significantly higher population growth rates in crop–wild hybrid *Raphanus* populations after three generations of natural selection (Campbell et al. 2014). Although strong, directional selection for early flowering or long leaves did not result in changes in lambda, these populations displayed significantly different demographic strategies. As a result, these highly weedy populations (i.e., early, control and large, wild and hybrid populations exhibited $\lambda \gg 1$) were equally successful using different germination or survival strategies.

Weed responses to strong anthropogenic selection pressure (i.e., potentially artificial but also unintentional, human selection), including herbicide resistance or altered germination schedules due to agricultural tilling schedules, account for a large proportion of documented cases of contemporary evolution in plants (Bone and Farres 2001; Delye et al. 2013; Heap 2015). Further, a number of examples of contemporary evolution provide support for surprising amounts of long-term crop allele introgression into weedy populations (e.g., Whitton et al. 1997; Hegde et al. 2006; Snow et al. 2010; Ellstrand et al. 2014). In contrast, some cases reveal surprisingly little introgression in wild relative populations planted near crops (Bartsch et al. 1999). Thus, empirical evidence for a strongly homogenizing role of gene flow from crops in the contemporary evolution of related wild or weedy populations is relatively common but not ubiquitous. This may be a consequence of the relative fitness of crop-like versus weed-like phenotypes within a weedy background. The evolution of increased success in weedy crop relatives after introduction provides convincing support that gene flow can introduce an adaptive, novel traits and thus increase population growth (Holt et al. 2004; Novack and Mack 2005). For instance, altered germination and survival of crop–wild hybrids were associated with higher relative fitness of hybrid radish in Texas, a newly invaded location (Hovick et al. 2012). Similarly, crop–wild hybridization in *Helianthus* has contributed to adaptive evolution in water stressed environments, by not only selecting for new leaf traits but also larger inflorescence size, a trait likely to change the demography of weedy sunflower population (Owart et al. 2014). Thus, rates of adaptive evolution that result in crop allele introgression depend on the rate of gene flow, the mode of inheritance of traits, and the relative fitness of heritable crop versus wild phenotypes in a weed population (Nuismer et al. 2012).

To assess the implications of weed evolution due to gene flow or selection, an LTRE perspective is useful as an LTRE approach can link change in population growth to changes in the vital rates (Fréville and Silvertown 2005). Fecundity contributed more to changes in population growth in hybrid than wild lineages. As well, large-leafed populations tended to be more fecund than control lineages and demonstrated higher population growth. In comparison, with relatively low seed production, early flowering lineages exhibited relatively low lambdas. As rates of fecundity impose a large influence on population growth rates (Appendix 2) and as hybrid populations possess greater genetic diversity than wild populations for large size (Campbell et al. 2009a,b), the above results suggest that hybrid biotypes could evolve higher population growth rates than wild biotypes, with additional selection for large size, that would improve their relative success colonizing new environments that are subject to an array of selection pressures.

Most risk assessments of crop–wild gene flow, including our own, consider fecundity of early-generation hybrid offspring as sufficient to assess the likelihood of persistent gene flow. Given that fecundity represented the demographic factor that contributed the most and the smallest contributions to $\Delta\lambda$s came from seed dormancy/mortality rates, these results suggest that this may be an adequate assessment approach, rather than a full-scale evaluation of the relative success of each demographic stage. However, germination success can seriously alter the relative fitness of genotypes (Hovick et al. 2012) and the contribution of each demographic stage to lambda can vary across years (Teitel et al. in press). Similar to our results, a LTRE conducted across 17 species of invasive and noninvasive plants revealed that invasive plant's large λs were mostly attributable to sexual reproduction (Burns et al. 2013), typical of successful invasion life-history strategy, where high fecundity allocation and plasticity is correlated with invasiveness (Daehler 2003; Morris and Doak 2004; Davidson et al. 2011). Relatively small contributions from seed dormancy suggest that attempts at suppressing seed banks may produce a less significant effect on λs. This is also reflected in the relatively low elasticities observed for seed dormancies (Appendix 2). However, *Raphanus* seed banks can be dynamic and remain dormant for several years (Teitel et al. in press). A longer-term assessment of seed bank viability as well as a more in-depth understanding of seed below-ground survival are needed to understand the full potential of seed dynamics for contributing to λ.

Acknowledgements

We thank J. Leonard, the UMBS staff, and many student researchers for their help in the greenhouse, field, and laboratory. The USDA (#2002-03715), NSF DDIG (DEB-0508615), and NSERC Discovery (#402305-2011) granting programs financially supported this research. The manuscript was improved by the constructive criticisms of two anonymous reviewers and A. Snow.

Data archiving statement

The data sets used in this work have been included in Appendix 1.

Literature cited

Arnold, M. L. 1997. Natural Hybridization and Evolution. Oxford University Press, Oxford, UK.

Bartsch, D., M. Lehnen, J. Clegg, M. Pohl-Orf, I. Schuphan, and N. C. Ellstrand 1999. Impact of gene flow from cultivated beet on genetic diversity of wild sea beet populations. Molecular Ecology 8:1733–1741.

Bone, E., and A. Farres 2001. Trends and rates of microevolution in plants. Genetica **112–113**:165–182.

Burgess, K. S., M. Morgan, L. Deverno, and B. C. Husband 2005. Asymmetrical introgression between two Morus species (*M. alba, M. rubra*) that differ in abundance. Molecular Ecology **14**:3471–3483.

Burns, J. H., E. A. Pardini, M. R. Schutzenhofer, Y. A. Chung, K. J. Seidler, and T. M. Knight 2013. Greater sexual reproduction contributes to differences in demography of invasive plants and their non-invasive relatives. Ecology **94**:995–1004.

Campbell, L.G. 2005. Can feral radishes become weeds? In Gressel J, ed. Crop Ferality and Volunteerism, pp. 193–208. CRC, Boca Raton, FL.

Campbell, L. G., A. A. Snow, and C. E. Ridley 2006. Weed evolution after crop gene introgression: greater survival and fecundity of hybrids in a new environment. Ecology Letters 9:1198–1209.

Campbell, L. G., A. A. Snow, and P. M. Sweeney 2009a. When divergent life histories hybridize: insights into adaptive life-history traits in an annual weed. New Phytologist 184:806–818.

Campbell, L. G., A. A. Snow, P. M. Sweeney, and J. M. Ketner 2009b. Rapid evolution in crop-weed hybrids under artificial selection for divergent life histories. Evolutionary Applications 2:172–186.

Campbell, L. G., Z. Teitel, M. Miriti, and A. A. Snow 2014. Context-specific enhanced invasiveness of Raphanus crop–wild hybrids: a test for associations between greater fecundity and population growth. Canadian Journal of Plant Science 94:1315–1324.

Caswell, H. 2001. Matrix Population Models: Construction, Analysis, and Interpretation, 2nd edn. Sinauer Associates, Sunderland, MA.

Caswell, H. 2007. Sensitivity analysis of transient population dynamics. Ecology Letters 10:1–15.

Daehler, C. C. 2003. Performance comparisons of co-occurring native and alien invasive plants: implications for conservation and restoration. Annual Review of Ecology Evolution and Systematics 34:183–211.

Darwin, C., and A. Wallace 1858. On the tendency of species to form varieties: and on the perpetuation of varieties and species by natural means of selection. As communicated by C. Lyell and J.D. Hooker. Journal of the Proceedings of the Linnean Society of London, Zoology 3:45–62.

Davidson, A. M., M. Jennions, and A. B. Nicotra 2011. Do invasive species show higher phenotypic plasticity than native species and if so, is it adaptive? A meta-analysis. Ecology Letters 14:419–431.

Delye, C., Y. Menchari, S. Michel, E. Cadet, and V. Le Corre 2013. A new insight into arable weed adaptive evolution: mutations endowing herbicide resistance also affect germination dynamics and seedling emergence. Annals of Botany 111:681–691.

Ehrlich, P. R., and P. H. Raven 1969. Differentiation of populations. Science 165:1228–1232.

Ellstrand, N. C., and K. A. Schierenbeck 2000. Hybridization as a stimulus for the evolution of invasiveness in plants? Proceedings of the National Academy of Sciences of the United States of America 97:7043–7050.

Ellstrand, N. C., P. Meirmans, J. Rong, D. Bartsch, A. Ghosh, T. J. deJong, P. Haccou et al. 2014. Introgression of crop alleles into wild or weedy populations. Annual Review of Ecology and Systematics 44:325–345.

Frankham, R. 2005. Genetics and extinction. Biological Conservation 126:131–140.

Fréville, H., and J. Silvertown 2005. Analysis of interspecific competition in perennial plants using life table response experiments. Plant Ecology 176:69–78.

Gomulkiewicz, R., R. D. Holt, and M. Barfield 1999. The effects of density dependence and immigration on local adaptation and niche evolution in a black-hole sink environment. Theoretical Population Biology 55:283–296.

Gould, S. J. 1989. Wonderful Life. Norton Co., Inc, New York.

Grant, P. R., and B. R. Grant 2002. Unpredicatable evolution in a 30-year study of Darwin's finches. Science 296:707–711.

Hairston, N. G. Jr, S. P. Ellner, M. A. Geber, T. Yoshida, and J. A. Fox 2005. Rapid evolution and the convergence of ecological and evolutionary time. Ecology Letters 8:1114–1127.

Hanski, I. 1999. Metapopulation Ecology. Oxford University Press, London.

Heap, I. 2015. The international survey of herbicide resistant weeds 2015. Available from http://www.weedscience.com (accessed on 24 February 2015).

Hegde, S. G., J. D. Nason, J. M. Clegg, and N. C. Ellstrand 2006. The evolution of California's wild radish has resulted in the extinction of its progenitors. Evolution 60:1187–1197.

Hendry, A. P. 2004. Selection against migrants contributes to the rapid evolution of ecologically dependent reproductive isolation. Evolutionary Ecology Research 6:1219–1236.

Hereford, J. 2009. A quantitative survey of local adaptation and fitness trade-offs. American Naturalist 173:579–588.

Hitchings, S. P., and T. J. C. Beebee 1998. Loss of genetic diversity and fitness in common toad (Bufo bufo) populations isolated by inimical habitat. Journal of Evolutionary Biology 11:269–283.

Holt, R. D., T. M. Knight, and M. Barfield 2004. Allee effects, immigration, and the evolution of species' niches. The American Naturalist 163:253–262.

Hooftman, D. A. P., J. G. B. Oostermeijer, M. M. J. Jacobs, and H. C. M. Den Nijs 2005. Demographic vital rates determine the performance advantage of crop-wild hybrids in lettuce. Journal of Applied Ecology 42:1086–1095.

Hooftman, D. A. P., M. J. DeJong, J. G. B. Oostermeijer, and H. C. M. Den Nijs 2007. Modelling the long-term consequences of crop-wild relative hybridization: a case study using four generations of hybrids. Journal of Applied Ecology 44:1035–1045.

Hooftman, D. A. P., A. J. Flavell, H. Jansen, J. C. M. den Nijs, N. H. Syed, A. P. Sørensen, P. Orozco-ter Wengel et al. 2011. Locus-dependent selection in crop-wild hybrids of lettuce under field conditions and its implication for GM crop development. Evolutionary Applications 4:648–659.

Hovick, S. M., and K. D. Whitney 2014. Hybridisation is associated with increased fecundity and size in invasive taxa: meta-analytic support for the hybridisation-invasion hypothesis. Ecology Letters 17:1464–1477.

Hovick, S. M., L. G. Campbell, A. A. Snow, and K. D. Whitney 2012. Hybridization alters early life-history traits and increases plant colonization success in a novel region. American Naturalist 179:192–203.

Hyatt, L. A., and S. Araki 2006. Comparative population dynamics of an invading species in its native and novel ranges. Biological Invasions 8:261–275.

Jordan, N., D. A. Mortensen, D. M. Prenzlow, and K. C. Cox 1995. Simulation analysis of crop-rotation effects on weed seedbanks. American Journal of Botany 82:390–398.

Kinnison, M. T., and G. H. Jr Nelson 2007. Eco-evolutionary conservation biology: contemporary evolution and the dynamics of persistence. Functional Ecology 21:444–454.

Kinnison, M. T., M. J. Unwin, and T. P. Quinn 2008. Eco-evolutionary vs. habitat contributions to invasion in salmon: experimental evaluation in the wild. Molecular Ecology 17:405–414.

Kirkpatrick, M., and N. H. Barton 1997. Evolution of a species range. American Naturalist 150:1–23.

Klinger, T., and N. C. Ellstrand 1994. Engineered genes in wild populations – fitness of weed-crop hybrids of Raphanus sativus. Ecological Applications 4:117–120.

de Kroon, H., J. van Groenendael, and J. Ehrlen 2000. Elasticities: a review of methods and model limitations. Ecology 81:607–618.

Lee, T. N., and A. A. Snow 1998. Pollinator preferences and the persistence of crop genes in wild radish populations (*Raphanus raphanistrum*, Brassicaceae). American Journal of Botany **85**:333–339.

Lenormand, T. 2002. Gene flow and the limits to natural selection. Trends in Ecology and Evolution **17**:183–189.

Mazer, S. J., A. A. Snow, and M. L. Stanton 1986. Fertilization dynamics and parental effects upon fruit development in *Raphanus raphanistrum* – consequences for seed size variation. American Journal of Botany **73**:500–511.

Mercer, K. L., D. L. Wyse, and R. G. Shaw 2006. Effects of competition on the fitness of wild and crop-wild hybrid sunflower from a diversity of wild populations and crop lines. Evolution **60**:2044–2055.

Mertens, S. K., F. van den Bosch, and J. A. P. Heesterbeek 2002. Weed populations and crop rotations: Exploring dynamics of a structured periodic system. Ecological Applications **12**:1125–1141.

Miriti, M. N., S. J. Wright, and H. F. Howe 2001. The effects of neighbors on the demography of a dominant desert shrub (*Ambrosia dumosa*). Ecological Monographs **71**:491–509.

Morris, W. F., and D. F. Doak 2004. Buffering of life histories against environmental stochasticity: accounting for a spurious correlation between the variabilities of vital rates and their contributions to fitness. American Naturalist **163**:579–590.

Novack, S. J., and R. N. Mack 2005. Genetic bottlenecks in alien species: influences on mating systems and introduction dynamics. In: D. F. Sax, J. J. Stachowicz, and S. G. Gaines, eds. Species Invasions: Insights into Ecology, Evolution and Biogeography. Sinauer, Sunderland.

Nuismer, S. L., A. MacPherson, and E. B. Rosenblum 2012. Crossing the threshold: gene flow, dominance and the critical level of standing genetic variation required for adaptation to novel environments. Journal of Evolutionary Biology **25**:2665–2671.

Owart, B. R., J. Corbi, J. M. Burke, and J. M. Dechaine 2014. Selection on crop-derived traits and QTL in sunflower (*Helianthus annuus*) crop-wild hybrids under water stress. PLoS One **9**:e102717.

Panetsos, C. A., and H. G. Baker 1967. The origin of variation in wild *Raphanus sativus* (Cruciferae) in California. Genetica **38**:243–274.

Perrins, J., M. Williamson, and A. Fitter 1992. Do annual weeds have predictable characters? Acta Oecologica, International Journal of Ecology **13**:517–533.

Pilson, D., and H. R. Prendeville 2004. Ecological effects of transgenic crops and the escape of transgenes into wild populations. Annual Review of Ecology, Evolution, and Systematics **35**:149–174.

Reznick, D. N., and H. Bryga 1987. Life-history evolution in guppies (*Poecilia reticulata*): 1. Phenotypic and genetic changes in an introduction experiment. Evolution **41**:1370–1385.

Rieseberg, L. H., O. Raymond, D. M. Rosenthal, Z. Lai, K. Livingstone, T. Nakazato, J. L. Durphy et al. 2003. Major ecological transitions in wild sunflowers facilitated by hybridization. Science **301**:1211–1216.

Roff, D. A. 1992. The Evolution of Life Histories: Theory and Analysis, Chapman and Hall, New York.

Simpson, G. G. 1944. Tempo and Mode in Evolution. Columbia University Press, New York.

Slatkin, M. 1987. Gene flow and the geographic structure of natural populations. Science **236**:787–792.

Snow, A. A., K. L. Uthus, and T. M. Culley 2001. Fitness of hybrids between weedy and cultivated radish: implications for weed evolution. Ecological Applications **11**:934–943.

Snow, A. A., D. Pilson, L. H. Rieseberg, M. J. Paulsen, N. Pleskac, M. R. Reagon, D. E. Wolf et al. 2003. A Bt transgene reduces herbivory and enhances fecundity in wild sunflowers. Ecological Applications **13**:279–286.

Snow, A. A., D. A. Andow, P. Gepts, E. M. Hallerman, A. Power, J. M. Tiedje, and L. L. Wolfenbarger 2005. Genetically engineered organisms and the environment: current status and recommendations. Ecological Applications **15**:377–404.

Snow, A. A., and L. G. Campbell 2005. Can feral radishes become weeds? In: J. Gressel, ed. Crop ferality and volunteerism. Boca Raton, FL, USA: CRC Press, 193–208.

Snow, A. A., T. M. Culley, L. G. Campbell, S. G. Hegde, and N. C. Ellstrand 2010. Long-term persistence of crop alleles in weed populations. New Phytologist **186**:537–548.

Stearns, S. C. 1992. The Evolution of Life Histories. Oxford University Press, New York.

Teitel, Z. 2014. The Effect of Climate Stress on Demographic Vital Rates and Seed Dormancy of a Hybridizing Weed. Department of Chemistry & Biology, Ryerson University, Toronto, ON.

Teitel, Z., A. E. Laursen, and L. G. Campbell. Germination rates of weedy radish populations (*Raphanus* spp.) altered by crop-wild hybridization, not human-mediated changes to soil moisture. Weed Research doi: 10.1111/wre.12194 [Epub ahead of print]

Warwick, S. I., and A. Francis 2005. The biology of Canadian weeds. 132. *Raphanus raphanistrum* L. Canadian Journal of Plant Science **85**:709–733.

Weiner, J., L. G. Campbell, J. Pino, and L. Echarte 2009. The allometry of reproduction within plant populations. Journal of Ecology **97**:1220–1233.

Whitney, K. D., R. A. Randell, and L. H. Rieseberg 2006. Adaptive introgression of herbivore resistance traits in the weedy sunflower *Helianthus annuus*. American Naturalist **167**:794–807.

Whitton, J., D. E. Wolf, D. M. Arias, A. A. Snow, and L. H. Rieseberg 1997. The persistence of cultivar alleles in wild populations of sunflowers five generations after hybridization. Theoretical and Applied Genetics **95**:33–40.

Yoshida, T., L. E. Jones, S. P. Ellner, G. F. Fussmann, and N. G. Jr Hairston 2003. Rapid evolution drives ecological dynamics in a predator–prey system. Nature **424**:303–306.

Appendix 1

Stage transitions and mortality over a growing season (2006) for artificially selected populations of *Raphanus raphanistrum* and advanced generation hybrids of *R. raphanistrum* × *R. sativus* grown in a common garden at the University of Michigan Biological Station, Pellston, MI, USA. Populations were selected for either early-flowering (Early) or long-leafed (Large) and compared to randomly mated control populations. Each selection treatment was replicated three times. Each value represents the number (proportion) of individuals that survived to a given

Biotype	Selection treatment	Selection replicate	Dormant	Died before flowering	Survived to flower	Average number of seeds per plant (SE)
Wild	Control	1	14	1	15	396.60 (119.02)
		2	16	1	13	141.88 (78.42)
		3	17	2	11	560.71 (278.98)
	Early	1	18	1	11	130.51 (50.88)
		2	10	2	18	358.29 (126.85)
		3	20	2	8	398.47 (289.79)
	Large	1	9	10	11	431.89 (240.28)
		2	8	9	13	288.22 (96.74)
		3	12	6	12	518.25 (287.66)
Hybrid	Control	1	12	5	13	185.34 (64.99)
		2	14	7	9	240.58 (122.85)
		3	11	4	15	173.94 (38.23)
	Early	1	9	4	17	64.37 (25.82)
		2	12	3	15	102.10 (37.37)
		3	12	8	10	176.29 (74.88)
	Large	1	14	9	7	192.84 (54.98)
		2	17	4	9	952.90 (319.28)
		3	15	7	8	408.79 (169.75)

demographic stage by October 2006 classified as dormant (did not emerge), died before flowering (germinated but did not flower), survived to flower (reproductive), and average number of seeds per plant. Populations were seeded at a density of 30 seeds in three replicate populations of each biotype by selection treatment by selection replicate combination.

Appendix 2

Estimates of elasticity in vital rates measured during a growing season (2006) in artificially selected populations of *Raphanus raphanistrum* and advanced generation hybrids of *R. raphanistrum* × *R. sativus* grown in a common garden at the University of Michigan Biological Station, Pellston, MI, USA.

Analytical methods

Elasticities, which describe how λ proportionally changes with changes to vital rates, were calculated for every individual vital rate and treatment combination in MATLAB by dividing each matrix by its dominant eigenvalue and weighting it by its sensitivity. Population growth rates and elasticities were calculated for individual replicates before averaging them across treatment combinations.

Elasticity analyses can be used to determine which life-history transitions have the greatest effect on population growth rates. This information has been used to develop weed management strategies for key 'choke points' that contribute most to population growth (e.g., (Jordan et al. 1995; Parker 2000; Mertens et al. 2002; Hyatt and Araki

Table A2. Elasticity (e, ±SE) of *Raphanus raphanistrum* (wild) and *R. raphanistrum* × *sativus* (hybrid) population growth rate (λ) to lower level demographic parameters after evolution in response to one of three artificial selection treatments.

Biotype and selection treatment	Elasticities of λ to demographic parameters			
	Dormancy	Germination	Flowering	Fecundity
Wild				
Control	0.036 ± 0.006	0.32 ± 0.002	0.32 ± 0.002	0.32 ± 0.002
Early	0.041 ± 0.01	0.32 ± 0.004	0.32 ± 0.004	0.32 ± 0.004
Large	0.020 ± 0.002	0.33 ± 0.0005	0.33 ± 0.0005	0.33 ± 0.0005
Hybrid				
Control	0.033 ± 0.003	0.32 ± 0.001	0.32 ± 0.001	0.32 ± 0.001
Early	0.037 ± 0.003	0.32 ± 0.001	0.32 ± 0.001	0.32 ± 0.001
Large	0.039 ± 0.004	0.32 ± 0.002	0.32 ± 0.002	0.32 ± 0.002

2006). Here, we use elasticity analyses to determine how changes in recruitment, survival, and fecundity could affect the relative invasiveness of early versus large genotypes (de Kroon et al. 2000; Caswell 2007).

Results

Based on elasticity measures, population growth was equally responsive to small changes in germination, survival to flowering, and fecundity for wild and hybrid genotypes under control, large leaf, and early flowering selection treatments and less responsive to small changes in seed bank dynamics (see Table A2 below). Whereas population growth was equally responsive to small changes in hybrid seed bank dynamics across selection treatments, population growth of early flowering wild plants was more responsive to changes in seed bank dynamics than control, which was more, in turn, more responsive than large-leaf selected treatments.

3

Proximity to agriculture is correlated with pesticide tolerance: evidence for the evolution of amphibian resistance to modern pesticides

Rickey D. Cothran,[1] Jenise M. Brown[1,*] and Rick A. Relyea[1]

1 Department of Biological Sciences and Pymatuning Laboratory of Ecology, University of Pittsburgh, Pittsburgh, PA, USA
* Present address: Department of Integrative Biology, University of South Florida, Tampa, FL, 33620, USA

Keywords
adaptation, ecotoxicology, life history evolution

Correspondence
Rickey D. Cothran, Department of Biological Sciences, University of Pittsburgh, 4249 Fifth Ave., Pittsburgh, PA 15260, USA.

e-mail: rdc28@pitt.edu

Abstract

Anthropogenic environmental change is a powerful and ubiquitous evolutionary force, so it is critical that we determine the extent to which organisms can evolve in response to anthropogenic environmental change and whether these evolutionary responses have associated costs. This issue is particularly relevant for species of conservation concern including many amphibians, which are experiencing global declines from many causes including widespread exposure to agrochemicals. We used a laboratory toxicity experiment to assess variation in sensitivity to two pesticides among wood frog (*Lithobates sylvaticus*) populations and a mesocosm experiment to ascertain whether resistance to pesticides is associated with decreased performance when animals experience competition and fear of predation. We discovered that wood frog populations closer to agriculture were more resistant to a common insecticide (chlorpyrifos), but not to a common herbicide (Roundup). We also found no evidence that this resistance carried a performance cost when facing competition and the fear of predation. To our knowledge, this is the first study demonstrating that organophosphate insecticide (the most commonly applied class of insecticides in the world) resistance increases with agricultural land use in an amphibian, which is consistent with an evolutionary response to agrochemicals.

Introduction

Global change often poses a major challenge for organisms because they must either move to regions that have more favorable environments or adapt to the novel conditions (Palumbi 2001; Meyers and Bull 2002). The use of agrochemicals, including pesticides, is one type of global change to which an increasing number of species are exposed as more land is being used for intensive agriculture (LeNoir et al. 1999; Hayes et al. 2010). Although we have made much progress in addressing the ecological consequences of pesticide exposure (Relyea and Hoverman 2006), the evolutionary consequences are poorly understood. The vast majority of evolutionary investigations are restricted to studies of target species, such as mosquitoes and crop pests, because evolved resistance poses economic and health concerns (Mallet 1989; Rosenheim et al. 1996). These

studies have demonstrated that invertebrate pest species often evolve resistance, but pesticide resistance sometimes carries a fitness cost that may reduce the health of populations even after exposure to pesticides has ceased (Carrière et al. 1994; Coustau and Chevillon 2000). However, fitness costs are not always detected (Arnaud and Haubruge 2002; Bielza et al. 2008; Lopes et al. 2008), and studies reporting no costs may be false negatives because costs of resistance may only emerge when the stress of the pesticides is combined with natural stressors (Coors and De Meester 2008; Hardstone et al. 2009).

There is a growing awareness that nontarget species often experience collateral damage from pesticides, often resulting in death or sublethal effects on behavior, physiology, or endocrinology (Weis et al. 2001; Hayes et al. 2010; Jansen et al. 2011a; Tuomainen and Candolin 2011). In nontarget species, however, we know very little about evolved resistance (Jansen et al. 2011b). Moreover, studies on vertebrate species are rare (but see Boyd et al. 1963; Vinson et al. 1963), and we have no information for some of the most commonly applied pesticides (e.g. organophosphate insecticides).

Of the many taxonomic groups that are affected by global change, amphibians are a group that is experiencing global declines, with 32% of species threatened and 43% of species experiencing declines (Stuart et al. 2004). The causes of these declines are diverse, including habitat loss, disease, and introduced species. In some locations, these declines appear to be related to pesticide exposure and it is becoming increasingly clear that these stressors are more lethal when combined (Wake 1991; LeNoir et al. 1999; Stuart et al. 2004; Hayes et al. 2010). Only a few studies have addressed the impacts of pesticides on amphibians from an evolutionary perspective. Recently, a phylogenetic signal of pesticide sensitivity in amphibians was found for the organochlorine pesticide endosulfan (Hammond et al. 2012). While this study demonstrates that characteristics common to amphibian families (e.g. conserved physiology within clades) can predict sensitivity to endosulfan, studies at the individual and population levels are necessary to understand contemporary responses to pesticides. Population-level studies have provided valuable insights on the tolerance of amphibians to nonpesticide toxicants (Persson et al. 2007; Brady 2012; Hopkins et al. 2012). Existing work on pesticides, restricted to the insecticide carbaryl, shows that amphibian species, populations, and individuals can vary in pesticide resistance and this resistance can carry a fitness cost (Bridges and Semlitsch 2000; Semlitsch et al. 2000; Bridges et al. 2001). However, no connection has been made between variation in resistance and patterns of land use. Only one study, using the insecticide DDT, has compared the resistance of amphibian populations from treated and untreated reference sites (Boyd et al. 1963). While the population from the pristine site was very sensitive, there was no clear mortality pattern for sites that were sprayed directly versus sites that probably experienced indirect exposure (e.g. drift or runoff). In addition, we have no information on whether amphibians can evolve resistance to major groups of pesticides that are commonly used today (e.g. organophosphate insecticides) or whether resistance varies across pesticides that have different modes of action.

We assessed whether wood frog populations vary in their resistance to the most commonly used insecticide (chlorpyrifos) and herbicide [Roundup Original MAX® (active ingredient: glyphosate)] in the agricultural sector (Grube et al. 2011), whether the variation in resistance is associated with variation in agricultural land use, and whether sensitivity to pesticides is associated with adaptive responses to competition and the threat of predation. We collected newly oviposited eggs from nine populations across a land-use gradient and reared them under common-garden conditions prior to using tadpoles in experiments. First, using a standard toxicology experiment, we tested the hypothesis that populations of wood frogs collected from ponds in areas with more agriculture were more tolerant to moderately lethal concentrations of chlorpyrifos and Roundup. Second, using an outdoor mesocosm study, we tested the hypothesis that more resistant populations of wood frogs would have reduced performance (measured as fitness components including survival, larval growth rate, and size at metamorphosis and time to metamorphosis) and that such costs would be more pronounced under stressful conditions (i.e. the presence of predators or high competition).

Methods

Animal collection and husbandry

Experiments were conducted at the University of Pittsburgh's Pymatuning Laboratory of Ecology during 2009 and 2010. Each year, we collected 9–10 recently-laid egg masses (composed of early-stage embryos) from each of nine wood frog populations. Wood frogs typically remain within 300 m of their natal pond and their genetic neighborhood is generally within 1 km of the breeding pond (Berven and Grudzien; Semlitsch 1998, 2000). In our study, the shortest distance between ponds was 4 km, so it is unlikely that animals from different ponds were from the same population. Ponds were chosen so that they varied in the amount of land nearby dedicated to the production of pasture/hay, row crops, and small grains (Fig. 1; see Figures S1 and S2 for a regional view of the ponds). For each pond, we considered both the distance to the nearest agricultural field and the proportion of land used for agriculture within a 500-m radius of the pond. Egg masses were hatched in covered 200-L plastic wading pools (4–5 egg masses per pool) filled with aged, untreated well water. Tadpoles were fed rabbit chow *ad libitum* until used in experiments.

Assessment of pesticide resistance

In the spring of 2010, we conducted a 48-h laboratory toxicity experiment using a completely randomized design to assess each population's sensitivity to chlorpyrifos and Roundup. We used published LC50 values and pilot studies to select nominal concentrations for each pesticide that would be moderately lethal to tadpoles (Sparling and Fellers 2007, 2009; Jones et al. 2009). Wood frogs from each of the nine populations were exposed to three treatments: (i) no-pesticide control, (ii) chlorpyrifos, and (iii) Roundup (active ingredient is glyphosate). For chlorpyrifos, we used a nominal concentration of 1.75 ppm. For glyphosate, we

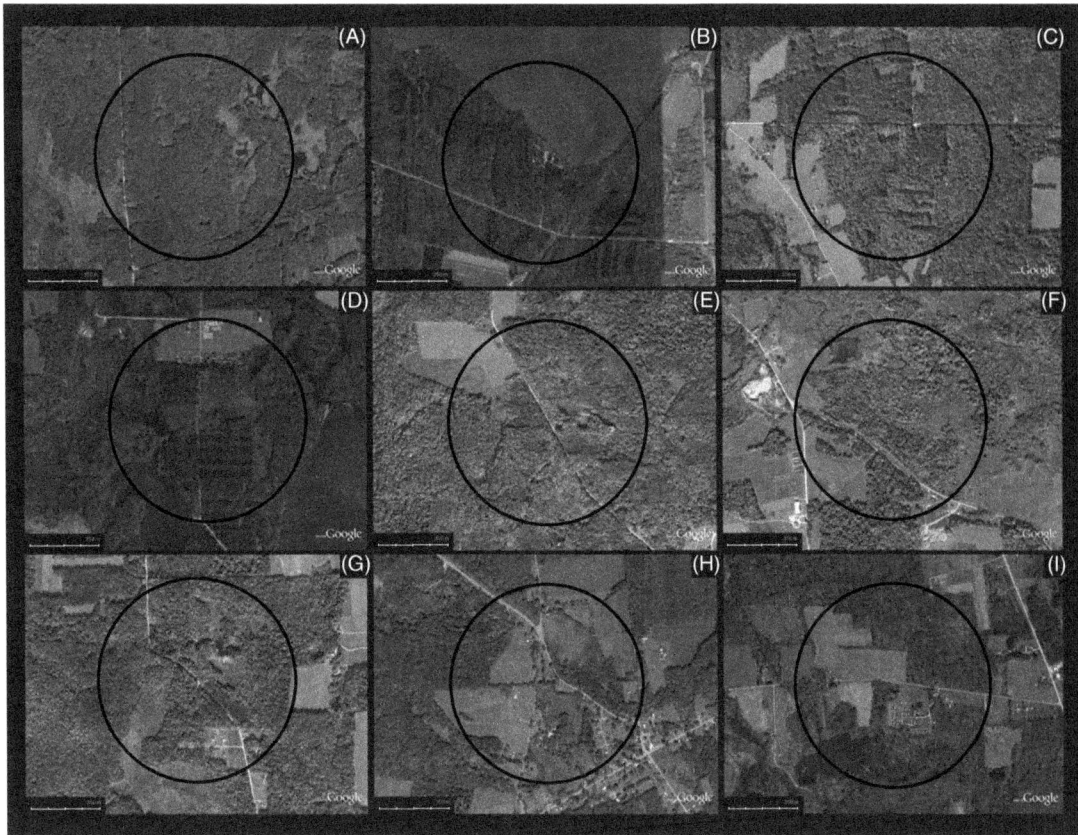

Figure 1 Arial maps showing surrounding land use for ponds. (A) Turkey Track, (B) Relyea, (C) Square, (D) Blackjack, (E) Road, (F) Log, (G) Bowl, (H) Graveyard, (I) Mallard. The circle represents a 500-m radius around each pond.

used a nominal concentration of 2.5 ppm for the first 24 h. This concentration did not cause much mortality over the first 24 h of the experiment so we increased the concentration to 2.75 ppm for the second 24 h. These are the most commonly used insecticide and herbicide in agriculture (Grube et al. 2011) and among the most commonly used pesticides in the study area (2002 estimated usage: chlorpyrifos = ≥ 0.5 kg/km^2; glyphosate = 0.6–2.6 kg/km^2; http://water.usgs.gov/nawqa/pnsp/). The 27 treatment combinations were replicated five times for a total of 135 experimental units.

Working pesticide solutions were prepared using technical grade chlorpyrifos (99.5% purity; Chem Service, West Chester, PA, USA) and Roundup Original MAX® (a formulation commonly used in agriculture). The water was carbon-filtered and UV-irradiated. Because chlorpyrifos is moderately insoluble in water, we used an ethanol carrier. Previous work has shown that the concentration of ethanol used (approximately 0.04% ethanol) does not affect tadpole survival; therefore, we did not include ethanol in the no-pesticide controls or the Roundup solutions (Jones et al. 2009). Solution samples were sent to the Mississippi State Chemical Laboratory

to ascertain actual concentrations. For chlorpyrifos, actual concentrations were 0.438 and 0.584 ppm for the day 1 dose and day 2 dose, respectively (detection limit = 0.01 ppm). For Roundup (i.e. glyphosate), actual concentrations were 3.218 and 3.675 ppm acid equivalents (a.e.) for the day 1 dose and day 2 dose, respectively (detection limit = 0.0075 ppm a.e.).

All embryos used in the experiment hatched within a 16-h period after the last population was collected from the wild. When we initiated the experiment, tadpoles had a mass (mean ± SD) of 51 ± 6 mg (range among populations = 40–60 mg) and were at or near Gosner stage 25 (Gosner 1960). Groups of 10 tadpoles were randomly assigned to 70-mL glass Petri dishes containing pesticide solutions. Individual tadpoles were not exposed to more than one pesticide treatment. Dishes were checked every 4 h and dead tadpoles were removed. After 24 h, the water was changed in all experimental units and the pesticides were reapplied. The experiment was terminated after 48 h. We used an analysis of variance (ANOVA) to examine the effects of pesticide, population, and their interaction on 48-h survival. Tukey's HSD test was used for multiple comparisons. Although population is usually used as a random factor in analyses of ecological data, we chose to use it as a fixed factor in our analyses. We were interested in whether the specific, representative populations that we chose based on surrounding land use differed in their mortality when exposed to pesticides. This required the inclusion of an interaction term to test the null that all populations had similar tolerances to pesticides, which required that both population and pesticide treatment be included as fixed effects in the model (Bennington and Thayne 1994; Hopkins et al. 2012).

Assessment of population responses to predator cues and competition

To assess whether pesticide resistance had costs, we raised the nine populations from hatchlings to metamorphosis in outdoor mesocosms under three different environmental conditions. We used a completely randomized design that crossed the nine populations with three environments: control, predator cues, and competition. We applied the predator cue and competition treatments to assess whether performance costs were only evident under stressful conditions. The 27 treatment combinations were replicated four times for a total of 108 experimental units. The experimental units were outdoor mesocosms constructed of wading pools filled with 90 L of well water, 100 g of dried oak leaves (*Quercus* spp.), and 1 g of rabbit chow to serve as structure and nutrients for periphyton growth. Water containing algae and zooplankton was collected from nearby ponds, predators were removed, and 500-mL aliquots were added to each pool. The local ponds used for collecting zooplankton were not used to collect amphibians for the study. Each pool also received a predator cage made of slotted drain pipe (8 × 10 cm) that was covered on both ends with fiberglass screen. Pools were covered with 60% shade cloth lids to prevent colonization by organisms and prevent the escape of metamorphosing frogs.

Treatments were assigned on May 11, 2009. For each population, initial tadpole mass was based on 20 randomly selected individuals that were not allocated to the experiment. For mesocosms assigned the predator treatment, we added a single larval dragonfly (*Anax junius*) to the predator cage and fed each predator three times a week using approximately 300 mg of wood frog tadpoles from a mixture of the nine populations. Cages in predator-free mesocosms were briefly lifted out of the water to equalize disturbance across mesocosms. To manipulate competition, we stocked 20 tadpoles in the control and predator cue treatments and 40 tadpoles in the competition treatment. Similar manipulations have been used to induce adaptive changes in tadpole traits (Van Buskirk and Relyea 1998; Relyea 2002).

On 4 June, we sampled 10 individuals from each pool to assess larval growth rates. Most tadpoles were at an intermediate developmental stage (mean Gosner stage ± SD = 33 ± 4; Gosner 1960; Werner and Anholt 1993; Jones et al. 2010). Mean individual mass was quantified for each pool, and all animals were returned to their pools. We calculated larval growth rate by subtracting the initial average mass of tadpoles from the average mass on 4 June and dividing by the number of days into the experiment. Growth rates were log-transformed.

On 5 June, we quantified tadpole behavior. Screen lids were removed 30 min prior to the observations. For each pool, we counted the number of tadpoles observed and the number moving during a 60-s period. Ten observations were recorded for each pool (five in the morning and five in the afternoon). From these data, we calculated the mean proportion of tadpoles observed and the mean proportion of tadpoles that were active in each pool.

Metamorphs were first observed on 10 June and we then checked for metamorphs daily. Metamorphs were removed from pools and kept in a plastic container with a small amount of water until full tail resorption (Gosner stage 46). Metamorphs were then euthanized in MS222 and preserved in 10% formalin. On 31 July, we concluded the experiment. Percent survival and mean metamorph mass were calculated for each pool.

We used a multivariate analysis of variance (MANOVA) to assess the effects of environment (control, with predator cues, and with increased competition) and population on tadpole performance (growth rate, time to metamorphosis, mass at metamorphosis, survival, activity, and refuge use), followed by ANOVAS and Tukey's HSD pairwise comparisons. Population was included as a fixed effect in the model because we were interested in whether the specific, representative populations chosen based on their surrounding land use differed in their responses to the environment treatments (Bennington and Thayne 1994).

Assessment of pesticide resistance across an agricultural land-use gradient and its associated costs

We used regressions to assess whether pesticide resistance is correlated with agricultural land use and whether resistance is correlated with the traits that differed among populations in the mesocosm experiment. We did this by calculating two measures of agricultural land: the proportion of land used within 500 m of each pond and the linear distance from a pond to the nearest agricultural field. To derive these measures, we used satellite images from Google Earth Pro and a National Land Cover Data (NLCD) map that uses satellite imagery with a 30-m resolution (http://landcover.usgs.gov/show_data.php?code=PA&state =Pennsylvania) (Vogelmann et al. 2001).

We calculated the proportion of land used for agriculture by defining an area encompassed by a 500-m radius from the center of each pond. We chose a 500-m radius because amphibians typically move <300 m from their natal pond (Semlitsch 1998, 2000) and the genetic neighborhood for wood frogs is generally within approximately 1 km (Berven and Grudzien 1990). Therefore, a 500-m radius is likely to cover the area that juvenile and adult animals would travel and experience pesticides. This provides a conservative measure of pesticide exposure because it does not consider exposures from drift or runoff. In addition, agricultural fields >500 m from small ponds do not have strong effects on aquatic systems (Declerck et al. 2006). We overlaid Google Earth images with a NLCD map of PA, USA and extracted land used for pasture/hay, row crops, and small grains using Photoshop CS5 Extended (Adobe Systems Inc., San Jose, CA, USA). We summed these land-use types and calculated the proportion of land used for agriculture. To improve linearity, the proportion of land used for agriculture was arcsine-transformed.

We measured the linear distance from the center of a pond to the closest agricultural field. We did not differentiate among different types of agriculture, because farmers in our area rotate crops from year to year. An important point is that the landscape around aquatic habitats may affect runoff, which could weaken distance from the point source and proportion of land used for agriculture as indicators of exposure to pesticides (Schriever and Liess 2007). However, the complex life cycle of amphibians means that they could come into contact with pesticides both in water, as larvae, and on land, as juveniles and adults. Thus, topographical and landscape features that reduce runoff into breeding ponds may not eliminate the risk of exposure. To improve linearity, the distance to agriculture was $\log_{10}$-transformed.

Inherent in this regression approach is an inability to pinpoint causation in the observed relationship. Experimental evolutionary approaches (e.g. quantitative genetic breeding designs and experimental evolution) would provide much needed data on the evolutionary implications of pesticide exposure to amphibians (Jansen et al. 2011b). Such approaches are difficult to employ in amphibians (due to complex life cycles and relatively long generation times), but would be invaluable in understanding the evolutionary implications of pesticide exposure to amphibians.

Results

Assessment of pesticide resistance

In the toxicology experiment, we found that survival in the controls was high ($\geq 98\%$), and populations exposed to the pesticides varied widely in their sensitivity (Fig. 2). In addition, patterns of sensitivity were not consistent

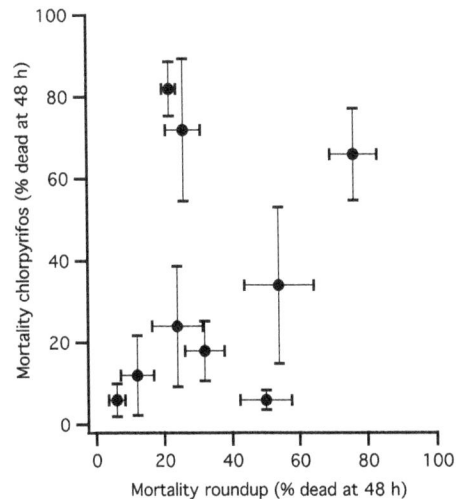

Figure 2 Variation among populations in sensitivity to pesticides. Results are 48-h mortality estimates for the insecticide chlorpyrifos and the herbicide Roundup Original MAX®. Data represent population means ± 1 SEM.

between chlorpyrifos and Roundup (population-by-pesticide interaction $F_{8,90} = 5.031$, $P < 0.001$). There was tremendous population variation in sensitivity to chlorpyrifos ($F_{8,45} = 6.17$, $P < 0.001$), with mortality ranging from 6% to 82%. Populations also varied widely in their sensitivity to Roundup ($F_{8,45} = 10.637$, $P < 0.001$), with mortality ranging from 6% to 76%. There was no correlation between a population's sensitivity to chlorpyrifos and Roundup (Pearson's $r = 0.398$, $P = 0.289$, $n = 9$). In addition, small differences in initial size had no effect on a population's sensitivity to either pesticide (both $P \geq 0.239$).

In considering the agricultural context of each population, we found that populations located closer to agriculture were more resistant to chlorpyrifos than populations located far from agriculture ($P = 0.026$, $R^2 = 0.531$; Fig. 3A). Additionally, populations with a higher proportion of land used for crops tended to be more resistant to chlorpyrifos than populations in more pristine areas ($P = 0.075$, $R^2 = 0.384$; Fig. 3C). Interestingly, sensitivity to Roundup was not correlated with either land-use variable (distance to agriculture: $P = 0.797$, proportion of land used for agriculture: $P = 0.874$; Fig. 3B,D).

Assessment of population responses to predator cues and competition

In the mesocosm experiment, we found that tadpole behavior and life history traits differed among populations ($F_{48,438} = 3.669$, $P < 0.001$) and among environments ($F_{12,138} = 29.553$, $P < 0.001$). However, all populations responded to these environments in the same direction with similar magnitudes (i.e. there was not a population-

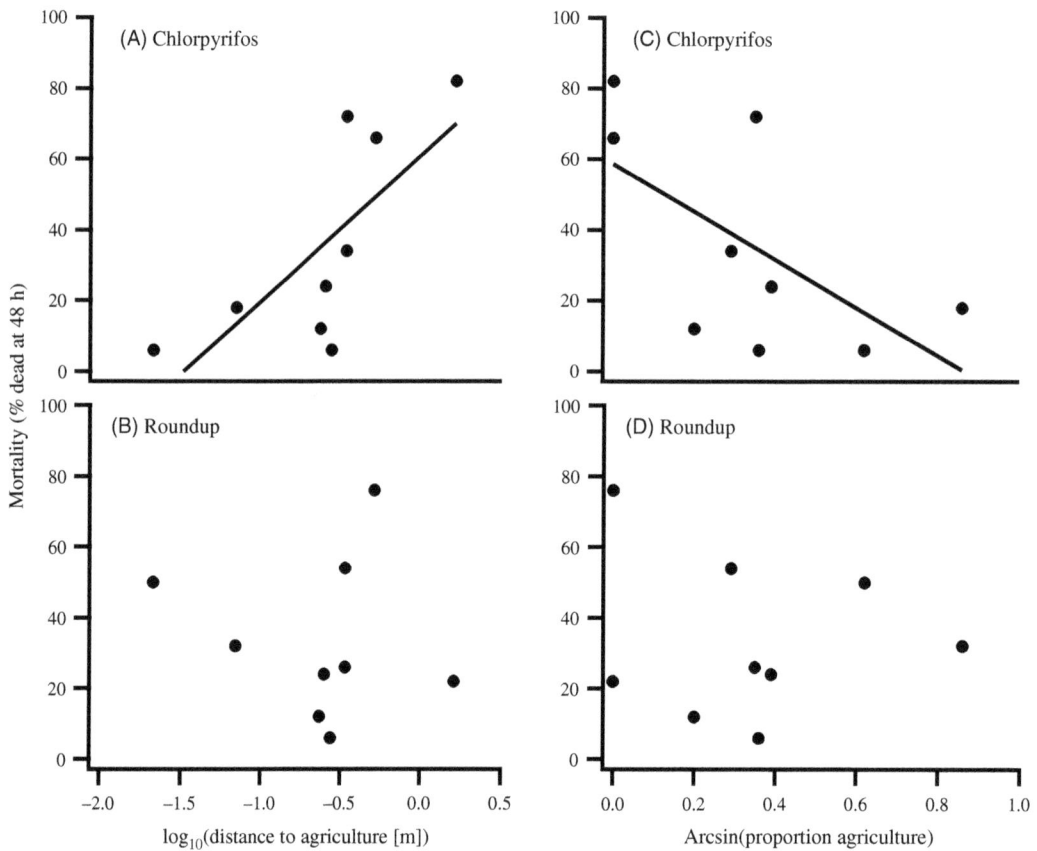

Figure 3 Relationship between the intensity of agricultural land use near ponds and a population's sensitivity to pesticides. Results are presented for (A, C) chlorpyrifos and (B, D) Roundup Original MAX®. Markers represent population means. Proportion of land used for agriculture was calculated within a 500-m radius of each pond. The fitted line for panel (A) was significant ($P = 0.026$), whereas the fitted line for panel (C) was marginally non-significant ($P = 0.075$).

by-environment interaction; $F_{96,438} = 0.9$, $P = 0.732$). The population differences were large for many of these traits (Figure S3). Competition and predator cues had strong effects on tadpole behavior and life history traits (Table 1). Relative to controls and average across populations, predator cues increased tadpole refuge use by 12%, decreased activity by 12%, and increased time to metamorphosis by 7%. Competition increased activity by 10%, decreased larval growth rate by 14%, decreased mass at metamorphosis

by 40%, and increased time to metamorphosis by 12% (Figure S4).

In the mesocosm experiment, we found no evidence that increased resistance was associated with fitness costs regardless of the environmental context. For Roundup, we found no relationship between a population's survival in the laboratory toxicity experiment and any of the wood frog traits expressed in the mesocosm experiment (all $P \geq 0.244$). For chlorpyrifos, we found no relationship

Table 1. ANOVA results from the outdoor mesocosm experiment. Growth rate and behavior (activity and refuge use) were recorded midway in the experiment. Time to and size at metamorphosis and survival were recorded at the end of the experiment.

Source	Growth rate	Time to metamorphosis	Mass at metamorphosis	Activity	Refuge use	Survival
Population	**52.714**	**6.362**	1.763	0.86	**2.218**	1.305
df = 8, 90	**<0.001**	**<0.001**	0.098	0.554	**0.036**	0.255
Environment	**120.789**	**87.234**	**225.414**	**35.962**	**6.817**	1.42
df = 8, 90	**<0.001**	**<0.001**	**<0.001**	**<0.001**	**0.002**	0.248
Population-by-environment	1.351	0.886	0.504	0.903	0.423	0.882
df = 16, 90	0.191	0.587	0.937	0.569	0.972	0.591

Bold values are statistically significant.

between a population's survival in the laboratory toxicity experiment and either the tadpoles' growth rate or the number of tadpoles observed (growth rate: $P = 0.853$; percentage of tadpoles observed: $P = 0.902$). However, we did find that more resistant populations metamorphosed up to 4 day quicker ($P = 0.01$, $R^2 = 0.637$; Fig. 4).

Discussion

We found that wood frog resistance to chlorpyrifos and Roundup was quite variable across populations. For chlorpyrifos, this variation was associated with proximity of the population to agriculture; populations closer to agriculture

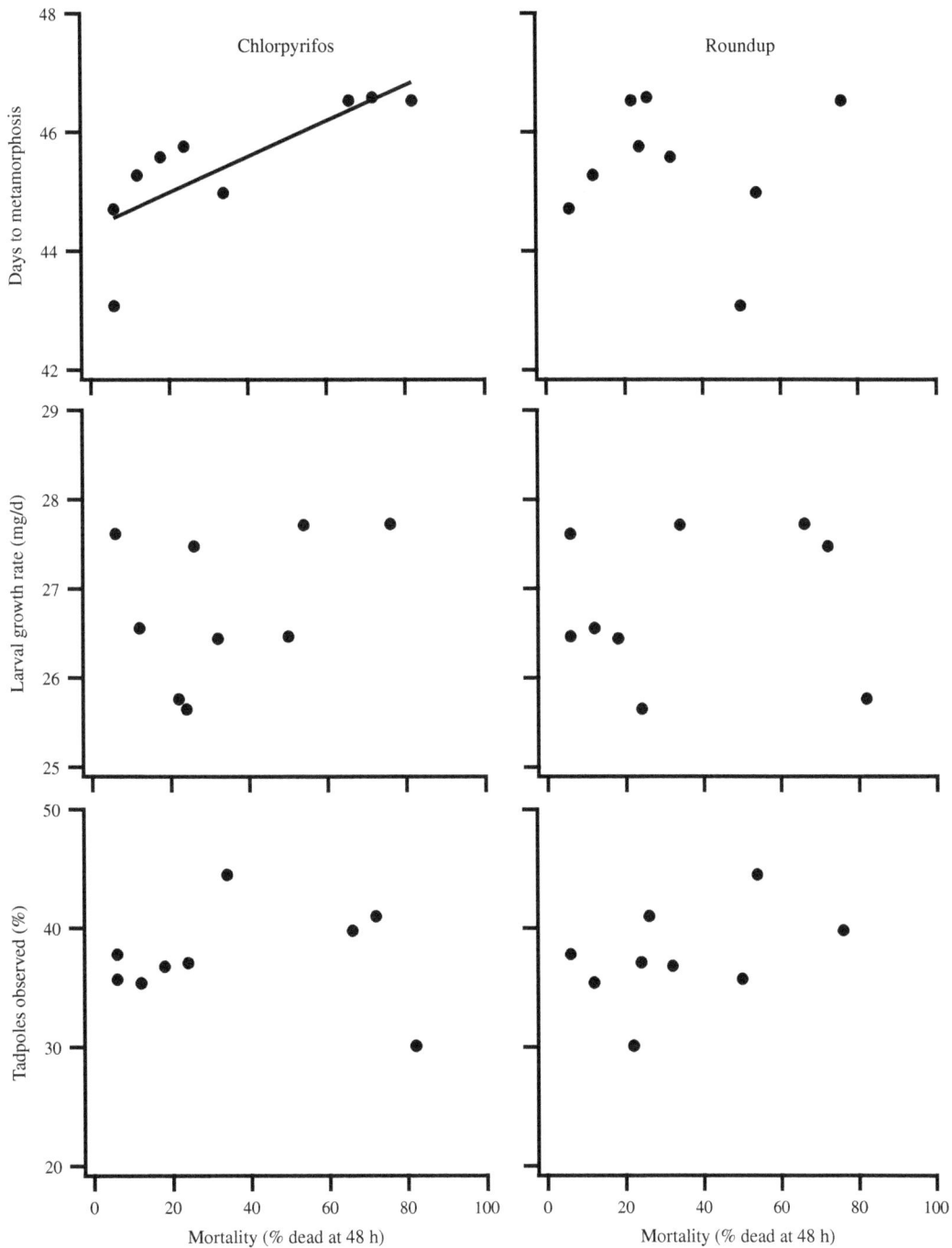

Figure 4 Relationship between a population's sensitivity to pesticides and life history and behavior. We present results for the three traits that differed among populations [days to metamorphosis, larval growth rate (mg/day), and proportion of tadpoles observed during behavior trials]. Data represent population means.

had higher survival than populations farther from agriculture. A similar pattern, although not statistically significant, was found when we used the proportion of land containing agriculture within a 500-m radius. In the mesocosm experiment, we found that predators induced lower activity and increased refuge use in wood frogs. This led to longer times to metamorphosis. We also found that higher competition, in the form of increased tadpole density, induced higher activity and decreased larval growth rate. This led to longer times to and smaller mass at metamorphosis. These results are consistent with previous studies that have addressed behavioral and life history responses to predator cues and increased competition (e.g. Relyea 2002). More importantly, there was population-level variation in these traits. However, we found no evidence that populations with greater pesticide resistance paid a performance cost for resistance in pesticide-free environments.

The chlorpyrifos results are consistent with an evolutionary response to insecticide exposure. Chlorpyrifos is the most commonly used insecticide in the agricultural sector with 3–4 million kg applied annually in the United States (Grube et al. 2011). Moreover, in our study area, chlorpyrifos is one of the most commonly used insecticides, and other insecticides that dominate the U.S. market (i.e. organophosphates) share the same mode of action as chlorpyrifos (Grube et al. 2011; http://water.usgs.gov/nawqa/pnsp/). The evolution of resistance could have occurred via selection imposed by chlorpyrifos or due to cross-resistance (i.e. when resistance to one pesticide confers resistance to other pesticides as well) to many carbamate and organophosphate insecticides that share the same mode of action. Cross-resistance to pesticides is commonly observed in pest species, and we have recently confirmed this to be the case in wood frogs (Georghiou 1972; Hua et al. 2013). Our results must be met with caution because we do not have pesticide concentration data for the ponds used in our study. It would be ideal to have long-term chlorpyrifos exposure data for each pond, to confirm past exposures that selected for tolerance and the current exposure risk faced by wood frogs. Given that we did not have this information, agricultural land use is our best predictor of exposure to pesticides and is correlated with sensitivity to pesticides and reduced genetic variation in aquatic invertebrates (Coors et al. 2009).

We also found variation among populations in sensitivity to Roundup, but it was not correlated with agricultural land use. There are several potential explanations. First, a variety of herbicides are used in our study area (see pesticide usage maps: http://water.usgs.gov/nawqa/pnsp/usage/maps/compound_listing.php?year=02) and, unlike insecticides that often share a single mode of action, these herbicides have diverse modes of action. Thus, any evolution of resistance to other herbicides is unlikely to confer cross-resistance to Roundup (Georghiou 1972). As a result, the relationship between agricultural land use and exposure to Roundup may be weak. Second, Roundup is the second most commonly used herbicide in the home and garden sector and the industry, commercial, and government sector and is commonly sprayed to control the growth of noxious plants (Giesy et al. 2000; Grube et al. 2011). Thus, populations far from agricultural land may still be exposed to Roundup due to nonagricultural exposures. Organophosphate pesticides are also commonly used in nonagricultural sectors; however, the agricultural sector by far uses the most herbicides and insecticides (herbicides: 83% of total use is agricultural; insecticides: 70%; Grube et al. 2011). Finally, we exposed tadpoles for a short period, at relatively high concentrations, and in an artificial environment, which may underestimate or overestimate each population's tolerance to pesticides (Relyea and Mills 2001; Suchail et al. 2001; Weis et al. 2001). Field experiments over relatively long periods would be valuable in confirming whether patterns of tolerance found in the current study hold when tadpoles are exposed to concentrations of pesticides found in natural ponds (Persson et al. 2007; Brady 2012).

We found no evidence that resistance to pesticides reduces an individual's performance in pesticide-free environments. We found that populations that were more resistant to chlorpyrifos metamorphosed faster than populations that were more sensitive to the pesticide; however, this effect was small and did not result in a smaller size at metamorphosis (which is much more important to fitness; Smith 1987). It is likely that the challenges (competitors and fear of predators) experienced by tadpoles in our mesocosm experiment were more benign than those in nature where tadpoles must contend with a suite of competitors and the combined stress of competition and predation. Therefore, it is possible that under more adverse conditions a performance cost to tolerance may have been uncovered (Coors and De Meester 2008; Hardstone et al. 2009). However, the presence of predator cues and more competitors was sufficient enough to cause changes in life history traits and behavior that are indicative of predator-induced and competitor-induced stress. Our results suggest that there never was a cost of resistance to chlorpyrifos or that costs were originally present, but populations evolved secondary mechanisms that reduce the costs. Reduced fitness costs of pesticide resistance can stem from compensatory mutations including allele replacement or the evolution of modifiers that reduce fitness costs (Guillemaud et al. 1998). Alternatively, fitness tradeoffs associated with local adaptation (e.g. to pesticide exposure) may not be as common as once thought (Hereford 2009). Such costs are expected to occur due to rampant pleiotropy (Fisher 1958). However, empirical estimates of pleiotropy are often weak and it is becoming increasingly clear that phenotypic evolution is often modular in nature, which would result in lower costs for expressing alleles that confer pesticide resistance (Wagner and Zhang 2011).

Here, we show that while pesticides can have a number of harmful effects on nontarget species, amphibian populations may be able to evolve resistance to pesticides without costs to how they respond to competitors and predators. However, researches on the genetic mechanisms that confer resistance to pesticides are necessary because it is possible that maternal effects and epigenetic mechanisms contributed to variation among wood frog populations in tolerance to pesticides (Morgan et al. 2007; Klerks et al. 2011). Interestingly, even if this is the case, maternal effects can themselves be adaptive and epigenetic inheritance can provide future generations the capacity to tolerate pesticide exposures (Mousseau and Fox 1998; Vandegehuchte and Janssen 2011). Our results appear to be the first example showing that amphibian populations near agriculture are more tolerant to pesticides, a result that is consistent with the evolution of pesticide resistance.

Acknowledgements

This research was funded by a NSF grant awarded to RAR. JB was supported on a NSF REU award. We thank B. Bailey, K. Chapman, J. Hua, J. Hammond, K. Henderson, D. Jones, A. Stiff, and A. Stoler for assistance on the project. We are grateful to two anonymous reviewers that provided constructive criticisms that improved the manuscript.

Literature cited

Arnaud, L., and E. Haubruge 2002. Insecticide resistance enhances male reproductive success in a beetle. Evolution 56:2435–2444.

Bennington, C., and W. Thayne 1994. Use and misuse of mixed model analysis of variance in ecological studies. Ecology 75:717–722.

Berven, K. A., and T. A. Grudzien 1990. Dispersal in the wood frog (Rana sylvatica): implications for genetic population structure. Evolution 44:2047–2056.

Bielza, P., V. Quinto, C. Gravalos, J. Abellan, and E. Fernandez 2008. Lack of fitness costs of insecticide resistance in the western flower thrips (Thysanoptera: Thripidae). Journal of Economic Entomology 101:499–503.

Boyd, C., S. Vinson, and D. Ferguson 1963. Possible DDT resistance in two species of frogs. Copeia 1963:426–429.

Brady, S. P. 2012. Road to evolution? Local adaptation to road adjacency in an amphibian (Ambystoma maculatum). Scientific Reports 2:235.

Bridges, C., and R. Semlitsch 2000. Variation in pesticide tolerance of tadpoles among and within species of ranidae and patterns of amphibian decline. Conservation Biology 14:1490–1499.

Bridges, C. M., R. D. Semlitsch, and A. Price 2001. Genetic variation in insecticide tolerance in a population of southern leopard frogs (Rana sphenocephala): implications for amphibian conservation. Copeia 2001:7–13.

Carrière, Y., J. Deland, D. Roff, and C. Vincent 1994. Life-history costs associated with the evolution of insecticide resistance. Proceedings of the Royal Society of London B: Biological Sciences 258:35–40.

Coors, A., and L. De Meester 2008. Synergistic, antagonistic and additive effects of multiple stressors: predation threat, parasitism and pesticide exposure in Daphnia magna. Journal of Applied Ecology 45:1820–1828.

Coors, A., J. Vanoverbeke, T. De Bie, and L. De Meester 2009. Land use, genetic diversity and toxicant tolerance in natural populations of Daphnia magna. Aquatic Toxicology 95:71–79.

Coustau, C., and C. Chevillon 2000. Resistance to xenobiotics and parasites: can we count the cost? Trends in Ecology and Evolution 15:378–383.

Declerck, S., T. De Bie, D. Erken, H. Hampel, S. Schrijvers, J. Van Wichelen, V. Gillard, et al. 2006. Ecological characteristics of small farmland ponds: associations with land use practices at multiple spatial scales. Biological Conservation 131:523–532.

Fisher, R. A. 1958. The Genetical Theory of Natural Selection: A Complete Variorum Edition. J. H. Bennett, ed. Oxford University Press, New York.

Georghiou, G. 1972. The evolution of resistance to pesticides. Annual Review of Ecology and Systematics 3:133–168.

Giesy, J. P., S. Dobson, and K. R. Solomon 2000. Ecotoxicological risk assessment for Roundup® herbicide. Reviews of Environmental Contamination and Toxicology 67:35–120.

Gosner, K. L. 1960. A simplified table for staging anuran embryos and larvae with notes on identification. Herpetologica 16:183–190.

Grube, A., D. Donaldson, and T. Kiely 2011. Pesticides Industry Sales and Usage: 2006 and 2007 Market Estimates. US Environmental Protection Agency, Washington, DC.

Guillemaud, T., T. Lenormand, D. Bourguet, C. Chevillon, N. Pasteur, and M. Raymond 1998. Evolution of resistance in Culex pipiens: allele replacement and changing environment. Evolution 52:443–453.

Hammond, J. I., D. K. Jones, P. R. Stephens, and R. A. Relyea 2012. Phylogeny meets ecotoxicology: evolutionary patterns of sensitivity to a common insecticide. Evolutionary Applications 5:593–606.

Hardstone, M. C., C. A. Leichter, and J. G. Scott 2009. Multiplicative interaction between the two major mechanisms of permethrin resistance, kdr and cytochrome P450-monooxygenase detoxification, in mosquitoes. Journal of Evolutionary Biology 22:416–423.

Hayes, T. B., P. Falso, S. Gallipeau, and M. Stice 2010. The cause of global amphibian declines: a developmental endocrinologist's perspective. Journal of Experimental Biology 213:921–933.

Hereford, J. 2009. A quantitative survey of local adaptation and fitness trade-offs. The American Naturalist 173:579–588.

Hopkins, G. R., S. S. French, and E. D. Brodie 2012. Potential for local adaptation in response to an anthropogenic agent of selection: effects of road deicing salts on amphibian embryonic survival and development. Evolutionary Applications, doi:10.1111/eva.12016.

Hua, J., R. Cothran, A. Stoler, and R. Relyea 2013. Cross tolerance in amphibians: wood frog mortality when exposed to three insecticides with a common mode of action. Environmental Toxicology and Chemistry 32:932–936.

Jansen, M., R. Stoks, A. Coors, W. van Doorslaer, and L. De Meester 2011a. Collateral damage: rapid exposure-induced evolution of pesticide resistance leads to increased susceptibility to parasites. Evolution 65:2681–2691.

Jansen, M., A. Coors, R. Stoks, and L. De Meester 2011b. Evolutionary ecotoxicology of pesticide resistance: a case study in Daphnia. Ecotoxicology 20:543–551.

Jones, D. K., J. I. Hammond, and R. A. Relyea 2009. Very highly toxic effects of endosulfan across nine species of tadpoles: lag effects and family-level sensitivity. Environmental Toxicology and Chemistry 28:1939–1945.

Jones, D. K., J. I. Hammond, and R. A. Relyea 2010. Roundup and amphibians: the importance of concentration, application time, and stratification. Environmental Toxicology and Chemistry 29:2016–2025.

Klerks, P. L., L. Xie, and J. S. Levinton 2011. Quantitative genetics approaches to study evolutionary processes in ecotoxicology; a perspective from research on the evolution of resistance. Ecotoxicology 20:513–523.

LeNoir, J., L. McConnell, G. Fellers, T. Cahill, and J. Seiber 1999. Summertime transport of current-use pesticides from California's Central Valley to the Sierra Nevada Mountain Range, USA. Environmental Toxicology and Chemistry 18:2715–2722.

Lopes, P. C., E. Sucena, M. E. Santos, and S. Magalhaes 2008. Rapid experimental evolution of pesticide resistance in *C. elegans* entails no costs and affects the mating system. PLoS ONE 3:e3741.

Mallet, J. 1989. The evolution of insecticide resistance: have the insects won? Trends in Ecology and Evolution 4:336–340.

Meyers, L., and J. Bull 2002. Fighting change with change: adaptive variation in an uncertain world. Trends in Ecology and Evolution 17:551–557.

Morgan, A., P. Kille, and S. Stürzenbaum 2007. Microevolution and ecotoxicology of metals in invertebrates. Environmental Science & Technology 41:1085–1096.

Mousseau, T. A., and C. W. Fox 1998. The adaptive significance of maternal effects. Trends in Ecology and Evolution 13:403–407.

Palumbi, S. 2001. Humans as the world's greatest evolutionary force. Science 293:1786–1790.

Persson, M., K. Rasanen, A. Laurila, and J. Merila 2007. Maternally determined adaptation to acidity in *Rana arvalis*: are laboratory and field estimates of embryonic stress tolerance congruent? Canadian Journal of Zoology 85:832–838.

Relyea, R. 2002. Competitor-induced plasticity in tadpoles: consequences, cues, and connections to predator-induced plasticity. Ecological Monographs 72:523–540.

Relyea, R., and J. Hoverman 2006. Assessing the ecology in ecotoxicology: a review and synthesis in freshwater systems. Ecology Letters 9:1157–1171.

Relyea, R., and N. Mills 2001. Predator-induced stress makes the pesticide carbaryl more deadly to gray treefrog tadpoles (*Hyla versicolor*). Proceedings of the National Academy of Sciences 98:2491–2496.

Rosenheim, J., M. Johnson, R. Mau, S. Welter, and B. Tabashnik 1996. Biochemical preadaptations, founder events, and the evolution of resistance in arthropods. Journal of Economic Entomology 89:263–273.

Schriever, C. A., and M. Liess 2007. Mapping ecological risk of agricultural pesticide runoff. Science of the Total Environment 384:264–279.

Semlitsch, R. 1998. Biological delineation of terrestrial buffer zones for pond-breeding salamanders. Conservation Biology 12:1113–1119.

Semlitsch, R. 2000. Principles for management of aquatic-breeding amphibians. The Journal of Wildlife Management 64:615–631.

Semlitsch, R., C. Bridges, and A. Welch 2000. Genetic variation and a fitness tradeoff in the tolerance of gray treefrog (*Hyla versicolor*) tadpoles to the insecticide carbaryl. Oecologia 125:179–185.

Smith, D. 1987. Adult recruitment in chorus frogs – effects of size and date at metamorphosis. Ecology 68:344–350.

Sparling, D., and G. Fellers 2007. Comparative toxicity of chlorpyrifos, diazinon, malathion and their oxon derivatives to larval *Rana boylii*. Environmental Pollution 147:535–539.

Sparling, D. W., and G. M. Fellers 2009. Toxicity of two insecticides to California, USA, anurans and its relevance to declining amphibian populations. Environmental Toxicology and Chemistry 28:1696–1703.

Stuart, S. N., J. S. Chanson, N. A. Cox, B. E. Young, A. S. L. Rodrigues, D. L. Fischman, and R. W. Waller 2004. Status and trends of amphibian declines and extinctions worldwide. Science 306:1783–1786.

Suchail, S., D. Guez, and L. P. Belzunces 2001. Discrepancy between acute and chronic toxicity induced by imidacloprid and its metabolites in *Apis mellifera*. Environmental Toxicology and Chemistry/SETAC 20:2482–2486.

Tuomainen, U., and U. Candolin 2011. Behavioural responses to human-induced environmental change. Biological Reviews 86:640–657.

Van Buskirk, J., and R. Relyea 1998. Selection for phenotypic plasticity in *Rana sylvatica* tadpoles. Biological Journal of the Linnean Society 65:301–328.

Vandegehuchte, M. B., and C. R. Janssen 2011. Epigenetics and its implications for ecotoxicology. Ecotoxicology 20:607–624.

Vinson, S., C. Boyd, and D. Ferguson 1963. Resistance to DDT in the mosquito fish, *Gambusia affinis*. Science 139:217.

Vogelmann, J., S. Howard, L. Yang, C. Larson, B. Wylie, and N. Van Driel 2001. Completion of the 1990s National Land Cover Data set for the conterminous United States from Landsat Thematic Mapper data and ancillary data sources. Photogrammetric Engineering and Remote Sensing 67:650–662.

Wagner, G. P., and J. Zhang 2011. The pleiotropic structure of the genotype-phenotype map: the evolvability of complex organisms. Nature Reviews Genetics 12:204–213.

Wake, D. 1991. Declining amphibian populations. Science 253:860.

Weis, J. S., G. Smith, T. Zhou, C. Santiago-Bass, and P. Weis 2001. Effects of contaminants on behavior: biochemical mechanisms and ecological consequences. BioScience 51:209–217.

Werner, E. E., and B. R. Anholt 1993. Ecological consequences of the trade-off between growth and mortality rates mediated by foraging activity. The American Naturalist 142:242–272.

4

Epidemiological and evolutionary management of plant resistance: optimizing the deployment of cultivar mixtures in time and space in agricultural landscapes

Frédéric Fabre,[1] Elsa Rousseau,[2,3,4,5,6] Ludovic Mailleret[2,3,4,5] and Benoît Moury[6]

1 UMR 1065 Unité Santé et Agroécologie du Vignoble, INRA, Villenave d'Ornon Cedex, France
2 Biocore Team, INRIA, Sophia Antipolis, France
3 UMR 1355 Institut Sophia Agrobiotech, INRA, Sophia Antipolis, France
4 UMR 7254 Institut Sophia Agrobiotech, Université Nice Sophia Antipolis, Sophia Antipolis, France
5 UMR 7254 Institut Sophia Agrobiotech, CNRS, Sophia Antipolis, France
6 UR 407 Pathologie Végétale, INRA, Montfavet, France

Keywords

evolutionary epidemiology, functional connectivity, heterogeneity of selection, landscape epidemiology, mosaic strategy, qualitative resistance, resistance durability, rotation strategy.

Correspondence

Frédéric Fabre, UMR 1065 Santé et Agroécologie du Vignoble (SAVE), INRA, 71, av. Edouard Bourlaux - CS 20032, 33882 Villenave d'Ornon Cedex, France.

e-mail: frederic.fabre@bordeaux.inra.fr

Abstract

The management of genes conferring resistance to plant–pathogens should make it possible to control epidemics (epidemiological perspective) and preserve resistance durability (evolutionary perspective). Resistant and susceptible cultivars must be strategically associated according to the principles of cultivar mixture (within a season) and rotation (between seasons). We explored these questions by modeling the evolutionary and epidemiological processes shaping the dynamics of a pathogen population in a landscape composed of a seasonal cultivated compartment and a reservoir compartment hosting pathogen year-round. Optimal deployment strategies depended mostly on the molecular basis of plant–pathogen interactions and on the agro-ecological context before resistance deployment, particularly epidemic intensity and landscape connectivity. Mixtures were much more efficient in landscapes in which between-field infections and infections originating from the reservoir were more prevalent than within-field infections. Resistance genes requiring two mutations of the pathogen avirulence gene to be broken down, rather than one, were particularly useful when infections from the reservoir predominated. Combining mixture and rotation principles were better than the use of the same mixture each season as (i) they controlled epidemics more effectively in situations in which within-field infections or infections from the reservoir were frequent and (ii) they fulfilled the epidemiological and evolutionary perspectives.

Introduction

Integrating the principles governing natural ecosystems into crop protection strategies should be a powerful way to face the challenge of doubling crop production in the next four decades while decreasing the environmental impact of agriculture (Tilman 1999). In natural ecosystems, the genetic diversity of hosts limits the spread of epidemics in a wide range of conditions (Ostfeld and Keesing 2012). In agroecosystems, functional diversity in disease resistance also decreases disease spread (Mundt 2002; Garrett et al. 2009). This approach has been particularly successful in the control of powdery mildew in barley and blast disease in rice (Wolfe 1985; Zhu et al. 2000). Limited efficacy has also been often reported (e.g., Garrett et al. 2009).

Disease resistance in plants often results from a molecular relationship governed by a gene-for-gene interaction (Flor 1971). For qualitative resistance genes (R genes) (i.e. genes that almost totally prevent plant infection), the interaction between the product of the R gene of the plant (which has at least two allelic forms: 'resistant' and 'susceptible') and the product of the avirulence gene of the pathogen (which has at least two allelic forms: 'non-adapted' and 'resistance-breaking') determines the resistance or susceptibility of the plant. Over the last decade, the molecular dissection of these interactions for viruses has revealed that (i) one or two nucleotide substitutions in virus avirulence genes are often sufficient to overcome resistance genes (Jenner et al. 2002; Kang et al. 2005; Fraile et al. 2011; Moury et al. 2011) and (ii) these substitutions have variable fitness costs in susceptible plants (Ayme et al. 2006; Janzac et al. 2010).

Pathogen adaptation is a key factor limiting the usefulness of mixtures of susceptible and resistant cultivars. In natural ecosystems, R genes may remain effective for long periods, because stable polymorphism is maintained by mechanisms such as negative frequency-dependent selection (Lewontin 1958; Brown and Tellier 2011; Zhan et al. 2015). Most of these mechanisms have been lost in modern agroecosystems. For example, cultivar selection by growers disrupts plant–pathogen coevolution (Bousset and Chèvre 2013). R genes thus often remain effective for only short periods, particularly for fungi and bacteria (McDonald and Linde 2002), but also for viruses (García-Arenal and McDonald 2003). This results in a classic 'boom-and-bust' cycle, in which cultivars carrying a new resistance gene are widely adopted by farmers, leading to a breakdown of resistance and replaced with new cultivars carrying another R gene. This cyclic system works well if sufficient R genes are available in the genetic resources for the plant species concerned. Unfortunately, R genes, particularly qualitative ones, are rare resources requiring careful, sustainable management.

The interplay between the spatial scale over which an epidemic spreads and that at which host heterogeneity is distributed also limits the usefulness of cultivar mixtures. One key reason for this is the need for control strategies to be applied at the same spatial scale as that of epidemic spread (Dybiec et al. 2004; Gilligan et al. 2007; Gilligan 2008). As many airborne and vectorborne diseases spread over long distance, epidemiologists and agriculture managers are increasingly focusing on the spatial scale of the landscape (Plantegenest et al. 2007; Real and Biek 2007; Parnell et al. 2009, 2010; Skelsey et al. 2010; Papaïx et al. 2011). At this scale, epidemics in individual fields follow two routes of infection, originating within the same field or in other fields (Park et al. 2001). Their relative importance depends on the spatial grain of the distribution of host heterogeneity in the landscape. This spatial grain is a component of landscape connectivity, 'the degree to which the landscape facilitates or impedes movement among resource patches' (Taylor et al. 1993). It remains unclear how landscape connectivity influences the interaction between uninfected and infected individuals (Meentemeyer et al. 2012), thereby affecting epidemics. The role of landscape connectivity in pathogen evolution must also be considered in the management of resistance

sustainability at the landscape scale (Papaïx et al. 2013; Mundt 2014).

We used a model coupling epidemiology and population genetics to study how mixtures of susceptible and resistant cultivars can control epidemics efficiently, over several seasons at the landscape scale, while preventing the emergence of adapted pathogens. Specifically, two management alternatives were considered (Zhan et al. 2015). In the first, the objective is to maximize crop yield (yield strategies). In the second, there are two objectives: to decrease yield loss due to pathogens and to slow the evolution of adapted pathogens (sustainable strategies). We investigated the role of landscape features (connectivity and epidemic intensity before the introduction of the resistant cultivar) and that of the molecular interactions between plant and pathogen. We found that the knowledge of the interaction between these factors is required to strategically associate susceptible and resistant cultivars in mixture and rotation.

Materials and methods

Model overview

The model couples plant epidemics and pathogen population genetics processes during a succession of cropping seasons. First, it describes during a cropping season the dynamics of epidemics in a landscape composed of susceptible (S) and resistant (R) plants and changes in the frequency of a resistance-breaking variant in the pathogen population. It deals with the case of any plant–pathogen for which the within- and between-host dynamics are clearly separated, typically a virus. Second, epidemics in successive cropping seasons are coupled to one another through the interaction of the crop with a reservoir compartment, containing diverse wild plant species that may serve as hosts for the pathogen during the crop-free period and as a source of inoculum for the initiation of infections in the next cropping period. In this landscape, we can consider three routes of infection: (i) between the reservoir and the fields, (ii) between fields, and (iii) within a field. The main variables and parameters of the model are listed in Table 1.

Within-season model of a landscape with only S fields

As a first step, the model was described verbally for a landscape with several fields in which only the S cultivar was sown ('S fields'). This situation also defines the baseline epidemiological situation before the deployment of the R plants. The variable modeled was $I_{S,y}(t)$, the number of plants infected in a S field at time t during year y. Its dynamic is governed by two processes: (i) the mean epidemic intensity in a field during a season (parameter Ω_{int}) and (ii) the relative importance of the three routes of

Table 1. Description of the parameters of the model of their range of variation and of the state variables of the model.

Parameters	Designation (Reference value)	Unit	Sensitivity analyses levels
Ω_{int}	Epidemic intensity before R deployment (in a landscape with only S plants)*,†	Unitless	4 levels: 0.1, 0.3, 0.5, 0.8
Ω_{pfl}	Landscape connectivity before R deployment (in a landscape with only S plants)‡,†	Unitless (vector)	4 levels: 1 (0.05, 0.05, 0.9), 2 (0.05, 0.9, 0.05), 3 (0.9, 0.05, 0.05), 4 (1/3, 1/3, 1/3)
λ	Characteristic of the pathogen reservoir§ (0.5)	Unitless	3 levels: 0.1, 0.5, 0.9
θ	Choice of R gene (defining the frequency of the resistance-breaking variant in S plants)¶	Unitless	5 levels: 10^{-8}, 10^{-6}, 10^{-4}, 10^{-2}, 0.5
n_y	Number of years of resistance deployment (15)	Year	2 levels: 10, 20
n_d	Duration of the cropping season (120)	Day	
n_f	Number of fields in the landscape (400)	Field	
n_p	Number of plants in a field (10^4)	Plant	

State variables	Designation		
$I_{S,y}$	Number of infected plants in a field with the S cultivar during year y	Plant	
$I_{R,y}$	Number of infected plants in a field with the R cultivar during year y	Plant	
$\alpha_{S,y}$	Rate of infection of the S cultivar from the reservoir during year y	Day^{-1}	
$\alpha_{R,y}$	Rate of infection of the R cultivar from the reservoir during year y	Day^{-1}	

*In a landscape with only the susceptible cultivar, the epidemic intensity is the mean frequency of S plants infected in a field during a cropping season.

†The parameters Ω_{int} and Ω_{pfl} define the epidemic dynamics in the landscape before the deployment of the resistant cultivar where epidemics repeat themselves identically every year.

‡$\Omega_{pfl} = (\Omega^1_{pfl}, \Omega^2_{pfl}, 1 - \Omega^1_{pfl} - \Omega^2_{pfl})$. In a landscape with only the susceptible cultivar, Ω^1_{pfl} measures the frequency of infection events originating from the reservoir, Ω^2_{pfl} the frequency of between-field infection events and $1 - \Omega^1_{pfl} - \Omega^2_{pfl}$ the frequency of within-field infection events. Ω_{pfl} defines the connectivity between the elements (fields, reservoir) of the landscape.

§λ represents the degree of decrease in the weighting of the reservoir pathogen load in an exponential moving average setting. Higher λ values result in the faster discounting of older reservoir pathogen loads. The levels of λ account for a wide range of the possible reservoir, with pathogen populations having a half-life of ≈6 months ($\lambda = 0.9$), ≈1 year ($\lambda = 0.5$) and ≈ 6 years ($\lambda = 0.1$).

¶The choice of qualitative R gene by plant breeders determines the number and fitness costs of the nucleotide substitutions that nonadapted pathogens must accumulate in their avirulence gene to overcome the R gene. In turn, these parameters determine the frequency of coexistence of resistance-breaking and nonadapted variants in S plants. The levels of θ account for resistance genes requiring 1 or 2 nucleotide substitutions in the avirulence gene of the pathogen to be broken down. The case of an RNA plant virus is addressed more specifically here. When only one (resp. two) mutation is required, assuming a mutation rate of 10^{-4} and given the distribution of fitness effects of single mutations (Carrasco et al. 2007), θ is likely to be in the range $[10^{-4}, 0.01]$ (resp. $[10^{-8}, 10^{-6}]$). $\theta = 0.5$ represents a situation in which one mutation with a very low (2×10^{-4}) fitness cost is required.

infection (vector of parameter Ω_{pfl}). The vector $\Omega_{pfl} = (\Omega^1_{pfl}, \Omega^2_{pfl}, 1 - \Omega^1_{pfl} - \Omega^2_{pfl})$ captures the infection profile of the landscape, which is dependent on the connectivity between its elements (fields, reservoir). More specifically, Ω^1_{pfl} is the proportion of infections originating from the reservoir, Ω^2_{pfl} is the proportion of between-field infections, and $1 - \Omega^1_{pfl} - \Omega^2_{pfl}$ is the proportion of within-field infections. These parameters are an intuitive and powerful way to characterize agricultural landscapes from an epidemiological point of view before resistance deployment ('R deployment') (Fabre et al. 2012a).

With only S fields, before R deployment, epidemics $I_{S,y}(t)$ repeat themselves identically from year to year during n_y successive cropping seasons. The virus load of the reservoir, which is proportional to $\alpha_{S,y}$, the rate of infection of the S cultivar from the reservoir during year y, also remains constant between years.

Within-season model of a landscape with S and R fields

The inclusion of a R cultivar leads to additionally consider the variable $I_{R,y}(t)$ of the number of plants infected in a field sown with the R cultivar ('R fields') at time t during year y. We considered a qualitative resistance. We therefore defined two pathogen variants: the nonadapted and the resistance-breaking variants, interacting in a gene-for-gene manner. The R cultivar can only be infected by the resistance-breaking variant. The S cultivar can be infected by both variants. We assumed that, on S plants, the nonadapted and resistance-breaking variants coexist in a mutation-selection balance. This equilibrium, characterized by parameter θ, results from the balance between the production of resistance-breaking variants through recurrent mutations of nonadapted variants and their counter-selection because of the fitness costs associated with the adaptive mutations. For a given mutation rate, θ depends on (i) the number of nucleotide substitutions required for the pathogen to break down the resistance conferred by the R gene and (ii) their impact on the competitiveness of the virus. θ can also be seen as characteristics of the R gene selected by breeders.

points, it displays discrete dynamics. Thereafter, n_y cropping seasons are considered, each lasting n_d days. The landscape consists of n_f fields, with a proportion φ_y of R fields during year y, with each field containing n_p plants.

The ODEs describing epidemics during cropping season y are

$$
\left\{
\begin{aligned}
&I_{S,y}(0) = I_{R,y}(0) = 0 \text{ for } y \in [1, n_y]\\
&\frac{dI_{S,y}}{dt} = (n_p - I_{S,y})\left[\alpha_{S,y} + \beta_C\left[((1-\varphi_y)n_f - 1)I_{S,y} + \varphi_y n_f I_{R,y}\right] + \beta_F I_{S,y}\right]\\
&\frac{dI_{R,y}}{dt} = (n_p - I_{R,y})\left[\alpha_{R,y} + \beta_C\left[(1-\varphi_y)n_f \theta I_{S,y} + (\varphi_y n_f - 1)I_{R,y}\right] + \beta_F I_{R,y}\right]
\end{aligned}
\right.
\tag{1}
$$

Between-season model in a landscape with both S and R fields

As soon as R fields occupy a proportion φ_y of the landscape during year y, the epidemic dynamics are no more of intensity Ω_{int} and profile $\mathbf{\Omega}_{pfl}$: they change from season to season. In particular, the rates $\alpha_{S,y}$ and $\alpha_{R,y}$, the rate of infection of the R cultivar from the reservoir during year y, will change each year, as a function of (i) the overall intensity of the epidemic in the R and the S fields in the previous year, as lower epidemic intensities decrease the pathogen load of the reservoir and (ii) the lifespan of the plants of the reservoir and the relative sizes of the cultivated and

Assuming that farmers sow healthy plants, the infections of S and R plants are initiated from the reservoir at rates $\alpha_{S,y}$ and $\alpha_{R,y}$, respectively. S plants may also be infected by pathogens from plants growing in the same field (rate β_F) or by pathogens from infected plants growing in other fields (rate β_C), regardless of the type of cultivar concerned. The same processes govern the infection of R plants, except that these plants can only be infected by pathogens from fields of S plants, at a rate discounted by θ.

The discrete equations describing the interseason dynamics of the pathogen load of the reservoir are based on an exponential moving average of weight λ:

$$
\left\{
\begin{aligned}
&\alpha_{S,1} = \alpha_E \quad \text{and} \quad \alpha_{R,1} = \theta\alpha_E,\\
&\alpha_{S,y} = \lambda\left[\alpha_E(A_{S,y-1} + A_{R,y-1})\right]/A_0 + (1-\lambda)\alpha_{S,y-1} \quad \text{for } y \in [2, n_y],\\
&\alpha_{R,y} = \lambda\left[\alpha_E(\theta A_{S,y-1} + A_{R,y-1})\right]/A_0 + (1-\lambda)\alpha_{R,y-1} \quad \text{for } y \in [2, n_y],
\end{aligned}
\right.
\tag{2}
$$

reservoir compartments. These two characteristics are summarized by a single parameter λ controlling the rate of renewal of the pathogen load in the reservoir, with high values of λ characteristic of a reservoir that is rapidly changing due to short host life spans and small reservoir size. Importantly, it is assumed that the plants of the reservoir are selectively neutral for the pathogen population. Thus, between seasons, the reservoir conserves the relative frequencies of the nonadapted and resistance-breaking variants arising from the crops. It is assumed that the frequency of the resistance-breaking variant in the reservoir is θ before the introduction of the R cultivar in the landscape.

Mathematical description of the model

We provide here a formal mathematical description of the model. The model is semidiscrete (Mailleret and Lemesle 2009). It mostly follows continuous dynamics described by ordinary differential equations (ODE), but, at given time

where $A_{S,y} = (1 - \varphi_y)n_f \int_0^{n_d} I_{S,y}(t)dt$, $A_{R,y} = \varphi_y n_f \int_0^{n_d} I_{R,y}(t)dt$ and $A_0 = n_f \int_0^{n_d} I_S^{\varphi_0}(t)dt$ [$I_S^{\varphi_0}(t)$ is defined below]. Thus, the lower the intensity of the overall epidemic in both R and S fields during year $y - 1$ (i.e. $A_{S,y-1} + A_{R,y-1}$) relative to the overall epidemic intensity of a landscape with only S fields (i.e. A_0), the larger the decrease in the pathogen load of the reservoir from season $y - 1$ to y. Similarly, higher values of λ correspond to a faster drop-off of pathogen levels from older epidemics.

Before R deployment, with only S fields, $dI_S^{\varphi_0}/dt = (n_p - I_S^{\varphi_0})(\alpha_E + \beta_C(n_f - 1)I_S^{\varphi_0} + \beta_F I_S^{\varphi_0})$ with $I_S^{\varphi_0}(0) = 0$. In this situation, the rates $(\alpha_E, \beta_C, \beta_F)$ defining the intensities of the three routes of infection are easier to interpret by considering the alternative parameters $(\Omega_{int}, \mathbf{\Omega}_{pfl})$ (Table S1, Fabre et al. 2012a).

Transformation techniques aiming to put the model in a dimensionless form showed that the results presented were independent of n_p and n_f, for high values of n_f, which is a reasonable assumption for agricultural landscapes. Moreover, using Ω_{int} and $\mathbf{\Omega}_{pfl}$ to define epidemics ensured that the results were independent of season length n_d. Epidemics characterized by Ω_{int} and Ω_{pfl} can occur over seasons of different lengths, provided that the epidemic parameters α_E, β_F, and β_C are rescaled appropriately.

Model analysis

Measuring the yield increase achieved with deployment strategies

A deployment strategy $\boldsymbol{\varphi} = \{\varphi_1, \varphi_2, \ldots, \varphi_{n_y}\}$ is the time series of the proportions of fields sown with R plants each year in the landscape during the time window of resistance deployment considered (n_y). For year y, the sum of the area under the disease progress curves $I_{S,y}(t)$ and $I_{R,y}(t)$ weighted by the proportions $(1 - \varphi_y)$ of S fields and φ_y of R fields is a proxy for yield losses due to the pathogen. Formally, these quantities are the integrals $A_{S,y}$ and $A_{R,y}$ defined above. The overall yield loss over n_y years is given by $D(\delta, \boldsymbol{\varphi}) = \sum_{y=1}^{n_y}[A_{S,y}(\delta, \varphi_y) + A_{R,y}(\delta, \varphi_y)]$, where $\delta = (\theta, \Omega_{int}, \Omega_{pfl}, \lambda, n_y)$ is a given set of model parameters.

The performance of the deployment strategy $\boldsymbol{\varphi}$ is measured by the relative damage $\Delta(\delta, \boldsymbol{\varphi}) = 100.D(\delta, \boldsymbol{\varphi})/(n_y A_0)$. Dividing $D(\delta, \boldsymbol{\varphi})$ by $n_y A_0$, the overall yield losses obtained in a landscape containing only S fields, provides an estimate of the damage obtained with $\boldsymbol{\varphi}$ relatively to the damage that would have been obtained without using the R cultivar. For example, a value of 80% indicates that the strategy $\boldsymbol{\varphi}$ decreases damage due to the pathogen by 20%.

Performance of R deployment strategies maximizing yield

The model was analyzed to explore how the yield increase obtained with optimal strategies of R deployment depended on five parameters of interest $\delta = (\theta, \Omega_{int}, \Omega_{pfl}, \lambda, n_y)$. We first considered strategies in which the same proportion of R fields was sown every year. These so-called constant-mixture yield strategies (CYS) are designed by determining the proportion φ_{YS}^C of R fields minimizing $D(\delta, \boldsymbol{\varphi})$.

We then determined to what extent the performance of CYS [measured by the relative damage $\Delta_{YS}^C(\delta, \varphi_{YS}^C)$] could be improved by modifying the proportion of R fields every 3 years. The resulting so-called variable-mixture yield strategies (VYS) were designed, for $n_y = 15$ only, by determining five proportions of R fields $\boldsymbol{\varphi}_{YS}^V = \{\varphi_{YS}^1, \varphi_{YS}^1, \ldots, \varphi_{YS}^5, \varphi_{YS}^5, \varphi_{YS}^5\}$ minimizing $D(\delta, \boldsymbol{\varphi})$, each proportion being applied during three successive cropping seasons. The performance of VYS was measured (i) as described above, by determining the relative damage $\Delta_{YS}^V(\delta, \boldsymbol{\varphi}_{YS}^V)$ and (ii) by determining the additional relative benefit of these strategies $A_{YS}^V(\delta) = \Delta_{YS}^C(\delta, \varphi_{YS}^C) - \Delta_{YS}^V(\delta, \boldsymbol{\varphi}_{YS}^V)$, corresponding to the percentage difference in

damage between the VYS and the reference CYS. A value of 10% indicates that the VYS is 10% points more beneficial than the CYS.

Performance of R deployment strategies maximizing both yield and resistance durability

We then focused on sustainable strategies designed to maximize yield while remaining sustainable. CYS and VYS strategies with the sole objective of maximizing yield are not necessarily sustainable. There is no reason that the frequency of the resistance-breaking variant will remain low.

We again first considered sustainable strategies in which the same proportion of R fields was sown in each year. These so-called constant-mixture sustainable strategies (CSS) were designed by determining the proportion φ_{SS}^C of R fields minimizing $D(\delta, \boldsymbol{\varphi})$, while ensuring that the mean proportion of infected R plants remained below the threshold value of 5% for all R fields in all years. We analyzed the difference in yield performance between CSS [measured by the relative damage $\Delta_{SS}^C(\delta, \varphi_{SS}^C)$] and CYS.

We then carried out the same analysis with variable-mixture sustainable strategies (VSS), defined by determining the five proportions of R fields $\boldsymbol{\varphi}_{SS}^V = \{\varphi_{SS}^1, \varphi_{SS}^1, \varphi_{SS}^1, \ldots, \varphi_{SS}^5, \varphi_{SS}^5, \varphi_{SS}^5\}$ minimizing $D(\delta, \boldsymbol{\varphi})$ while ensuring that the mean proportion of infected R plants remained below 5%, for all fields and all years. These strategies were identified only for $n_y = 15$, each proportion being applied during three successive cropping seasons. We analyzed the difference in yield performance between VSS [measured by the relative damage $\Delta_{SS}^V(\delta, \boldsymbol{\varphi}_{SS}^V)$] and CYS.

Implementation of model analyses

The model and analyses were implemented in R software (http://www.r-project.org/). The model was solved with the 'lsoda' function (library 'deSolve'). Optimal strategies were identified with a Nelder–Mead algorithm (detailed in Appendix S1 and Figure S1). The numerical exploration of the model was performed by combining global sensitivity analysis, one-at-a-time analysis and hierarchical clustering methods (Appendix S2 and Table S1).

Results

Analysis of deployment strategies maximizing yield
Sensitivity analyses

Sensitivity analyses of the relative damage obtained with constant-mixture yield strategies (CYS) indicated that three of the five parameters of interest considered were important (Figure S2A) given the large range of production situations explored (Table 1). The sum of the main indices (revealing the individual effect of factors) of epidemic intensity, landscape connectivity and choice of R gene accounted for 82% of the variance for relative damage. The variance explained rose to 95% when we added their second- and third-order interactions. The most influential factors were the intensity of the epidemic before the deployment of R plants (accounting for 42% of the variance), followed by the choice of R gene (31%). The third factor, landscape connectivity, accounted for almost 10% of the variance. By contrast, the characteristics of the reservoir and the duration of resistance deployment had only a marginal impact on the performance of the strategy.

Effect of the main factors
The effects of the choice of R gene and epidemic intensity were studied jointly for three contrasting patterns of landscape connectivity. Below, the range of the parameter θ is designated by a binary alternative. This parameter takes into account the number of mutations of the avirulence gene of the pathogen required to break down the resistance mediated by the R gene, and the fitness costs of these mutations. Given the probability distribution of the nonlethal fitness effects of single mutations in plant viruses (Carrasco et al. 2007), values of θ in the range $[10^{-4}, 0.01]$ are indicated a 'R gene *typically* requiring 1 mutation to be broken down' while values in the range $[10^{-8}, 10^{-6}]$ indicated a 'R gene *typically* requiring two mutations to be broken down' (Table 1, Fabre et al. 2012a).

The case of a landscape in which epidemics are mostly driven by within-field infections is illustrated in Fig. 1. In this landscape, before the deployment of the resistant cultivar, 90% of the infections were within-field infections (Fig. 1A). This is the worst-case scenario for the control of epidemics with an R gene. CYS yield marked disease control only for the lower range of epidemic intensities. At best, damage is decreased by more than 90% only when epidemic intensities are <0.35 before the deployment of resistance (Fig. 1B,C). Choosing R genes harder to break down only slightly decreased the damage (Fig. 1C vs B).

In this landscape, the adoption of a variable-mixture yield strategy (VYS), in which the proportion of R fields is changed every 3 years, provided the greatest additional relative benefit. This was true regardless of the choice of R gene and, to a lesser extent, of epidemic intensity. When a margin of improvement exists for CYS (typically when their relative damages are >10%), the mean additional relative benefit was 8% and the highest values (up to 18%) were observed for intermediate epidemic intensities (Table S2). They were obtained only with highly periodic strategies, in which the optimal proportions of R fields varied from almost 1 to almost 0 (Fig. 1F,G). When CYS are already efficient (relative damage <10%), there is likely to

be little gain with VYS. Alternation patterns for the optimal proportions of R fields are less contrasted (Fig. 1D,E).

The situation was very different for landscapes in which epidemics were driven mostly by between-field infections. In such landscapes, before the deployment of the resistant cultivar, 90% of infections are between-field infections (Fig. 2A). The range of epidemic intensities for which high levels of disease control (i.e., relative damage < 10%) were obtained with CYS was much larger and included all production situations in which epidemic intensity was below 0.5 before the deployment of the R cultivar (Fig. 2B), and even below 0.6 for resistance genes typically requiring two mutations to be broken down (Fig. 2C). However, for higher epidemic intensities, relative damage increased steeply, whatever the resistance gene.

Here, VYS were no more effective than CYS. The additional relative benefit accrued never exceeded 0.5% even when relative damages of CYS were > 10% (Table S2). The optimal proportions of R fields with VYS followed a smooth pattern (Fig. 2D–G). As for CYS, they were based on almost equal proportions of R and S fields, particularly for lower levels of disease control (i.e. relative damage > 10%) (Fig. 2F,G).

Finally, the case of a landscape in which epidemics are driven mostly by infections initiated from the reservoir is illustrated in Fig. 3. In such landscapes, before the deployment of the resistant cultivar, 90% of infections originate from the reservoir (Fig. 3A). Here, there is a substantial difference between R genes typically requiring only one mutation to be broken down (Fig. 3B) and R genes typically requiring two mutations to be broken down (Fig. 3C). When two mutations are required, high levels of disease control (i.e. relative damage < 10%) were achieved for the entire range of epidemic intensities explored. In most cases, pathogen damage was reduced by more than 99%, the optimal strategy being to use only the R cultivar. No additional benefit was obtained from the use of VYS. When a single mutation is typically required to break down the R gene (Fig. 3B), the performance of CYS depends on the fitness cost of the mutation. High fitness costs ($\theta = 10^{-4}$ in Fig. 3B) are associated with a high degree of disease control for epidemic intensities <0.7. This threshold falls to 0.45 for intermediate fitness costs ($\theta = 0.01$ in Fig. 3B). For both fitness costs, VYS provided substantial additional benefit (mean of 11%; highest value 16%) when the relative damage of CYS was >10% (Table S2). These benefits were also obtained with highly periodic strategies (Figure S3, panels C and D).

The additional benefit obtained with VYS depended also of the rate of renewal of the pathogen load in the reservoir (Table S2). In slowly changing reservoir, they were always very low (<0.5%) in all landscapes (Table S2, $\lambda = 0.1$). In rapidly changing reservoir, they were roughly doubled

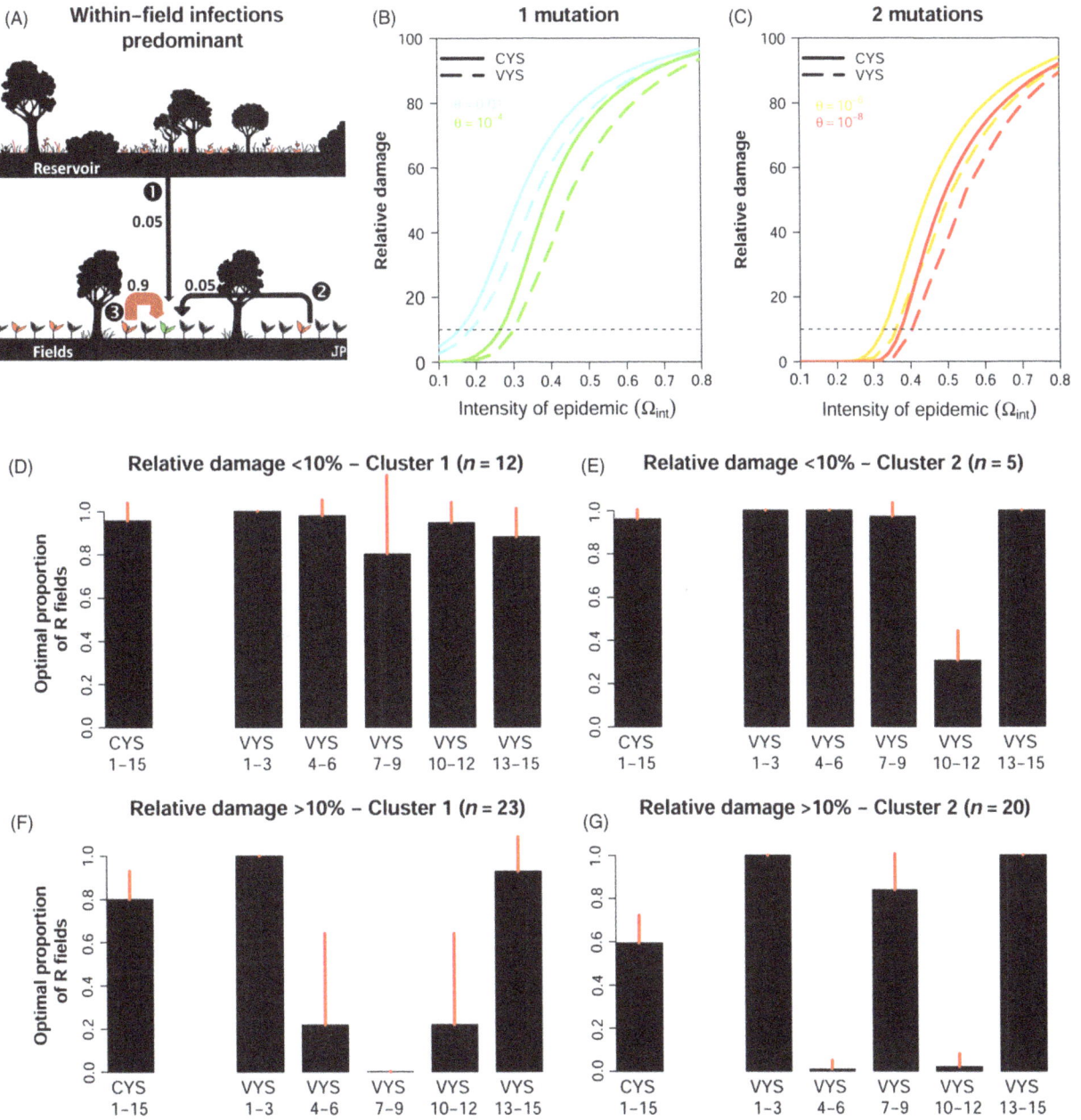

Figure 1 Comparison of the damage reduction achieved with constant-mixture and variable-mixture yield strategies in a landscape in which epidemic dynamics are driven mostly by within-field infections. (A) In this landscape, before the deployment of the resistant cultivar, 90% of the infections are within-field infections (arrow ❸). The other two infection routes, from the reservoir (arrow ❶) and between-field (arrow ❷), each accounts for 5% of the infection events. (B, C) The effects of resistant cultivar choice (θ) and of epidemic intensity before the deployment of the resistant cultivar (Ω_{int}) on the relative damage obtained with optimal constant-mixture yield strategies (CYS) and optimal variable-mixture yield strategies (VYS). Values of θ in the range [10^{-4}, 0.01] (resp. [10^{-8}, 10^{-6}]) correspond to resistance genes typically requiring one (resp. two) mutation to be broken down, depending on the fitness costs associated with these mutations. The dotted line indicates relative damage of 10%, above this arbitrary threshold a substantial margin of improvement exists for CYS. (D, E). Clustering in two groups of the time series of the optimal proportion of R fields in VYS when the relative damage obtained with CYS is <10% (17 of the 60 parameter combinations displayed in graphs B and C). Bars show the mean (±standard deviation) of the proportion of R fields to sown in years 1–3, 4–6, 7–9, 10–12, and 13–15, and the proportion to sown with the CYS. (F, G) As for (D, E), but for relative damage obtained with CYS ≥10% (43 of 60 cases).

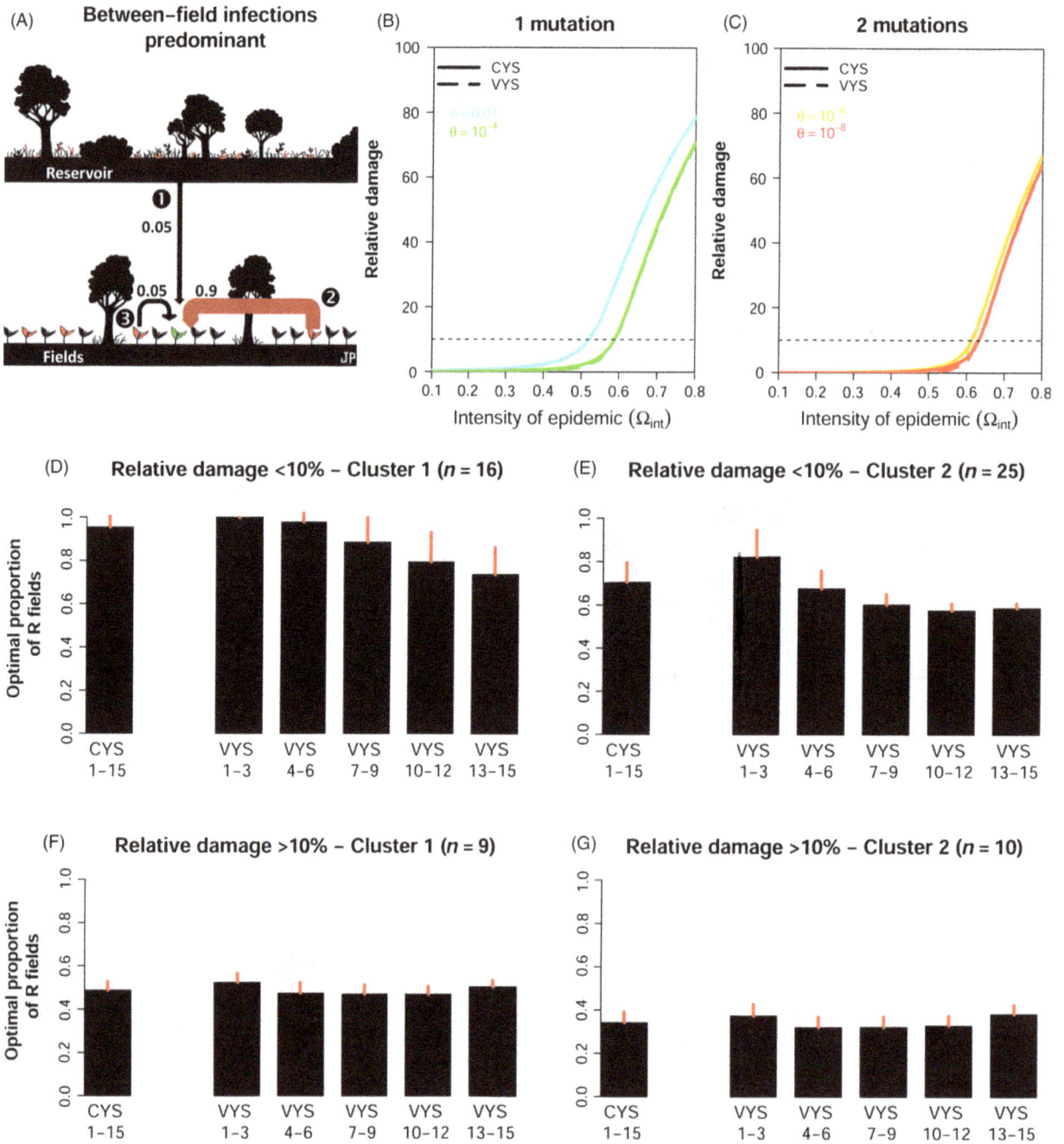

Figure 2 Comparison of damage reduction achieved with constant-mixture and variable-mixture yield strategies in a landscape in which epidemic dynamics are driven mostly by between-field infections. (A) In this landscape, before R deployment, 90% of infections are between-field infections (arrow ❷). (B, C) Effects of the choice of resistant cultivar (θ) and of epidemic intensity before the deployment of the resistant cultivar (Ω_{int}) on the relative damage obtained with constant-mixture yield strategies (CYS) and variable-mixture yield strategies (VYS). (D, E). Clustering into two groups of the time series of the optimal proportion of R fields in VYS when the relative damage obtained with CYS is <10% (41 of the 60 parameter combinations displayed in graphs B and C). (F, G) As for (D, E), but for relative damage obtained with CYS ≥10% (19 of 60 cases). Please refer to Fig. 1 for further details.

(from 8% to 17% for landscapes in which within-field infections predominated, from to 11% to 19% for landscapes in which infections from the reservoir predominated). Additional benefits still remained very low (<1%) for landscapes in which between-field infections predominated (Table S2, $\lambda = 0.9$).

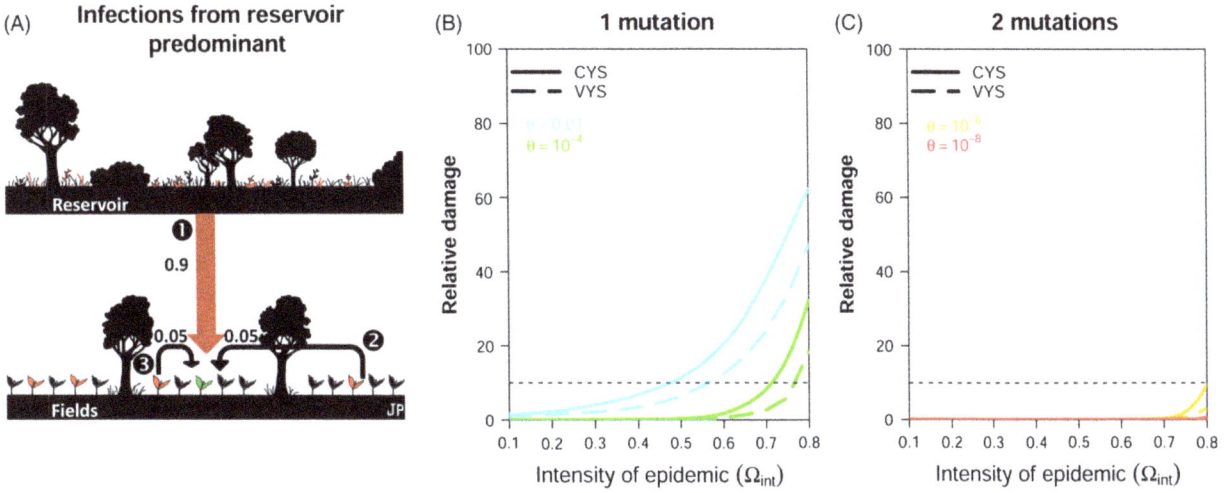

Figure 3 Comparison of damage reduction achieved with constant-mixture and variable-mixture yield strategies in a landscape in which epidemic dynamics are driven mostly by infections from the reservoir. (A) In this landscape, before R deployment, 90% of infections originate from the reservoir (arrow ❶). (B, C) Effects of the choice of resistant cultivar (θ) and of epidemic intensity before the deployment of the resistant cultivar (Ω_{int}) on the relative damage obtained with constant-mixture yield strategies (CYS) and variable-mixture yield strategies (VYS).

Analysis of deployment strategies maximizing yield while preserving resistance durability

Sensitivity analyses highlighted the same three factors as for CYS. Epidemic intensity, landscape connectivity, and choice of the R gene together accounted for 78% of the variance for the relative damage obtained with constant-mixture sustainable strategies (CSS). The most influential factors were the choice of R gene (accounting for 40% of the variance) and epidemic intensity before R deployment

(30%). The connectivity of the landscape alone accounted for 8% of the variance.

The yield performance of CSS was compared to that of CYS in Fig. 4 for a R gene typically requiring one mutation to be broken down. It defined thresholds for epidemic intensity below which relative damages were identical for both strategies (i.e., the solid and dashed lines merge in Fig. 4). These thresholds were low (epidemic intensities <0.25) for landscapes in which within-field infections predominated (Fig. 4A), but much higher (>0.55)

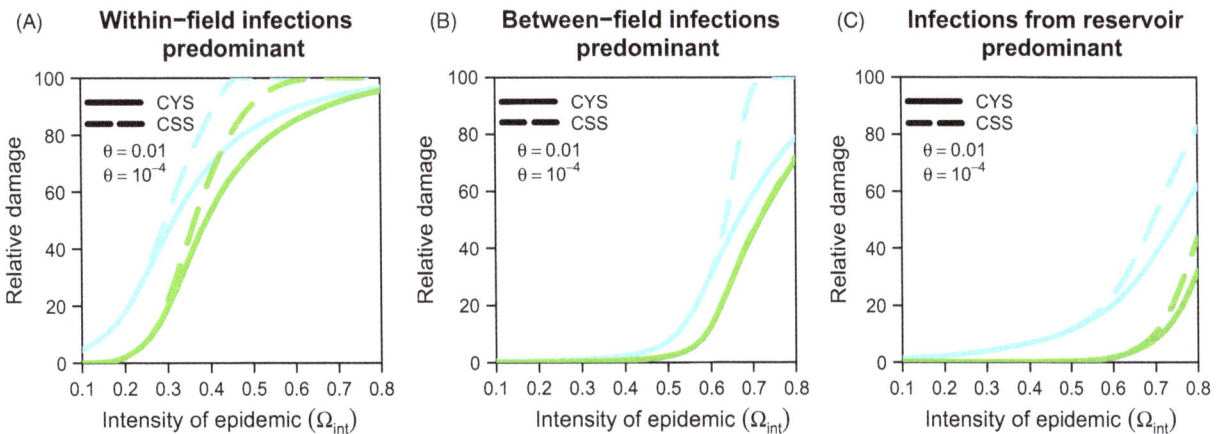

Figure 4 Comparison of the damage reduction achieved with constant-mixture yield strategies and constant-mixture sustainable strategies in three contrasting patterns of landscape connectivity. (A) Effects of epidemic intensity before R deployment (Ω_{int}) on the relative damage obtained with constant-mixture yield strategies (CYS) and constant-mixture sustainable strategies (CSS) in a landscape in which 90% of the infections are within-field infections [$\Omega_{pfl} = (0.05, 0.05, 0.9)$]. The range of θ illustrates a resistance gene typically requiring 1 nucleotide substitution in the avirulence gene of the pathogen to be broken down. (B) As for (A) but for a landscape in which 90% of infections are between-field infections [$\Omega_{pfl} = (0.05, 0.9, 0.05)$]. (C) As for (A) but for a landscape in which 90% of the infections are infections from the reservoir [$\Omega_{pfl} = (0.9, 0.05, 0.05)$].

elsewhere (Fig. 4B,C). In situations in which within-field infections predominated, relative damages for CSS were always 100% for epidemic intensities >0.6 (Fig. 4A). It is therefore not possible to use the R gene and to ensure that the mean proportion of infected R plants is below 5% in all fields and all years. On average, the additional relative cost incurred when using a CSS rather than a CYS was 8% in landscapes in which within-field infections prevailed and 3% when the other two routes of infection predominated. The use of variable-mixture sustainable strategies (VSS) systematically improved performance in landscapes in which (i) within-field infections prevailed (the mean additional relative cost decreasing from 8% to 4%) and (ii) infections originating from the reservoir prevailed (the mean additional relative cost of 3% being converted into a mean additional relative benefit of 1%). However, VSS did not decrease the range of epidemic intensities in situations in which the adoption of CSS was impossible (not illustrated).

For R genes typically requiring two mutations to be broken down, the picture was much simpler (Figure S4). CSS substantially increased the damage due to the pathogen when the dynamics of the epidemic were driven principally by within-field infections (8% on average) (Figure S4A). No such losses occurred with the other two patterns of landscape connectivity (Figure S4B,C).

Discussion

We analyzed a model explicitly linking the epidemiological and population genetics processes occurring at nested levels of biological organization: (i) within-host population genetics and evolutionary processes and (ii) between-host epidemiological processes. If we assume that the initial pathogen population has never been exposed to a resistant cultivar, then the model describes, at the landscape scale, the dynamics of emergence of a resistance-breaking pathogen and the associated epidemiological dynamics during a succession of cropping seasons in which the resistant cultivar can be sown in variable proportions. We investigated the factors determining the levels of disease control obtained with deployment strategies using the principles of cultivar mixture only (constant-mixture strategies) or using cultivar mixture and rotation (variable-mixture strategies). Two management alternatives were also considered, only minimizing pathogen damage in the short term (yield strategies) or, following Gilligan (2008), reconciling the conflicting goals of reducing damage in the short term while preserving resistance durability in the longer term (sustainable strategies).

In our simulations, deployment of the R cultivar controlled epidemics efficiently. Assuming that the range of parameter variations explored (Table 1) is representative of

production situations worldwide, CSS reduced damage by more than 90% in 65% of them. Interestingly, R genes conferring resistance to viruses were found to be durable in 68% of the deployment 'stories' complied worldwide by García-Arenal and McDonald (2003). The choice of R gene was found to be a key factor for effective disease control. The dedicated parameter θ can be estimated *in planta* with high-throughput sequencing techniques when the mutations breaking down the resistance are known (Fabre et al. 2012b). This is now the case for many viral pathosystems (Kang et al. 2005; Moury et al. 2011). Resistance-breaking mutants, now easily obtained in laboratory conditions, are representative of the resistance-breaking isolates found in field conditions (e.g., Moury et al. 2004; Ayme et al. 2006; Hajimorad et al. 2011).

The effect of landscape on plant–pathogen evolution remains unclear, in both cultivated (Mundt 2014) and natural (Meentemeyer et al. 2012) ecosystems. Several studies have considered the management of epidemics at this scale (e.g., Gilligan et al. 2007; Parnell et al. 2009, 2010; Skelsey et al. 2010; Filipe et al. 2012), mostly without taking pathogen evolution into account (but see Papaïx et al. 2013), but only a few have focused specifically on landscape connectivity. Sullivan et al. (2011) found that establishing conservation habitat corridors in fragmented landscapes also increased the incidence of vectorborne plant parasites. Parnell et al. (2009, 2010) found that the dynamics of Asiatic citrus canker and the optimal strategies for its eradication depended on the spatial configuration of its host in the landscape. Skelsey et al. (2010) compared the effect of S and partially R potato cultivar mixtures at three levels on potato late blight epidemics. For their specific pathosystem, within-field mixtures provided the best control. Papaïx et al. (2014) investigated, in a more general context, but without considering pathogen evolution, how the basic components of landscape functional connectivity (proportion and aggregation of S and R cultivars, pathogen fitness on these cultivars, dispersal ability) affected epidemics for an airborne foliar disease. They demonstrated major effects of (i) the proportion of fields sown with a qualitative R cultivar and (ii) their aggregation within the landscape. In our model, the parameter Ω_{pfl} defines functional connectivity in a landscape containing only the S cultivar before R deployment. It results from interactions between the physical characteristics of the host habitats (except for host density, see below) and the dispersal capacities of the pathogen (Meentemeyer et al. 2012). This parameter has a major impact on the dynamics of R breakdown. Over the range of production situations explored (Table 1), CYS reduced the damage by more than 90% in 32% of situations in which within-field infection was the prevailing route of infection, 59% of situations in which between-field infection predominated and 68% of situations in which infections originating from the reservoir prevailed. In our setting, overall host density in the area determined epidemic intensity before resistance deployment (parameter Ω_{int}). As previously reported (e.g. Skelsey et al. 2010, 2013), reducing the crop acreage and the size of the wild host habitat decreases Ω_{int} and leads to the efficient control of epidemics.

The efficiency of R deployment strategies is favored by frequent between-field infections. This is because the numerous between-field infection events occurring in these landscapes include some in which a nonadapted pathogen from a field of susceptible plants is introduced into a field of resistant plants, resulting in unsuccessful infection. These failed infections are particularly probable if (i) the counter-selection of resistance-breaking pathogens in the S cultivar is strong and (ii) the proportions of S and R fields are close to 0.5. This effect is known as the dilution effect (Mundt 2002). In practice, the frequency of between-field infections relative to within-field infections can be increased by designing landscapes with, for a given area, a larger number of smaller (genetically homogeneous) fields (i.e. increasing the number of plant genotype units and decreasing the plant genotype unit area). Using an epidemiological model (without pathogen evolution), Mundt and Brophy (1988) showed that, for a given overall host area, increasing the number of genotype units increases the effectiveness of cultivar mixtures and, consequently, disease control. For given field number and size, between-field infections are also favored by random, rather than aggregated, patterns of S and R cultivars conferring complete resistance (Papaïx et al. 2014).

Yield and sustainable strategies were the most effective in landscapes in which infection from the reservoir was the prevailing route of infection. This situation can be described as pathogen spillover (Daszak et al. 2000). The gain was found to be greater for higher epidemic intensities than for landscapes in which between-field infections prevailed (Figs 2 vs 3 for $\Omega_{int} > 0.5$). In the model, the pathogen load of the reservoir drove the rate of infection of crops from this compartment and was itself dependent on the intensity of epidemics on crops (eqn 2). Deploying a R cultivar initially reduces the intensity of the epidemic through a dilution effect, which, in turn, reduces the pathogen load of the reservoir. This 'serial inoculum dilution' mechanism is effective provided that the frequency of the resistance-breaking pathogen variant in the reservoir remains low (i.e. as long as only few R plants are infected). This mechanism is also effective because we assumed an initial low frequency of the resistance-breaking variant in the reservoir, equal to the mutation-selection balance. This is likely if the R gene is newly introduced from far-off environment but also if the R gene exists in the local wild hosts

at low frequency, a condition favored by a high genetic diversity of wild hosts. In practice, the relative proportion of infections originating from the reservoir could be increased by manipulating the wild host community, as shown by Power and Mitchell (2004) for a virus of annual wild and cultivated grasses. Theoretically, it would be of interest to determine which functional connectivity parameters optimize disease control. The existence of an optimum is suggested by the peaking of pathogen dispersal between habitat patches at intermediate scales of habitat heterogeneity relative to the dispersal capacity of pathogens (Skelsey et al. 2013).

One of our key findings concerns variable-mixture strategies. Resistance management strategies to date have focused on the use of the same mixture every year (van den Bosch and Gilligan 2003; Ohtsuki and Sasaki 2006; Sapoukhina et al. 2009; Skelsey et al. 2010; Fabre et al. 2012a; Bourget et al. 2013; Lo Iacono et al. 2013; Papaïx et al. 2013). Such strategies are inevitable for perennial crops (Sapoukhina et al. 2009), but temporal shifts in the mixture of S and R fields are possible for annual crops. Variable-mixture strategies provided a substantial additional benefit over constant-mixture strategies (i) when within-field infections predominated, regardless of the gene introduced (Fig. 1B,C), and (ii) when infections from the reservoir predominated, for R genes typically requiring only one mutation to be broken down (Fig. 3B). By contrast, variable-mixture strategies were useless when between-field infections predominated (Fig. 2B,C) as the dilution effect is maximal when the proportions of S and R fields are similar. Variable-mixture strategies can also be used to circumvent the conflict between efficient epidemic control and R durability. Most optimal variable-mixture strategies combined the principles of cultivar mixture (during a season) and mixture alternation (i.e. between seasons). The best strategies for managing insecticide (Mani 1989) and antibiotic (Masterton 2010) resistances also involved maximizing treatment heterogeneity (REX consortium, 2013), by varying them in time and space. Almost nearly 20% of optimal variable-mixture strategies involve switching the proportion of R fields from one extreme to the other. Such periodic strategies are particularly interesting in terms of their acceptability to farmers as they are similar to crop rotations, a popular agronomic practice worldwide.

The added value of linking within- and between-host dynamics in nested models for studies of disease emergence is recognized (van den Bosch and Gilligan 2003; Mideo et al. 2008). In our model, a single parameter summarizes the within-host dynamics: the mutation-selection balance between nonadapted and resistance-breaking pathogen variants in S hosts. Despite this simplicity, interactions between within-host and between-host processes accounted for 8% and 13% of the relative damage obtained with CYS and CSS, respectively (Figure S2). There was an interesting interaction between landscape connectivity and the choice of R gene, which determined the mutation-selection balance. R genes typically requiring two mutations, rather than just one, are particularly effective when infections originating from the reservoir predominate (Fig. 3B vs C). Careful R gene choice was less crucial elsewhere (Figs 1 and 2 panels B and C). It follows that associating R gene choice with control methods (either chemical, cultural, or biological) decreasing epidemic intensity and dedicated landscape planning could be the corner stone to achieve integrated and sustainable disease management.

Several hypotheses may have important effects on model outputs. We assumed that the fitness effects of mutations are constant, implying that resistance-breaking mutants revert to their initial frequencies after removal of the R gene. This has been observed (or inferred) for some viruses (e.g., ToMV – *Tm1* gene, Harrison 2002; PVY – *Pvr4* gene, Janzac et al. 2010; tobamovirus – *L* gene, Fraile et al. 2011), but is not general rule for plant–pathogen (Torres-Barcelo et al. 2010; Mundt 2014). Like most models dealing with the evolution of resistance to xenobiotics (REX consortium, 2010), we ignored genetic drift. Within plants, bottlenecks occur at many steps of the virus cycle (Fabre et al. 2012b; Gutiérrez et al. 2012). Between plants, narrow bottlenecks occur during transmission (Moury et al. 2007; Betancourt et al. 2008), overwintering (Kiyosawa 1989) or due to environmental stochasticity (Lo Iacono et al. 2013). These bottlenecks may contribute to delay the emergence of resistance-breaking variants. At a larger scale, we assumed that the relative frequencies of the nonadapted and resistance-breaking variants arising from crops remained unchanged in the reservoir compartment. This is clearly a baseline hypothesis, but little evidence is available to affirm or refute it. Natural populations of wild host plants are highly patchy, with many diverse genotypes of the same species, and different environmental conditions between populations (Zhan et al. 2015). Wild hosts are also the main source of disease resistance in crops, and they contain a large diversity of R genes and alleles. All these factors make strong directional selection of one particular pathogen variant over another unlikely. More generally, this point highlights the need for further research at the agro-ecological interface and setting up experiments at this scale.

Acknowledgements

The drawing of the landscape was realized by Juliette Poidatz. Authors thank two anonymous reviewers for their constructive comments on an earlier draft of the manuscript. This work was supported by the project 'VirAphid' (ANR-10-STRA-0001) of the Agence Nationale de la Recherche (ANR) and by Institut National de la Recherche Agronomique (INRA) (project 'TakeControl' of the metaprogramme SMaCH). Numerical simulations were performed with the cluster Avakas of the university of Bordeaux.

Literature cited

Ayme, V., S. Souche, C. Caranta, M. Jacquemond, J. Chadoeuf, A. Palloix, and B. Moury 2006. Different mutations in the genome-linked protein VPg of Potato Virus Y confer virulence on the *pvr2³* resistance in pepper. Molecular Plant-Microbe Interactions **19**: 557–563.

Betancourt, M., A. Fereres, A. Fraile, and F. García-Arenal 2008. Estimation of the effective number of founders that initiate an infection after aphid transmission of a multipartite plant virus. Journal of Virology **82**:12416–12421.

van den Bosch, F., and C. A. Gilligan 2003. Measures of durability of resistance. Phytopathology **93**:616–625.

Bourget, R., L. Chaumont, and N. Sapoukhina 2013. Timing of pathogen adaptation to a multicomponent treatment. PLoS One **8**:e71926.

Bousset, L., and A. M. Chèvre 2013. Stable epidemic control in crops based on evolutionary principles: adjusting the metapopulation concept to agro-ecosystems. Agriculture, Ecosystems & Environment **165**:118–129.

Brown, J. K. M., and A. Tellier 2011. Plant-parasite coevolution: bridging the gap between genetics and ecology. Annual Review of Phytopathology **49**:345–367.

Carrasco, P., F. de la Iglesia, and S. F. Elena 2007. Distribution of fitness and virulence effects caused by single-nucleotide substitutions in tobacco etch virus. Journal of Virology **81**:12979–12984.

Daszak, P., A. A. Cunningham, and A. D. Hyatt 2000. Wildlife ecology – emerging infectious diseases of wildlife – threats to biodiversity and human health. Science **287**:443–449.

Dybiec, B., A. Kleczkowski, and C. A. Gilligan 2004. Controlling disease spread on networks with incomplete knowledge. Physical Review E **70**:066145.

Fabre, F., E. Rousseau, L. Mailleret, and B. Moury 2012a. Durable strategies to deploy plant resistance in agricultural landscapes. New Phytologist **193**:1064–1075.

Fabre, F., J. Montarry, J. Coville, R. Senoussi, V. Simon, and B. Moury 2012b. Modelling the evolutionary dynamics of viruses within their hosts: a case study using high-throughput sequencing. PLoS Pathogens **8**:e1002654.

Filipe, J. A. N., R. C. Cobb, R. K. Meentemeyer, C. A. Lee, Y. S. Valachovic, A. R. Cook, D. M. Rizzo et al. 2012. Landscape epidemiology and control of pathogens with cryptic and long-distance dispersal: sudden oak death in northern Californian forests. PLoS Computational Biology **8**:e1002328.

Flor, H. H. 1971. Current status of the gene-for-gene concept. Annual Review of Phytopathology **9**:275–296.

Fraile, A., I. Pagán, G. Anastasio, E. Sáez, and F. García-Arenal 2011. Rapid genetic diversification and high fitness penalties associated with pathogenicity evolution in a plant virus. Molecular Biology and Evolution **28**:1425–1437.

García-Arenal, F., and B. A. McDonald 2003. An analysis of the durability of resistance to plant viruses. Phytopathology **93**:941–952.

Garrett, K. A., L. N. Zúñiga, E. Roncal, G. A. Forbes, C. C. Mundt, Z. Su, and R. J. Nelson 2009. Intraspecific functional diversity in hosts and its effect on disease risk across a climatic gradient. Ecological Applications 19:1868–1883.

Gilligan, C. A. 2008. Sustainable agriculture and plant diseases: an epidemiological perspective. Philosophical Transactions of the Royal Society of London. Series B, Biological Sciences 363:741–759.

Gilligan, C. A., J. E. Truscott, and A. J. Stacey 2007. Impact of scale on the effectiveness of disease control strategies for epidemics with cryptic infection in a dynamical landscape: an example for a crop disease. Journal of the Royal Society Interface 4:925–934.

Gutiérrez, S., Y. Michalakis, and S. Blanc 2012. Virus population bottlenecks during within-host progression and host-to-host transmission. Current Opinion in Virology 2:546–555.

Hajimorad, M. R., R. H. Wen, A. L. Eggenberger, J. H. Hill, and M. A. S. Maroof 2011. Experimental adaptation of an RNA virus mimics natural evolution. Journal of Virology 85:2557–2564.

Harrison, B. D. 2002. Virus variation in relation to resistance-breaking in plants. Euphytica 124:181–192.

Janzac, B., J. Montarry, A. Palloix, O. Navaud, and B. Moury 2010. A point mutation in the polymerase of Potato Virus Y confers virulence toward the Pvr4 resistance of pepper and a high competitiveness cost in susceptible cultivar. Molecular Plant-Microbe Interactions 23:823–830.

Jenner, C. E., X. Wang, F. Pronz, and J. A. Walsh 2002. A fitness cost for turnip mosaic virus to overcome host resistance. Virus Research 86:1–6.

Kang, B. C., I. Yeam, and M. M. Jahn 2005. Genetics of plant virus resistance. Annual Review of Phytopathology 43:581–621.

Kiyosawa, S. 1989. Breakdown of blast resistance in relation to general strategies of resistance gene deployment to prolong effectiveness of resistance in plants. In: K. J. Leonard, and W. E. Fry, eds. Plant Disease Epidemiology, vol. 2, pp. 251–283. McGraw-Hill, New York.

Lewontin, R. C. 1958. A general method for investigating the equilibrium of gene frequency in a population. Genetics 43:419–434.

Lo Iacono, G., F. van den Bosch, and C. A. Gilligan 2013. Durable resistance to crop pathogens: an epidemiological framework to predict risk under uncertainty. PLoS Computational Biology 9:e1002870.

Mailleret, L., and V. Lemesle 2009. A note on semi-discrete modelling in the life sciences. Philosophical Transactions. Series A, Mathematical, Physical, and Engineering Sciences 367:4779–4799.

Mani, G. S. 1989. Evolution of resistance with sequential application of insecticides in time and space. Proceedings of the Royal Society of London. Series B, Biological Sciences 238:245–276.

Masterton, R. G. 2010. Antibiotic heterogeneity. International Journal of Antimicrobial Agents 36:S15–S18.

McDonald, B. A., and C. Linde 2002. Pathogen population genetics, evolutionary potential, and durable resistance. Annual Review of Phytopathology 40:349–379.

Meentemeyer, R. K., S. E. Haas, and T. Václavík 2012. Landscape epidemiology of emerging infectious diseases in natural and human-altered ecosystems. Annual Review of Phytopathology 50:379–402.

Mideo, N., S. Alizon, and T. Day 2008. Linking within- and between-host dynamics in the evolutionary epidemiology of infectious diseases. Trends in Ecology and Evolution 23:511–517.

Moury, B., C. Morel, E. Johansen, L. Guilbaud, S. Souche, V. Ayme, C. Caranta et al. 2004. Mutations in potato virus Y genome-linked protein determine virulence toward recessive resistances in Capsicum annuum and Lycopersicon hirsutum. Molecular Plant-Microbe Interactions 17:322–329.

Moury, B., F. Fabre, and R. Senoussi 2007. Estimation of the number of virus particles transmitted by an insect vector. Proceedings of the National Academy of Sciences of the United States of America 104:17891–17896.

Moury, B., A. Fereres, F. García-Arenal, and H. Lecoq 2011. Sustainable management of plant resistance to viruses. In: C. Caranta, M. A. Aranda, M. Tepfer, and J. J. Lopez-Moya, eds. Recent Advances in Plant Virology, pp. 219–236. Caister Academic Press.

Mundt, C. C. 2002. Use of multiline cultivars and cultivar mixtures for disease management. Annual Review of Phytopathology 40:381–410.

Mundt, C. C. 2014. Durable resistance: a key to sustainable management of pathogens and pests. Infection, Genetics and Evolution 14:446–455.

Mundt, C. C., and L. S. Brophy 1988. Influence of number of host genotype units on the effectiveness of host mixtures for disease control: a modelling approach. Phytopathology 78:1087–1094.

Ohtsuki, A., and A. Sasaki 2006. Epidemiology and disease-control under gene-for-gene plant-pathogen interaction. Journal of Theoretical Biology 238:780–794.

Ostfeld, R. S., and F. Keesing 2012. Effects of host diversity on infectious disease. Annual Review of Ecology, Evolution, and Systematics 43:157–182.

Papaïx, J., H. Goyeau, P. Du Cheyron, H. Monod, and C. Lannou 2011. Influence of cultivated landscape composition on variety resistance: an assessment based on wheat leaf rust epidemics. New Phytologist 191:1095–1107.

Papaïx, J., O. David, C. Lannou, and H. Monod 2013. Dynamics of adaptation in spatially heterogeneous metapopulations. PLoS One 8: e54697.

Papaïx, J., S. Touzeau, H. Monod, and C. Lannou 2014. Can epidemic control be achieved by altering landscape connectivity in agricultural systems? Ecological Modelling 284:35–47.

Park, A. W., S. Gubbins, and C. A. Gilligan 2001. Invasion and persistence of plant parasites in a spatially structured host population. Oikos 94:162–174.

Parnell, S., T. R. Gottwald, F. van den Bosch, and C. A. Gilligan 2009. Optimal strategies for the eradication of Asiatic citrus canker in heterogeneous host landscapes. Phytopathology 99:1370–1376.

Parnell, S., T. R. Gottwald, C. A. Gilligan, N. J. Cunniffe, and F. van den Bosch 2010. The effect of landscape pattern on the optimal eradication zone of an invading epidemic. Phytopathology 100:638–644.

Plantegenest, M., C. Le May, and F. Fabre 2007. Landscape epidemiology of plant diseases. Journal of the Royal Society Interface 4:963–972.

Power, A., and C. E. Mitchell 2004. Pathogen spillover in disease epidemics. The American Naturalist 164:S79–S89.

Real, L. A., and R. Biek 2007. Spatial dynamics and genetics of infectious diseases on heterogeneous landscapes. Journal of the Royal Society Interface 4:935–948.

REX consortium. 2010. The skill and style to model the evolution of resistance to pesticides and drugs. Evolutionary Applications 3: 375–390.

REX consortium. 2013. Heterogeneity of selection and the evolution of resistance. Trends in Ecology and Evolution 28: 110–118.

Sapoukhina, N., C. E. Durel, and B. Le Cam 2009. Spatial deployment of gene-for-gene resistance governs evolution and spread of pathogen populations. Theoretical Ecology 2:229–238.

Skelsey, P., W. A. H. Rossing, G. J. T. Kessel, and W. van der Werf 2010. Invasion of Phytophthora Infestans at the landscape level: how do spatial scale and weather modulate the consequences of spatial heterogeneity in host resistance? Phytopathology 100:1146–1161.

Skelsey, P., K. A. With, and K. A. Garrett 2013. Pest and disease management: why we shouldn't go against the grain. PLoS One **8**:e75892.

Sullivan, L. L., B. L. Johnson, L. A. Brudvig, and N. M. Haddad 2011. Can dispersal mode predict corridor effects on plant parasites? Ecology **92**:1559–1564.

Taylor, P. D., L. Fahrig, K. Henein, and K. G. Merriam. 1993. Connectivity is a vital element of landscape structure. Oikos **68**:571–573.

Tilman, D. 1999. Global environmental impacts of agricultural expansion: the need for sustainable and efficient practices. Proceedings of the National Academy of Sciences **96**:5995–6000.

Torres-Barcelo, C., J. A. Daros, and S. F. Elena 2010. Compensatory molecular evolution of HC-Pro, an RNA-silencing suppressor from a plant RNA virus. Molecular Biology and Evolution **27**:543–551.

Wolfe, M. S. 1985. The current status and prospects of multiline cultivars and variety mixtures for disease resistance. Annual Review of Phytopathology **23**:251–273.

Zhan, J., P. H. Thrall, J. Papaïx, L. Xie, and J. J. Burdon 2015. Playing on a pathogen's weakness: using evolution to guide sustainable plant disease control strategies. Annual Review of Phytopathology **53**:19–43.

Zhu, Y. Y., H. R. Chen, J. H. Fan, Y. Y. Wang, Y. Li, J. B. Chen, J. X. Fan et al. 2000. Genetic diversity and disease control in rice. Nature **406**:718–722.

5

What, if anything, are hybrids: enduring truths and challenges associated with population structure and gene flow

Zachariah Gompert[1] and C. Alex Buerkle[2]

1 Department of Biology, Utah State University, Logan, UT, USA
2 Department of Botany, University of Wyoming, Laramie, WY, USA

Keywords
admixture, conservation biology, genetic ancestry, hybridization, population genetics.

Correspondence
Zachariah Gompert, Department of Biology, Utah State University, Logan, UT 84322, USA.

e-mail: zach.gompert@usu.edu

Abstract

Hybridization is a potent evolutionary process that can affect the origin, maintenance, and loss of biodiversity. Because of its ecological and evolutionary consequences, an understanding of hybridization is important for basic and applied sciences, including conservation biology and agriculture. Herein, we review and discuss ideas that are relevant to the recognition of hybrids and hybridization. We supplement this discussion with simulations. The ideas we present have a long history, particularly in botany, and clarifying them should have practical consequences for managing hybridization and gene flow in plants. One of our primary goals is to illustrate what we can and cannot infer about hybrids and hybridization from molecular data; in other words, we ask when genetic analyses commonly used to study hybridization might mislead us about the history or nature of gene flow and selection. We focus on patterns of variation when hybridization is recent and populations are polymorphic, which are particularly informative for applied issues, such as contemporary hybridization following recent ecological change. We show that hybridization is not a singular process, but instead a collection of related processes with variable outcomes and consequences. Thus, it will often be inappropriate to generalize about the threats or benefits of hybridization from individual studies, and at minimum, it will be important to avoid categorical thinking about what hybridization and hybrids are. We recommend potential sampling and analytical approaches that should help us confront these complexities of hybridization.

Introduction

Sexual reproduction that involves mating with other individuals (outcrossing rather than selfing) and meiotic recombination mix alleles among different genomic backgrounds. Physical dispersal of individuals before reproduction moves alleles farther from where they originated by mutation and is referred to as gene flow. At some point, crosses can occur between individuals that are unrelated enough that we refer to these as hybrids. Although hybridization has sometimes been viewed as an unimportant dead end, there is a long history of interest in hybridization as a potent creative and destructive evolutionary process (e.g. Stebbins 1950; Ellstrand 1992; Rieseberg and Wendel 1993; Buerkle et al. 2003; Arnold 2006). Numerous cases where hybridization and introgression have had substantial ecological or evolutionary consequences in plants are known. For example, hybridization between the sunflower species *Helinathus annuus* and *Helinathus petiolaris* resulted in multiple distinct hybrid species (Rieseberg et al. 1990, 1995, 2003a), and hybridization in *Populus* affects community composition and ecosystem processes (Driebe and Whitham 2000; Martinsen et al. 2000; Whitham et al. 2006; Floate et al. 2016). Hybridization is particularly common among oak species, where it may spread or generate adaptive genetic variation and where it has been proposed as a key component of natural and human-induced invasions (Petit et al. 2004; Moran et al. 2012).

The consequences of hybridization are directly relevant to aspects of conservation biology and agriculture. Hybridization, whether natural or human induced, can affect the origin, maintenance, and loss of biodiversity (Rhymer and Simberloff 1996; Wolf et al. 2001; Buerkle et al. 2003; Zalapa et al. 2010; Muhlfeld et al. 2014). Hybridization in plants could help endemic species survive periods of climate change (Becker et al. 2013) or result in extinction, when, for example, native species are assimilated by non-native species or experience demographic decline due to outbreeding depression (Ellstrand 1992; Levin et al. 1996; Balao et al. 2015; Gómez et al. 2015). Introgressive hybridization also occurs between crops and their wild relatives, and this too can have beneficial or detrimental consequences for biodiversity (Linder et al. 1998; Ellstrand et al. 2013; Hufford et al. 2013; Warschefsky et al. 2014). Of particular interest is the potential for crop–wild hybridization to allow modified or engineered genes to escape into the wild, which could negatively affect native species or increase public distrust of genetically modified crops (Ellstrand 2001; Stewart et al. 2003; Chapman and Burke 2006; Garnier et al. 2014). Another practical issue is whether and under what conditions hybrid populations or taxa warrant conservation efforts. Hybrids were not granted protection under the US Endangered Species Act, but this was questioned in a federal rule proposed in 1996 (this rule was never adopted; Allendorf et al. 2001, 2013). The proposed federal rule used the term 'intercross' rather than 'hybrid' to avoid a negative connotation of the latter (Allendorf et al. 2013) and we suspect that some people would view even natural hybrids as less worthy of protection than 'pure' species (e.g. the decision to conserve eastern wolves has in part been based on species or hybrid status; Rutledge et al. 2015). Clearly, the potential outcomes and practical consequences of hybridization are multifarious, and thus, different cases of hybridization will need to be treated differently.

Confronting this complexity requires careful consideration of what hybridization is, and when distinguishing among different processes is necessary and possible. The recognition of hybrids between named taxa is relatively uncontroversial, but it is somewhat poorly resolved as to what distance of a cross constitutes hybridization, and what therefore qualifies as a hybrid (Harrison 1993; Arnold 2006; Allendorf et al. 2013). Similarly, different histories of gene flow and selection, such as primary divergence versus secondary contact, have been referred to as hybridization (Barton and Hewitt 1985). However, discriminating among these different histories could be necessary from a management perspective, if, for example, we are to treat cases of natural and human-induced hybridization differently as suggested by Allendorf et al. (2001). Unfortunately, different histories of hybridization can generate very similar or identical patterns of genetic and phenotypic variation (e.g. Barton and Hewitt 1985; Kruuk et al. 1999; Barton and de Cara 2009). This means we might not always be able

to distinguish different histories even when doing so would be useful.

In this article, we review and discuss ideas that are relevant to recognition of hybrids and supplement these with simulations to illustrate important contrasts. We acknowledge that is atypical to have a paper contain review, synthesis of concepts and novel simulations, but we think the combination can be useful. The issues we address have a relatively long history, some of which is underappreciated, and clarifying these ideas should have practical consequences for managing hybridization and gene flow in plants. A reexamination of some of these points is worthwhile too because recent population genomic studies have led to a greater appreciation of variation within species and genomic heterogeneity in differentiation between species or populations (e.g. Martin and Orgogozo 2013; Gompert et al. 2014; Mandeville et al. 2015). Additionally, we have learned more about models and approaches that can be used to describe patterns of variation in hybrids (Patterson et al. 2012; Gompert and Buerkle 2013). Along these lines, it is important to recognize what we can and cannot infer about hybrids and hybridization from molecular data; in other words, we must be aware that genetic data provide incomplete information about hybridization. Our simulations and discussion focus on patterns of variation when hybridization is recent and populations are polymorphic; this contrasts with the bulk of theoretical work that concerns long-term equilibrium outcomes of hybridization and often is most applicable when hybridizing taxa exhibit fixed differences. This distinction increases the novelty of our results and makes them particularly informative for applied issues and contemporary hybridization following recent ecological change. In the following, we first address the question of what constitutes hybridization and then turn to the definition of hybrids. We combine literature review and new simulations to answer these questions and conclude each section with recommendations for applied studies of hybridization and gene flow in plants.

What, if anything, is hybridization

Hybridization has been variously defined as interbreeding between different species or subspecies, distinct populations or cultivars, or any individuals with heritable phenotypic differences (Stebbins 1950; Barton and Hewitt 1985; Harrison 1993; Allendorf et al. 2001; Arnold 2006). However, such distinctions downplay the continuous nature of genetic and phenotypic differentiation and distract from the fact that gene flow can have similar consequences anywhere along this continuum (Mayr 1963; Mallet et al. 2007; Martin and Orgogozo 2013). For example, because of population genetic structure and local adaptation within species, intraspecific gene flow can have positive, negative, or negligible effects on populations that are similar to those of interspecific gene flow (e.g. Ellstrand 1992; Kremer et al. 2012; Nosil et al. 2012; Roe et al. 2014). Moreover, the consequences of interspecific gene flow frequently depend on the specific individuals involved, because of polymorphisms within and among conspecific populations (Sweigart et al. 2007; Escobar et al. 2008; Good et al. 2008; Gompert et al. 2013). In other words, it is the evolutionary and ecological consequences of gene flow that should be considered when defining hybridization. Importantly, the consequences of gene flow do not depend on taxonomy or a specific definition of species, but rather on the nature of differences between groups. Of course, such differences also represent a continuum, and thus, an unambiguous and objective definition of hybridization as something distinct from gene flow is not likely possible. With that said, we think it is useful to reserve the term hybridization for cases where outcrossing and gene flow occur between populations that differ, at least quantitatively, at multiple heritable characters or genetic loci that affect fitness. Thus, we argue that the distinction between gene flow and hybridization is fuzzy and quantitative, rather than discrete and qualitative. While such a view could complicate management decisions, we think it more accurately captures patterns of variation in nature.

Different histories or geographies of gene flow and selection have often been referred to as hybridization. For example, several authors have argued that both primary divergence with gene flow and gene flow following secondary contact (i.e. gene flow after a prolonged period of geographic separation with very little or no gene flow) constitute hybridization (Barton and Hewitt 1985). We think that the case for secondary contact is uncontroversial, but that informed opinions might differ about whether primary divergence includes hybridization. Certainly, primary divergence is not the common conception of hybridization in conservation biology (Allendorf et al. 2001, 2013). Likewise, hybrid zones maintained primarily by exogenous (environment dependent) versus endogenous (environment-independent) selection have been classified and treated similarly. However, management efforts could benefit from distinguishing among these different histories and processes. We might be more inclined to intervene when secondary contact occurs after an anthropogenic disturbance than when primary divergence occurs, even if the latter takes place in a disturbed area.

An equally important question is whether and under what conditions we can in fact discriminate among these different cases. On the one hand, theory shows that over the long-term, primary divergence and secondary contact with exogenous or endogenous selection have similar equilibrium conditions and result in similar geographic patterns of genetic and phenotypic variation (Endler 1977;

Barton and Hewitt 1985; Kruuk et al. 1999; Navarro and Barton 2003; Barton and de Vladar 2009; Barton 2013; Flaxman et al. 2014). However, it is also true that well-documented examples of these different cases are known. For example, convergent clines in flowering time in sunflowers are best explained by primary divergence driven by exogenous selection (Blackman et al. 2011; Kawakami et al. 2011), whereas hybridization between *H. annuus* and *H. petiolaris*, which are not sister species, can be attributed to secondary contact (Rieseberg 1991). Additionally, the bulk of evidence suggests that many classic hybrid zones are tension zones maintained by endogenous selection (reviewed in Barton and Hewitt 1985). Consistent with this, Dobzhansky–Muller incompatibilities (DMIs) have been documented in several plant taxa, such as *Mimulus* and *Solanum* (Sweigart et al. 2007; Moyle and Nakazato 2010).

Here, we ask when genetic analyses commonly used to study hybridization might mislead us about the history or nature of gene flow and selection. We are particularly interested in cases where being misled could affect decisions in applied science. We consider primary divergence versus secondary contact, and neutral evolution versus selection on a quantitative trait along an environmental gradient or reduced hybrid fitness due to intrinsic epistatic incompatibilities (i.e. DMIs). We simulate genetic data under each of these conditions and then summarize the results by (i) examining allele frequency and trait clines, (ii) summarizing genetic variation with principal component analysis (PCA), and (iii) estimating admixture proportions. Our goal is not an exhaustive evaluation of these methods, but rather to provide illustrative examples of the potential to be misled by genetic data. We then turn to the related problem of finite sampling. In particular, we show that sparse population sampling when organisms are continuously distributed can lead to false inferences about population structure. That is to say, clinal variation can appear more demic and even suggestive of hybrid speciation. Importantly, and in contrast to most theoretical work on hybridization or hybrid zones, our simulations incorporate shared polymorphism across populations (or species), rather than focusing on genetic markers with fixed differences. This is realistic in general and better reflects the current generation of molecular data (e.g. SNPs identified and scored through genotyping-by-sequencing or exome sequencing).

Simulations and analyses

We used individual-based, genetically explicit simulations to generate pseudo-data under different demographic and evolutionary histories. Simulations were conducted using the program nemo version 2.3.44 (Guillaume and Rougemont 2006). Generations were discrete, and each generation consisted of the following ordered events: breeding, dispersal, viability selection (some histories), and aging. Patches were arranged according to a 1-D stepping-stone model with dispersal allowed only between adjacent patches (dispersal off the outer-edges of the patch vector was allowed). We assumed logistic growth within each patch with a carrying capacity of 5000 individuals and a mean fecundity of two. Genomes consisted of a single chromosome with a recombinational map length of one Morgan. We tracked 200 neutral bi-allelic SNPs in all simulations, and 10 quantitative trait SNPs or DMI SNPs in relevant subsets of the simulations. In all cases, mutation rates were 0.0001 per locus per generation and SNPs were distributed according to a random uniform distribution along the recombinational map of the chromosome (this included neutral and non-neutral SNPs). Simulations lasted 2000 generations.

Starting allele frequencies were generated for neutral markers, quantitative trait SNPs and DMI SNPs to mimic secondary contact or primary divergence (Figure S1). Ancestral allele frequencies were first generated for neutral SNPs by sampling from a beta distribution with α and β equal to 20 (this distribution has a mean of 0.5 and a standard deviation of 0.08). We then obtained initial allele frequencies for the two taxa experiencing secondary contact by sampling from $beta(\alpha = \pi \frac{1-F}{F}, \beta = (1 - \pi) \frac{1-F}{F})$, where π is the ancestral allele frequency for the SNP and F corresponds to F_{ST} (Balding and Nichols 1995; Falush et al. 2003), which was set to 0.3 (i.e. substantial population genetic differentiation). We assigned one set of allele frequencies to patches 1–5 and a different set of allele frequencies to patches 6–10. We used the same procedure to generate initial neutral allele frequencies for primary divergence, except the same allele frequencies were assigned to all 10 patches. We initialized quantitative SNPs by assuming the two taxa were perfectly adapted to alternative ends of the patch vector (secondary contact; mean phenotypes of −0.5 and 0.5 were used for patches 1–5 and 6–10, respectively), or by setting the mean phenotype in each patch equal to 0 (primary divergence). We initialized DMI SNPs with different taxa fixed for different sets of derived alleles, such that no fitness reduction occurred within taxa but hybrids would experience reduced fitness (secondary contact), or with all populations fixed for the ancestral allele.

We then simulated five replicate data sets with the following conditions: neutral evolution following secondary contact (no DMIs and no effect of the quantitative trait on fitness), exogenous selection along an environmental gradient with primary divergence, exogenous selection along an environmental gradient following secondary contact, exogenous selection at a sharp ecotone with primary divergence, exogenous selection at a sharp ecotone following secondary

contact, endogenous selection caused by DMIs with primary divergence, and endogenous selection caused by DMIs following secondary contact (summarized in Table 1). We repeated all simulations with migration rates of 0.01 and 0.001. Exogenous selection was based on a single quantitative trait that was under stabilizing selection in each patch; we used a Gaussian fitness function with mean μ and variance 0.5. μ varied from −0.5 to 0.5 in steps of 0.1 (most patches) or 0.2 (patches 5 and 6) between patches for the environmental gradient, and was set to −0.5 (patches 1–5) or 0.5 (patches 6–10) for the sharp ecotone. This means that an individual perfectly adapted to one end of the patch vector would have relative fitness of 0.37 at the other end. DMIs were modeled as negative fitness effects between derived alleles at pairs of SNPs. Considering a single locus pair, we assumed the double homozygote for different derived alleles had a fitness of 0.6, and an individual heterozygous at one locus and homozygous for derived alleles at the other had a fitness of 0.8; all other genotypes had a fitness of 1.0. We assumed fitness was absolute (not relative) and multiplicative across DMIs.

Additional data were simulated to evaluate the effect of limited sampling on inference. Our primary motivations were to determine whether sampling gaps would provide false evidence of discrete population clusters or a lack of hybrids when the underlying population structure was continuous (i.e. with isolation by distance). Here, we assumed neutral primary divergence in a 1-D stepping-stone model with 50 patches, each with a carrying capacity of 2500 individuals and a dispersal rate between neighboring patches of 0.001 (our focus on neutral primary divergence reflects our interest in isolation by distance). We initialized neutral allele frequencies as described above. We analyzed either samples from all 50 patches (50 or 5 individuals each),

Table 1. . Summary of conditions for simulations conducted with nemo (five replicates each).

Geography	Selection	Migration rate
Secondary contact	None	0.001
Primary divergence	Exogeneous, smooth gradient	0.001
Secondary contact	Exogeneous, smooth gradient	0.001
Primary divergence	Exogeneous, sharp ecotone	0.001
Secondary contact	Exogeneous, sharp ecotone	0.001
Primary divergence	Endogenous (DMIs)	0.001
Secondary contact	Endogenous (DMIs)	0.001
Secondary contact	None	0.01
Primary divergence	Exogeneous, smooth gradient	0.01
Secondary contact	Exogeneous, smooth gradient	0.01
Primary divergence	Exogeneous, sharp ecotone	0.01
Secondary contact	Exogeneous, sharp ecotone	0.01
Primary divergence	Endogenous (DMIs)	0.01
Secondary contact	Endogenous (DMIs)	0.01

DMI, Dobzhansky–Muller incompatibilities.

from sets of four patches at the edges and center of the patch vector (50 individuals each), and from the 12 center patches (50 individuals each).

We used three common analytical approaches to quantify and summarize patterns of genetic variation from the simulations: (i) character and allele frequency clines, (ii) ordination via PCA, and (iii) inference of admixture proportions. We plotted geographic clines in allele frequencies at all neutral SNPs and for the quantitative trait (with the exceptions of DMI simulations, which did not include a quantitative trait). Allele frequencies were polarized such that the rarer allele in the first patch was shown. We conducted PCA on the centered genotype data from 50 individuals from each patch in each simulation (i.e. 500 individuals total for most simulations) using the prcomp function in R (R Development Core Team 2015). Genotypes were coded as 0, 1, or 2 copies of one allele at each locus. We estimated admixture proportions using these same genetic data. We used the program admixture version 1.23 (Alexander et al. 2009) for this, which fits the same model as the admixture model in structure Pritchard et al. (2000), but uses maximum likelihood rather than Bayesian inference. We used the block-relaxation method for parameter estimation with a tolerance of 0.0001 and the Quasi-Newton algorithm for convergence acceleration.

Our analyses show that time since the onset of secondary contact or primary divergence has a profound effect on patterns of genetic variation (Figs 1–3), even over the relatively short temporal scale of our simulations (2000 generations). By the end of the simulations, allele frequency clines were somewhat similar for both histories (i.e. secondary contact and primary divergence), despite clear differences earlier on. PCA and admixture proportions gave similar results. Thus, there may be a relatively narrow window of time during which can distinguish between these histories based on patterns of genetic or phenotypic data. With that said, time here is measured in generations, which could represent vastly different amounts of absolute time for species with different life histories and reproductive strategies (e.g. annual plants versus long-lived, clonal trees).

At the end of the simulations (2000 generations), allele frequency clines and population structure were weak overall, particularly when the migration rate was 0.01 (Figures S2–S4). Phenotypic clines were much more pronounced and followed the environmental gradient or ecotone when exogenous selection occurred. This contrast is not surprising even though the neutral SNPs and quantitative trait SNPs were linked on a single chromosome, because without greater allele frequency differences among populations, limited linkage disequilibrium (LD) is expected. A lower migration rate slowed the decay of differences following secondary contact, but also resulted

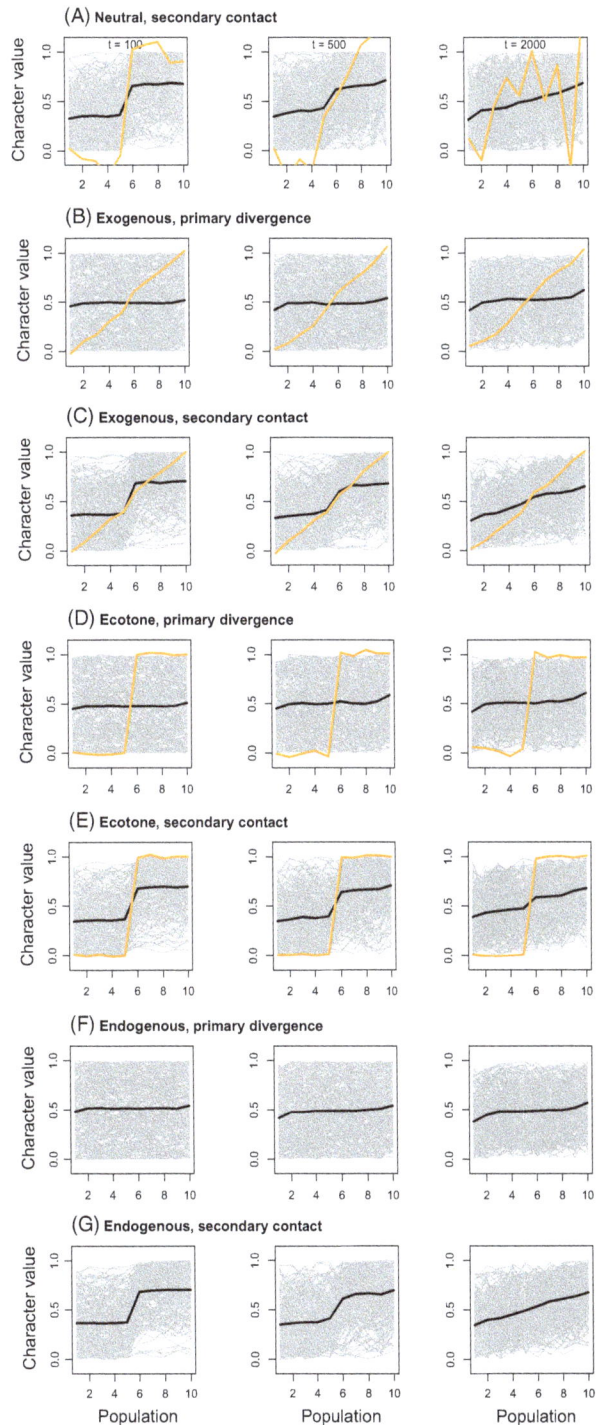

Figure 1 Plots show neutral allele frequency (gray) and quantitative trait (orange) clines from simulated data with a migration rate of 0.001. The mean allele frequency cline with SNPs polarized such that the allele plotted was rarer in patch 1 than patch 10 is depicted with a black line. Clines after 100, 500, and 2000 generations are shown. Results from a single simulation are shown, but replicate simulations produced qualitatively similar results. Clines from simulations with a higher migration rate of 0.01 are shown in Figure S2.

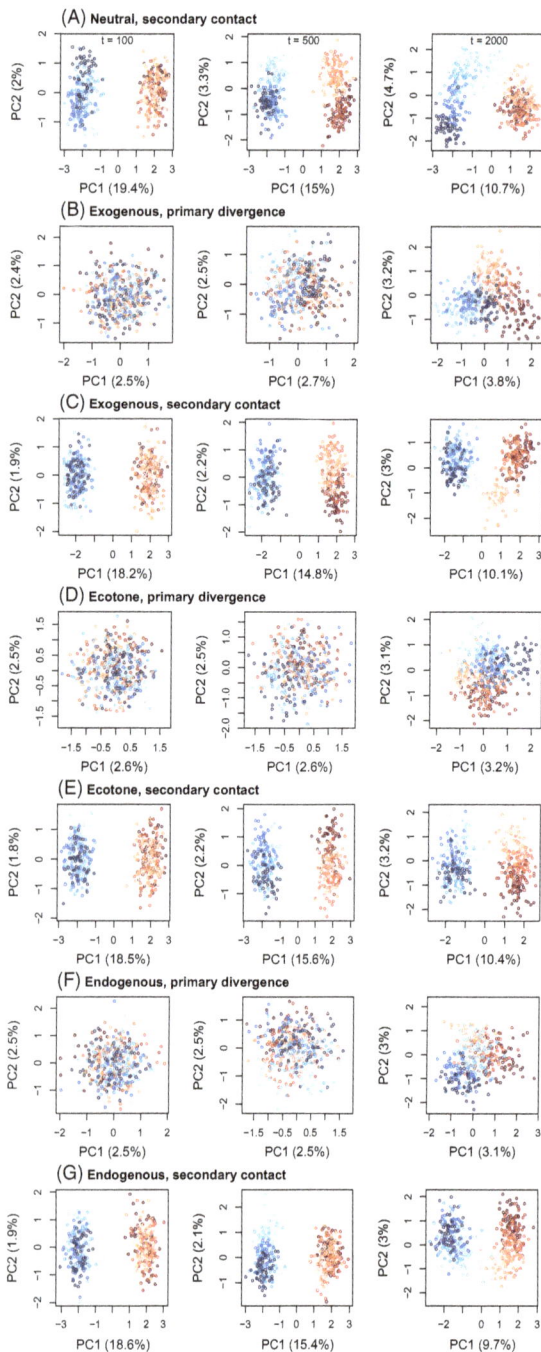

Figure 2 Scatterplots summarize patterns of genotypic variation for simulated data based on principal component analysis (PCA). Points denote individuals and are colored based on patch (dark red and dark blue for patches 1 and 10, with lighter shades indicating patches closer to the center). Results are shown for a migration rate of 0.001 and 100, 500, or 2000 generations. Results from a single simulation are shown, but replicate simulations produced qualitatively similar results. Clines from simulations with a higher migration rate of 0.01 are shown in Figure S3.

in smaller-scale isolation by distance, including sharp phenotypic clines under the neutral secondary contact model, which could be incorrectly attributed to selection. Consistent with previous studies focused on equilibrium dynamics (Kruuk et al. 1999), we found that patterns of variation generated by exogenous and endogenous selection can also be difficult to distinguish earlier in the evolutionary process.

Neutral simulations that included 50 patches resulted in weak population structure overall, and this pattern was robust to sampling a smaller number of individuals per patch (5 vs 50; Fig. 4). However, other sampling approaches resulted in greater distortions of the true population structure. Sampling only center and edge patches resulted in three distinct genotypic clusters, which could be incorrectly interpreted as evidence of an isolated hybrid lineage or even hybrid species (e.g. Gompert et al. 2014). Even sampling only the central patches exaggerates levels of population structure. Together these results highlight the importance of broad geographic sampling to accurately recover clinal variation (also see Witherspoon et al. 2006; Schwartz and McKelvey 2009), as opposed to more limited sampling of putative hybrids and isolated 'pure' parental populations.

Recommendations

Our illustrative simulations are consistent with other theoretical work on hybridization (e.g. Barton and Hewitt 1985; Kruuk et al. 1999; Barton and de Vladar 2009) and show that it will often be difficult to discriminate among different histories of selection and gene flow from genetic data. However, we show that even though primary divergence and secondary contact are thought of as hybridization and result in similar long-term or equilibrium patterns of genetic variation (Barton and Hewitt 1985), recent primary divergence and secondary contact generate different patterns of variation. These differences occur because time is required for LD to buildup between neutral and selected variants with primary divergence (Barton and de Vladar 2009; Flaxman et al. 2014), whereas allele frequency differences between geographically isolated populations will generate LD upon secondary contact. This also means that, during the early stages of hybridization, secondary contact might often lead to segregation of greater functional (and nonfunctional) variation than primary divergence. On the other hand, early stages of primary divergence might be limited to sharp phenotypic and genetic differences for strongly selected characters (e.g. Poelstra et al. 2014; Soria-Carrasco et al. 2014), with less segregating variation for other traits or genes in hybrids. We thus recommend that conservation and management practitioners treat recent primary divergence and secondary contact distinctly, as

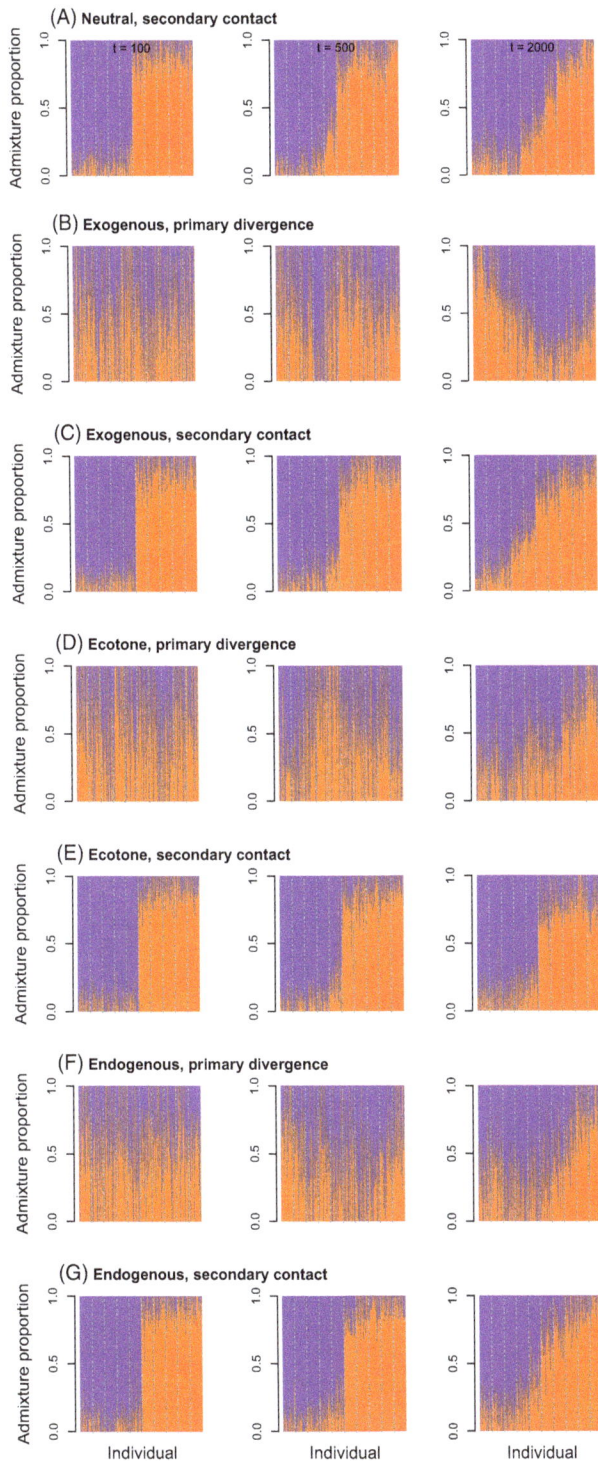

Figure 3 Barplots show maximum likelihood estimates of admixture proportions. Different colors denote ancestry from different hypothetical source populations. Here, we give results for a migration rate of 0.001 and 100, 500, or 2000 generations from a single set of simulations. Replicate simulations produced qualitatively similar results. Admixture from simulations with a higher migration rate of 0.01 is shown in Figure S4.

these processes can be distinguished and have different consequences. Once hybridization has occurred for a greater amount of time, patterns will become similar, and additional data, such as the phylogenetic relationship between or geographic distribution of hybridizing species, will be needed to parse these histories. Our results also show that widespread geographic sampling is important to accurately describe population structure and patterns of hybridization. As argued by practitioners of landscape genetics (e.g. Schwartz and McKelvey 2009), this means that structured, sensible sampling is preferable to sparse opportunistic sampling, or sampling focused on ends of a continuum.

Additional information will likely be gained from studies of hybridization that parse different types of genetic variants rather than treating them all in a single analysis (e.g. Gompert et al. 2014). For example, our simulations and discussion have considered genetic polymorphism, but we have focused on common rather than rare genetic variants. Rare variants, that is genetic variants with minor allele frequencies <1%, have become more accessible with current sequencing methods and could further help discriminate among different histories and provide information about recent evolutionary dynamics (Gravel et al. 2011; Mathieson and McVean 2012; Nelson et al. 2012). In particular, rare variants are often spatially restricted and can be informative about the dispersal of individuals from neighboring populations (Slatkin 1985; Barton and Bengtsson 1986; Gompert et al. 2014) and thus might provide better measures of contemporary gene flow among plant populations of conservation concern. Although more difficult to identify, genetic variants affecting important phenotypes or those linked to such variants could provide additional information if they are strongly structured by the environment (e.g. contrast phenotypic and neutral clines in Fig. 1). When one or a few genes of large effect determine functional phenotypes, it might be useful to examine patterns of genetic variation at these loci. However, when phenotypic variation is due to many variants with smaller effects, statistical approaches that combine information across genetic loci will be more useful (Berg and Coop 2014). Complementary methods that attempt to identify genetic variants potentially affected by selection in hybrids could also be used (e.g. Payseur et al. 2004; Gompert and Buerkle 2009, 2011). Thus, studies of hybridization between crops and wild species or native and non-native plants, as well as gene flow in plants with fragmented populations, would benefit from an increased emphasis on the spread of functional genetic variation via hybridization (e.g. Rieseberg and Willis 2007; Hufford et al. 2013). Such information is needed to determine the fitness consequences of hybridization and thus to decide when hybridization should be valued, allowed or prevented.

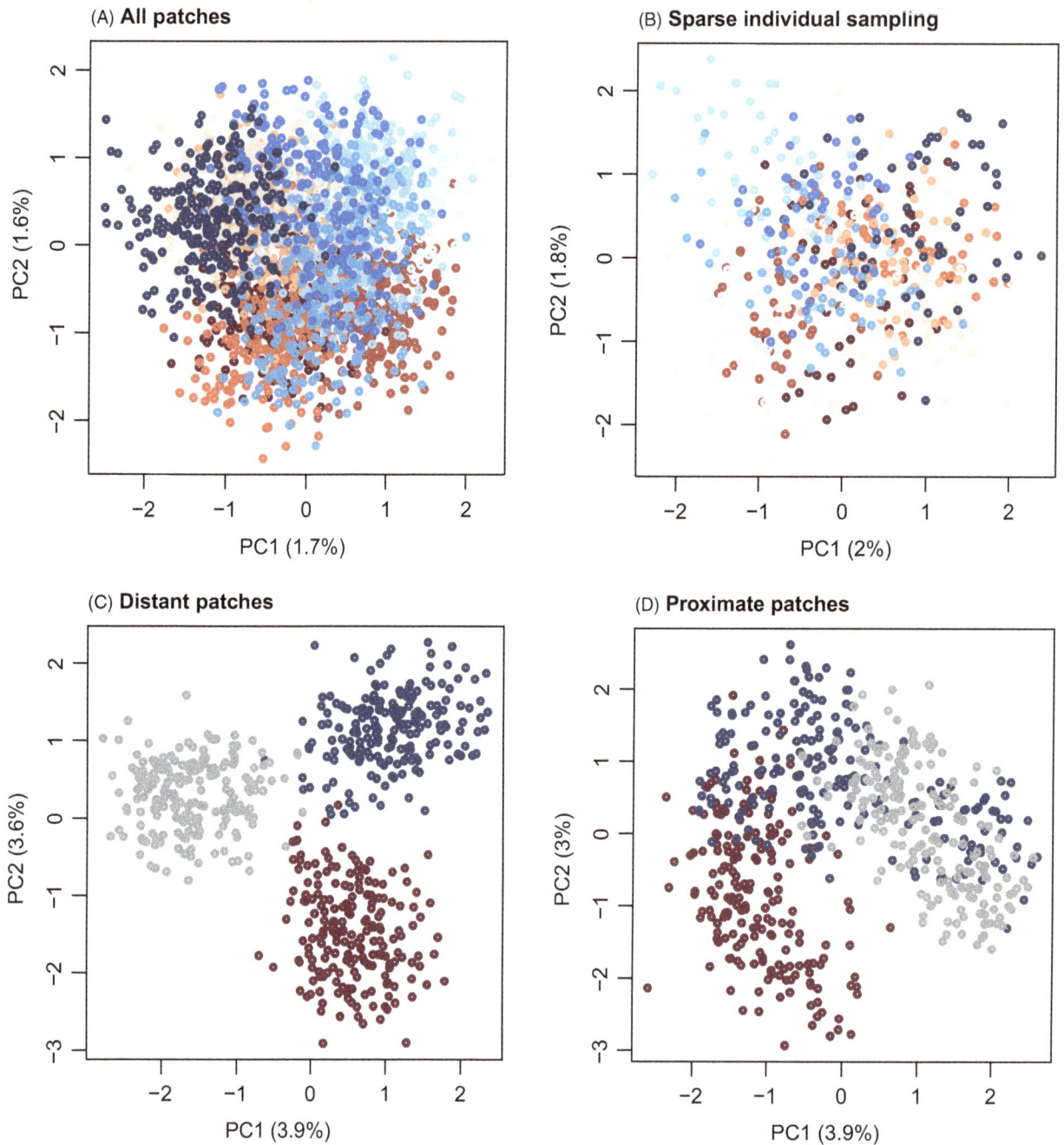

Figure 4 Principal component analysis (PCA) plots illustrate the effect of subsampling on summaries of genetic variation. Points denote individuals and are colored based on patch. Dark red, dark blue, and gray are used to denote peripheral and central patches when a subset of patches were sampled; otherwise dark red and blue indicate patches on opposite ends, with lighter colors used for more central patches. In panes (A) and (B), 50 or 5 individuals were included from each patch. In pane (C), 50 individuals were included from patches 1–4, 24–27 and 47–50, and in pane (D), 50 individuals were sampled from patches 20–31. Results are shown for a migration rate of 0.001 and 100, 500, or 2000 generations.

What, if anything, are hybrids

As noted above, there is a long history of recognizing phenotypically intermediate individuals as putative hybrids between differentiated parental populations or species, including the use of multivariate phenotypic analysis (Alston and Turner 1962; Hatheway 1962; Freeman et al. 1991) and gaining understanding of evolutionary relationships through crossing studies (e.g. Heiser 1947, 1956; Rieseberg 2000). The advent of molecular markers gave rise to the use of genetic information as the basis of inference of ancestry and the recognition of hybrids (e.g. Harrison and Arnold 1982; Vanlerberghe et al. 1986; Barton and Gale 1993). A variety of statistical models exist to support the recognition of hybrids and their distinction from individuals from parental populations (including species), both using population genetic (Boecklen and Howard 1997; Barton 2000; Pritchard et al. 2000; Anderson and Thompson 2002; Falush et al. 2003) and tree-based models (Durand et al. 2011; Patterson et al. 2012). These various models and their implementation in software allow quantitative, model-based recognition of hybrids, given sufficient, informative genetic data (Anderson and Thompson 2002; Falush et al. 2003; Vaha and Primmer 2006). Yet, the papers that describe these models often include explicit cautionary statements regarding the difficulty of distinguishing among different hybrid genealogies, as well as distinguishing hybrids from parentals (e.g. Barton 2000; Anderson and Thompson 2002). Aside from the problem of alleles shared between parental taxa and the resulting imperfect information about ancestry from allelic state, hybrids can be difficult to recognize simply because genetic recombination and sexual reproduction in different genealogies can lead to the same, ambiguous combination of alleles in genotypes. While the genetic variation that results from hybridization is known, it is not clear that as biologists we appreciate the extent to which different hybrid genealogies can lead to the same genetic composition. To illustrate the overlapping expectations for ancestry and genotypic composition of hybrids, we present a simple set of simulations in this section (reprising related simulations and results in Fitzpatrick 2012; Gompert et al. 2014; Lindtke et al. 2014), and their continuous variation along multiple dimensions of hybridization. These illustrations lead to the conclusions that it can be misleading to think about ancestry categories of hybrids and that hybrids will often be genetically and functionally diverse.

The fractional contribution of two (or more) parental taxa to the ancestry of hybrids is a common measure of hybridity and ancestry and is typically referred to as a hybrid index (Barton and Gale 1993; Boecklen and Howard 1997; Buerkle 2005) or admixture proportion (Pritchard et al. 2000; Falush et al. 2003). In the simple case of putative hybridization between two parental taxa, the hybrid index or admixture proportion (q) corresponds to variation along a single axis, with parental ancestry at each end and hybrids intermediate. Summarizing admixture in this way is very common, but it also disregards important information about the history of admixture (Barton 2000; Anderson and Thompson 2002; Fitzpatrick 2012; Lindtke et al. 2012; Gompert et al. 2014; Lindtke et al. 2014). For example, F_1 individuals will have a hybrid index of 0.5, but this is also the expected (mean) hybrid index of any

$F_2 \cdots F_n$ hybrid individuals, which do not have one of the parental taxa as a parent after the first generation of hybridization (i.e. they have experienced no backcrossing). Consequently, whereas a hybrid index does quantify a continuum of genetic hybridity and is preferable to a categorical analysis, it cannot discriminate among very different genealogies, including the differences in ancestry between an F_2 and an F_{20}. Additional information can be obtained from a second dimension of admixture, the fraction of loci that combine ancestry from the two parental taxa, which has been referred to as interspecific heterozygosity or interpopulation ancestry (denoted Q_{12} here; Barton 2000; Fitzpatrick and Shaffer 2007; Fitzpatrick 2012; Lindtke et al. 2012; Gompert et al. 2014; Lindtke et al. 2014). Some software models this parameter explicitly from genetic data in hybrids and source populations (e.g. `HIest` and `entropy`; models for interpopulation ancestry are described in Fitzpatrick 2012; Gompert et al. 2014), but the most commonly used software for admixture analysis does not (`structure`; Pritchard et al. 2000; Falush et al. 2003). The combination of admixture proportion (q) and interpopulation ancestry (Q_{12}) contains additional information about admixture histories and thus is a general tool for summarizing the genomic composition of hybrids. For one, it allows identification of individuals that had a parental taxon as an immediate parent (including F_1 and any backcrossed hybrids), as these have maximal Q_{12} for a given q.

Simulations and analyses

As has been done in previous studies (Fitzpatrick 2012; Gompert et al. 2014; Lindtke et al. 2014), we performed individual-based simulations of hybridization. In the first set, we repeatedly modeled two generations of hybridization that included parental, F_1, F_2, and backcross (BC) individuals. In a second set, we used replicates to generate expectations for the ancestry of F_2, F_5 and F_{20} individuals. The simulations were of finite populations of 50 individuals that contribute to the parentage of any set of progeny (F_1, F_2, etc.). Diploid meiotic recombination and segregation were modeled, with 1000 marker loci distributed across 10 chromosomes, and a single, randomly located crossover per chromosome in each gamete. Thus, we were able to track ancestry with complete knowledge. To superimpose allelic states (including shared alleles between parental taxa and polymorphism within), we utilized an F-model for shared ancestry of parental taxa and the genetic drift they experienced relative to the common ancestor (as above, and in Balding and Nichols 1995; Falush et al. 2003), with a beta distribution of allele frequencies in the ancestral population with parameters α and β equal to 0.8 (this distribution has a mean allele frequency of 0.5 and a standard deviation of 0.31). We set $F_{ST} = 0.5$ and only considered a random subset of 1000 marker loci with a minor allele frequency >0.05 in the sampled individuals (i.e. what are typically referred to as 'polymorphic' loci or common variants). We arbitrarily sampled 20 individuals of each of the parental taxa, 20 F_2, and 10 each F_5, F_{20}, and BC to each parental taxon. The simulations were performed in R (version 3.2.2; R Development Core Team 2015) and the script to perform the simulations is in the Supporting information.

Our simulations tracked both the ancestry and allelic state of loci, and we present summaries of both (Fig. 5). Because we simulated admixture, we had perfect knowledge of the admixture proportions and interpopulation ancestry rather than needing to infer them. If one were to infer ancestries based on models and software (e.g. Gompert et al. 2014), there would be more uncertainty and variance around the true values shown here (uncertainty in ancestry is inversely proportional to allele frequency differences between the parental taxa, that is, to the extent that allelic state is informative about ancestry). With the level of allele frequency difference between our parental populations ($F_{ST} = 0.5$), recognition of parental individuals and distinguishing them from all hybrids was unambiguous

with PCA (Fig. 5, PCA performed in R; R Development Core Team 2015; it would be more difficult to distinguish parental and hybrid individuals based on allelic state if parental populations were more similar genetically). In terms of ancestry, F_1 individuals were distinguishable from more advanced generation F_n hybrids (F_2, F_5, and F_{20}) and backcrossed individuals on the basis of their maximal interpopulation ancestry. Likewise, BC individuals are recognizable on the basis of their maximal interpopulation ancestry for a given admixture proportion. Distinguishing among different generations of backcrossing (e.g. whether F_1, or F_2 was hybrid parent) would not be possible based only on the information contained in Q_{12} and q (knowledge of chromosomal blocks of ancestry would be helpful; Gompert and Buerkle 2013). Segregation even in the F_1 is highly variable and ancestry in later generation hybrid F_n parents is expected to overlap with that of the F_1. More generally and as noted in previous research, discriminating between genealogies beyond the first two generations of admixture is difficult (Barton 2000; Anderson and Thompson 2002) without additional information. This is illustrated in these simulations by the overlapping expectations for Q_{12} and q across the individuals in the F_2, F_5, and F_{20} generations. While drift would cause Q_{12} to decline over further

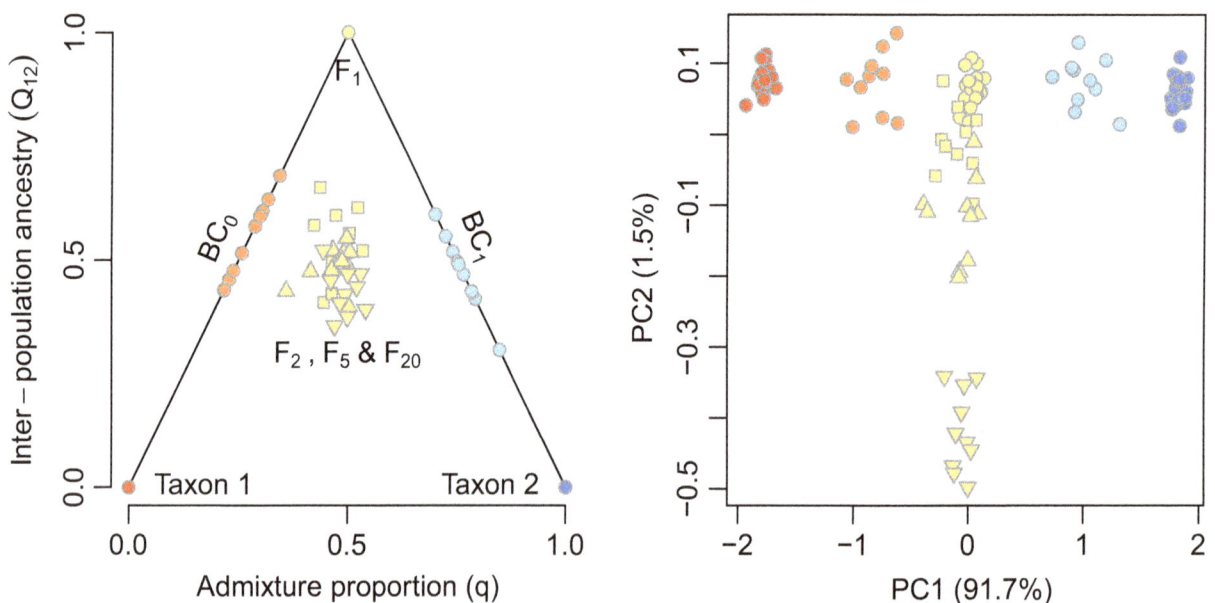

Figure 5 Ancestry for simulated individuals from parental taxa (Taxon 1 and 2) and hybrids vary in admixture proportion (q) and the fraction of loci at which individuals have ancestry from both parental taxa (Q_{12}, interpopulation ancestry; left pane of plot). Hybrids that are progeny from a cross involving one (BC) or both (F_1) parental taxa have maximal interpopulation ancestry for a given admixture proportion (on the edges of the triangle). In contrast, progeny from crosses between hybrid individuals ($F_2 \cdots F_n$) has less than maximal interpopulation ancestry for a given admixture proportion. Principal component analysis (PCA) of genetic covariances among individuals in the simulated population (right pane) shows that genetic differences between the parental species (ancestry variation) constitute the dominant axis of genetic variation (colors as in left pane). $F_1 \cdots F_n$ are genetically intermediate on PC1, and across all hybrids, PC1 mirrors the admixture proportion. F_{20} individuals (downward-pointing triangles) are distinguishable genetically from earlier F_n hybrids and in general PC2 is associated with genetic variation among F_n generations.

generations and ultimately lead to the fixation of ancestry states in finite populations over time (Stam 1980; Chapman and Thompson 2002, 2003; MacLeod et al. 2005; Buerkle and Rieseberg 2008), 20 generations are insufficient to have a detectable effect in a simulated population of 50 individuals.

If hybridization is restricted to two generations, with sufficient genomic sampling, it can be possible to identify different parental, F_1, F_2, and BC categories. Here, we have considered perfect knowledge of ancestry and previous work has addressed sources of uncertainty that would lessen the prospects for clear expectations for ancestry (Anderson and Thompson 2002; Vaha and Primmer 2006; Burgarella et al. 2009; Fitzpatrick 2012). But beyond the issues of whether hybridization involves only two generations of hybridization and uncertainty in empirical ancestry estimates, analysis of ancestry categories is problematic because these classes mask the fact that they will contain genetic and functional phenotypic variation. Perhaps we stand to be the most misled in the case of F_1 individuals, where allelic polymorphisms in the parental populations will result in genetically variable F_1 individuals, contrary to the expectation for a single genotype of F_1 resulting from typical experimental crosses between homozygous parents. This genetic variation is evident among F_1 individuals in our simulations (Fig. 5). This variation would be even greater if one considered F_1 individuals from different geographic locations, where allele frequencies in parental taxa are likely to differ even more (e.g. Gompert et al. 2014; Mandeville et al. 2015). Somewhat similarly, categorical treatment of ancestry in hybrids leads to overlapping expectations for Q_{12} and q in F_2, F_5, and F_{20} individuals, but analysis of their allelic states shows that F_{20} individuals differ genetically from early F_n generations (PCA in Fig. 5). These genetic differences could be responsible for functional phenotypic differences and comparable functional differences might arise from more subtle genotypic differences between F_n generations.

Overall, the use of the term *hybrid classes* or *categories*, and methods for their inference, could obscure important variation that exists within classes. Instead, Rieseberg and Carney (1998) suggested it is worthwhile to focus on the fitness of individual genotypes, rather than hybrid classes. Certainly, our simple model illustrates that it can be nonsensical to refer to the 'fitness of hybrids', as the genetic and ancestry composition of hybrids can be highly variable and hybrids would be expected to vary substantially for phenotypes. Given the expected genetic and phenotypic variability within hybrids, and the potential for transgressive phenotypes (Rieseberg et al. 1999a, 2003b), discussion of hybrid fitness should be in the context of the typical complexity of tying phenotype (including fitness) to genotype in natural populations, which is particularly difficult in variable environments and in variable genetic backgrounds (Weiss 2008; Rockman 2012).

Recommendations

In both applied and basic science settings, knowledge of the existence and attributes of hybrids can provide a foundation for learning about species interactions and maintenance (Arnold 2006; Allendorf et al. 2013). For example, a predominance of BC hybrids would lead to genetic exchange between parental taxa and a potential local erosion of species differences, whereas if hybrids are restricted to relatively abundant F_n individuals, these will affect the demography of parental taxa through wasted reproductive effort on F_1 hybrid progeny and possibly through competition. Our simple model reflects our understanding that the ancestry and genetic composition of hybrids vary along multiple axes and treatment of hybrids as a singular entity would disregard potentially important variation. Thus, management decisions might need to consider the types of hybrids generated and could even accommodate different actions for different hybrids within the same biological system.

Furthermore, hybrids beyond the F_1 will also vary in ancestry along their chromosomes, both in tracts of ancestry that have not yet recombined, and as a result of drift and selection leading individual loci to deviate from the average ancestry in the genome (reviewed in Gompert and Buerkle 2013). For this reason, hybrids have been of interest to evolutionary biologists who are interested in the genetics of species boundaries (Rieseberg et al. 1999b; Rieseberg and Buerkle 2002; Buerkle and Lexer 2008; Payseur 2010; Gompert et al. 2012). Incomplete reproductive isolation and hybridization have provided support for the 'genic view' of speciation and species boundaries (Wu 2001; Abbott et al. 2013). Additionally, recent studies in a variety of taxa have drawn attention to variability in the genetic outcomes of hybridization that followed secondary contact between the same pairs of species in multiple locations or contexts (Rieseberg 2006; Nolte et al. 2009; Teeter et al. 2010; Lepais and Gerber 2011; Lagache et al. 2013; Gompert et al. 2014; Mandeville et al. 2015). For both applied and basic evolutionary biology, this variability in outcomes means that it can be difficult to formulate categorical statements about the composition, importance, and likely conservation threats of hybrids. The empirical abundance of parental taxa and hybrids at one site may or may not be informative about other locations where the taxa co-occur (e.g. Aldridge and Campbell 2009; Mandeville et al. 2015). Likewise, as noted above, genetic variation in parents and hybrids makes it difficult to make categorical statements about the genotypes, phenotypes, and fitness of hybrids (e.g. Sweigart et al. 2007; Good et al. 2008). This challenge is not a matter of uncertainty that arises from analytical approaches and software, but is inherent to the process of hybridization, as we have illustrated with the simulations in this paper.

Overall, these complexities mean it will be difficult to know the consequences of hybridization without detailed study (Allendorf et al. 2013), which could include estimation of multiple dimensions of ancestry (Gompert and Buerkle 2013), sampling multiple geographic locations and contexts (e.g. Hamilton et al. 2013; Haselhorst and Buerkle 2013; Gompert et al. 2014; Mandeville et al. 2015), and characterization of the demography of parental and hybrid individuals in populations (e.g. Carney et al. 2000; Fitzpatrick and Shaffer 2007).

Synthesis and conclusions

A common theme of our results and discussion is that hybridization is not a singular process, but rather a collection of related processes with variable outcomes and consequences. In support of this, as noted above, empirical studies have often documented variation in outcomes of hybridization in different locations or contexts, in terms of the genomic composition of hybrids, patterns of introgression, and the ecological consequences of hybridization (e.g. Yanchukov et al. 2006; Lepais et al. 2009; Nolte and Tautz 2010; Teeter et al. 2010; Nice et al. 2013; Gompert et al. 2014; Mandeville et al. 2015). Indeed, consistent outcomes of hybridization appear to be mostly limited to taxa that exhibit limited intraspecific variation for loci affecting fitness and where endogenous selection dominates (e.g. Buerkle and Rieseberg 2001). Such variability limits our ability to predict the outcome of specific instances of hybridization and thus is relevant for our understanding of evolutionary biology in general and has practical consequences for management. For example, invasion by a non-native species could result in extirpation of a native species in one area but not in another, or transgene escape from a crop could occur readily into some wild populations but not others. Thus, it might be difficult to make valid general statements about the threats or benefits of hybridization, even for individual species. Likewise, extrapolation from single empirical examples (i.e. studies or sites) could be problematic. Clearly, such problems will be exacerbated when species exhibit substantial isolation by distance or local adaptation and fail to function as cohesive entities. We conclude by noting that while we have focused on hybridization between pairs of diploid populations or species, the points we have made should also apply when hybridization generates polyploids or involves multiple species. However, in such cases, even more factors could affect the ecological and evolutionary

dynamics, rendering the outcomes of these instances of hybridization even less predictable.

Acknowledgements

We thank Norman Ellstrand and Loren Rieseberg for the invitation to contribute to this special issue. Computing, storage, and other resources from the Division of Research Computing in the Office of Research and Graduate Studies at Utah State University are gratefully acknowledged.

Literature cited

Abbott, R., D. Albach, S. Ansell et al. 2013. Hybridization and speciation. Journal of Evolutionary Biology 26:229–246.

Aldridge, G., and D. Campbell 2009. Genetic and morphological patterns show variation in frequency of hybrids between *Ipomopsis* (*Polemoniaceae*) zones of sympatry. Heredity 102:257–265.

Alexander, D. H., J. Novembre, and K. Lange 2009. Fast model-based estimation of ancestry in unrelated individuals. Genome Research 19:1655–1664.

Allendorf, F., R. Leary, P. Spruell, and J. Wenburg 2001. The problems with hybrids: setting conservation guidelines. Trends in Ecology & Evolution 16:613–622.

Allendorf, F. W., G. Luikart, and S. N. Aitken 2013. Conservation and the Genetics of Populations, 2nd edn. Wiley-Blackwell, Hoboken, NJ.

Alston, R. E., and B. L. Turner 1962. New techniques in analysis of complex natural hybridization. Proceedings of the National Academy of Sciences of the USA 48:130–137.

Anderson, E. C., and E. A. Thompson 2002. A model-based method for identifying species hybrids using multilocus genetic data. Genetics 160:1217–1229.

Arnold, M. L. 2006. Evolution Through Genetic Exchange. Oxford University Press, Oxford, UK.

Balao, F., R. Casimiro-Soriguer, J. L. García-Castaño, A. Terrab, and S. Talavera 2015. Big thistle eats the little thistle: does unidirectional introgressive hybridization endanger the conservation of *Onopordum hinojense*? The New Phytologist 206:448–458.

Balding, D. J., and R. A. Nichols 1995. A method for quantifying differentiation between populations at multi-allelic loci and its implications for investigating identity and paternity. Genetica 96:3–12.

Barton, N. H. 2000. Estimating multilocus linkage disequilibria. Heredity 84:373–389.

Barton, N. 2013. Does hybridization influence speciation? Journal of Evolutionary Biology 26:267–269.

Barton, N., and B. Bengtsson 1986. The barrier to genetic exchange between hybridizing populations. Heredity 57:357–376.

Barton, N. H., and M. A. R. de Cara 2009. The evolution of strong reproductive isolation. Evolution 63:1171–1190.

Barton, N. H., and K. S. Gale 1993. Genetic analysis of hybrid zones. In R. G. Harrison, ed. Hybrid Zones and the Evolutionary Process, pp. 13–45. Oxford University Press, New York, NY.

Barton, N. H., and G. M. Hewitt 1985. Analysis of hybrid zones. Annual Review of Ecology and Systematics 16:113–148.

Barton, N. H., and H. P. de Vladar 2009. Statistical mechanics and the evolution of polygenic quantitative traits. Genetics 181:997–1011.

Becker, M., N. Gruenheit, M. Steel et al. 2013. Hybridization may facilitate in situ survival of endemic species through periods of climate change. Nature Climate Change **3**:1039–1043.

Berg, J. J., and G. Coop 2014. A population genetic signal of polygenic adaptation. PLoS Genetics **10**:e1004412.

Blackman, B. K., S. D. Michaels, and L. H. Rieseberg 2011. Connecting the sun to flowering in sunflower adaptation. Molecular Ecology **20**:3503–3512.

Boecklen, W. J., and D. J. Howard 1997. Genetic analysis of hybrid zones: numbers of markers and power of resolution. Ecology **78**:2611–2616.

Buerkle, C. A. 2005. Maximum-likelihood estimation of a hybrid index based on molecular markers. Molecular Ecology Notes **5**:684–687.

Buerkle, C. A., and C. Lexer 2008. Admixture as the basis for genetic mapping. Trends in Ecology & Evolution **23**:686–694.

Buerkle, C. A., and L. H. Rieseberg 2001. Low intraspecific variation for genomic isolation between hybridizing sunflower species. Evolution **55**:684–691.

Buerkle, C. A., and L. H. Rieseberg 2008. The rate of genome stabilization in homoploid hybrid species. Evolution **62**:266–275.

Buerkle, C. A., D. E. Wolf, and L. H. Rieseberg 2003. The origin and extinction of species through hybridization. In C. A. Brigham, and M. W. Schwartz, eds. Population Viability in Plants: Conservation, Management, and Modeling of Rare Plants, pp. 117–141. Springer Verlag, New York, NY.

Burgarella, C., Z. Lorenzo, R. Jabbour-Zahab et al. 2009. Detection of hybrids in nature: application to oaks (*Quercus suber* and *Q. ilex*). Heredity **102**:442–452.

Carney, S. E., K. A. Gardner, and L. H. Rieseberg 2000. Evolutionary changes over the fifty-year history of a hybrid population of sunflowers (*Helianthus*). Evolution **54**:462–474.

Chapman, M. A., and J. M. Burke 2006. Letting the gene out of the bottle: the population genetics of genetically modified crops. The New Phytologist **170**:429–443.

Chapman, N. H., and E. A. Thompson 2002. The effect of population history on the lengths of ancestral chromosome segments. Genetics **162**:449–458.

Chapman, N. H., and E. A. Thompson 2003. A model for the length of tracts of identity by descent in finite random mating populations. Theoretical Population Biology **64**:141–150.

Driebe, E., and T. Whitham 2000. Cottonwood hybridization affects tannin and nitrogen content of leaf litter and alters decomposition. Oecologia **123**:99–107.

Durand, E. Y., N. Patterson, D. Reich, and M. Slatkin 2011. Testing for ancient admixture between closely related populations. Molecular Biology and Evolution **28**:2239–2252.

Ellstrand, N. C. 1992. Gene flow by pollen: implications for plant conservation genetics. Oikos **63**:77–86.

Ellstrand, N. C. 2001. When transgenes wander, should we worry? Plant Physiology **125**:1543–1545.

Ellstrand, N. C., P. Meirmans, J. Rong et al. 2013. Introgression of crop alleles into wild or weedy populations. Annual Review of Ecology, Evolution, and Systematics **44**:325–345.

Endler, J. A. 1977. Geographic Variation, Speciation, and Clines. Princeton University Press, Princeton, NJ.

Escobar, J. S., A. Nicot, and P. David 2008. The different sources of variation in inbreeding depression, heterosis and outbreeding depression in a metapopulation of *Physa acuta*. Genetics **180**:1593–1608.

Falush, D., M. Stephens, and J. K. Pritchard 2003. Inference of population structure using multilocus genotype data: linked loci and correlated allele frequencies. Genetics **164**:1567–1587.

Fitzpatrick, B. 2012. Estimating ancestry and heterozygosity of hybrids using molecular markers. BMC Evolutionary Biology **12**:131.

Fitzpatrick, B. M., and H. B. Shaffer 2007. Hybrid vigor between native and introduced salamanders raises new challenges for conservation. Proceedings of the National Academy of Sciences of the USA **104**:15793–15798.

Flaxman, S. M., A. C. Wacholder, J. L. Feder, and P. Nosil 2014. Theoretical models of the influence of genomic architecture on the dynamics of speciation. Molecular Ecology **23**:4074–4088.

Floate, K. D., J. Godbout, M. K. Lau, N. Isabel, and T. G. Whitham 2016. Plant-herbivore interactions in a trispecific hybrid swarm of *Populus*: assessing support for hypotheses of hybrid bridges, evolutionary novelty and genetic similarity. The New Phytologist **209**:832–844, 2015-19480.

Freeman, D. C., W. A. Turner, E. D. McArthur, and J. H. Graham 1991. Characterization of a narrow hybrid zone between two subspecies of big sagebrush (*Artemisia tridentata*: Asteraceae). American Journal of Botany **78**:805–815.

Garnier, A., H. Darmency, Y. Tricault, A. M. Chèvre, and J. Lecomte 2014. A stochastic cellular model with uncertainty analysis to assess the risk of transgene invasion after crop-wild hybridization: oilseed rape and wild radish as a case study. Ecological Modelling **276**:85–94.

Gómez, J. M., A. González-Megías, J. Lorite, M. Abdelaziz, and F. Perfectti 2015. The silent extinction: climate change and the potential hybridization-mediated extinction of endemic high-mountain plants. Biodiversity and Conservation **24**:1843–1857.

Gompert, Z., and C. A. Buerkle 2009. A powerful regression-based method for admixture mapping of isolation across the genome of hybrids. Molecular Ecology **18**:1207–1224.

Gompert, Z., and C. A. Buerkle 2011. Bayesian estimation of genomic clines. Molecular Ecology **20**:2111–2127.

Gompert, Z., and C. A. Buerkle 2013. Analyses of genetic ancestry enable key insights for molecular ecology. Molecular Ecology **22**:5278–5294.

Gompert, Z., T. L. Parchman, and C. A. Buerkle 2012. Genomics of isolation in hybrids. Philosophical Transactions of the Royal Society of London B: Biological Sciences **367**:439–450.

Gompert, Z., L. K. Lucas, C. C. Nice, J. A. Fordyce, C. A. Buerkle, and M. L. Forister 2013. Geographically multifarious phenotypic divergence during speciation. Ecology and Evolution **3**:595–613.

Gompert, Z., L. K. Lucas, C. A. Buerkle, M. L. Forister, J. A. Fordyce, and C. C. Nice 2014. Admixture and the organization of genetic diversity in a butterfly species complex revealed through common and rare genetic variants. Molecular Ecology **23**:4555–4573.

Good, J. M., M. A. Handel, and M. W. Nachman 2008. Asymmetry and polymorphism of hybrid male sterility during the early stages of speciation in house mice. Evolution **62**:50–65.

Gravel, S., B. M. Henn, R. N. Gutenkunst et al. 2011. Demographic history and rare allele sharing among human populations. Proceedings of the National Academy of Sciences of the USA **108**:11983–11988.

Guillaume, F., and J. Rougemont 2006. Nemo: an evolutionary and population genetics programming framework. Bioinformatics **22**:2556–2557.

Hamilton, J. A., C. Lexer, and S. N. Aitken 2013. Genomic and phenotypic architecture of a spruce hybrid zone (*Picea sitchensis* × *P. glauca*). Molecular Ecology **22**:827–841.

Harrison, R. G. 1993. Hybrids and Hybrid Zones: Historical Perspective, Chap. 1, pp. 3–12. Oxford University Press, New York, NY.

Harrison, R. G., and J. Arnold 1982. A narrow hybrid zone between closely related cricket species. Evolution **36**:535–552.

Haselhorst, M. S. H., and C. A. Buerkle 2013. Population genetic structure of *Picea engelmannii*, *P. glauca* and their previously unrecognized hybrids in the central Rocky Mountains. Tree Genetics & Genomes **9**:669–681.

Hatheway, W. H. 1962. A weighted hybrid index. Evolution **16**:1–10.

Heiser, C. B. Jr 1947. Hybridization between the sunflower species *Helianthus annuus* and *H. petiolaris*. Evolution **1**:249–262.

Heiser, C. B. 1956. Biosystematics of *Helianthus debilis*. Madroño **13**:145–167.

Hufford, M. B., P. Lubinksy, T. Pyhäjärvi, M. T. Devengenzo, N. C. Ellstrand, and J. Ross-Ibarra 2013. The genomic signature of crop-wild introgression in maize. PLoS Genetics **9**:e1003477.

Kawakami, T., T. J. Morgan, J. B. Nippert et al. 2011. Natural selection drives clinal life history patterns in the perennial sunflower species, *Helianthus maximiliani*. Molecular Ecology **20**:2318–2328.

Kremer, A., O. Ronce, J. J. Robledo-Arnuncio et al. 2012. Long-distance gene flow and adaptation of forest trees to rapid climate change. Ecology Letters **15**:378–392.

Kruuk, L. E. B., S. J. E. Baird, K. S. Gale, and N. H. Barton 1999. A comparison of multilocus clines maintained by environmental adaptation or by selection against hybrids. Genetics **153**:1959–1971.

Lagache, L., E. K. Klein, E. Guichoux, and R. J. Petit 2013. Fine-scale environmental control of hybridization in oaks. Molecular Ecology **22**:423–436.

Lepais, O., and S. Gerber 2011. Reproductive patterns shape introgression dynamics and species succession within the European white oak species complex. Evolution **65**:156–170.

Lepais, O., R. Petit, E. Guichoux et al. 2009. Species relative abundance and direction of introgression in oaks. Molecular Ecology **18**:2228–2242.

Levin, D. A., J. K. Francisco-Ortega, and R. K. Jansen 1996. Hybridization and the extinction of rare plant species. Conservation Biology **10**:10–16.

Linder, C. R., I. Taha, G. J. Seiler, A. A. Snow, and L. H. Rieseberg 1998. Long-term introgression of crop genes into wild sunflower populations. Theoretical and Applied Genetics **96**:339–347.

Lindtke, D., C. A. Buerkle, T. Barbará et al. 2012. Recombinant hybrids retain heterozygosity at many loci: new insights into the genomics of reproductive isolation in *Populus*. Molecular Ecology **21**:5042–5058.

Lindtke, D., Z. Gompert, C. Lexer, and C. A. Buerkle 2014. Unexpected ancestry of *Populus* seedlings from a hybrid zone implies a large role for postzygotic selection in the maintenance of species. Molecular Ecology **23**:4316–4330.

MacLeod, A. K., C. S. Haley, J. A. Woolliams, and P. Stam 2005. Marker densities and the mapping of ancestral junctions. Genetical Research **85**:69–79.

Mallet, J., M. Beltran, W. Neukirchen, and M. Linares 2007. Natural hybridization in Heliconiine butterflies: the species boundary as a continuum. BMC Evolutionary Biology **7**:28.

Mandeville, E. G., T. L. Parchman, D. B. McDonald, and C. A. Buerkle 2015. Highly variable reproductive isolation among pairs of catostomus species. Molecular Ecology **24**:1856–1872.

Martin, A., and V. Orgogozo 2013. The loci of repeated evolution: a catalog of genetic hotspots of phenotypic variation. Evolution **67**:1235–1250.

Martinsen, G. D., K. D. Floate, A. M. Waltz, G. M. Wimp, and T. G. Whitham 2000. Positive interactions between leafrollers and other arthropods enhance biodiversity on hybrid cottonwoods. Oecologia **123**:82–89.

Mathieson, I., and G. McVean 2012. Differential confounding of rare and common variants in spatially structured populations. Nature Genetics **44**:243–246.

Mayr, E. 1963. Animal Species and Evolution. Harvard University Press, Cambridge, MA.

Moran, E. V., J. Willis, and J. S. Clark 2012. Genetic evidence for hybridization in red oaks (*Quercus* sect. *Lobatae*, *Fagaceae*). American Journal of Botany **99**:92–100.

Moyle, L. C., and T. Nakazato 2010. Hybrid incompatibility "snowballs" between *Solanum* species. Science **329**:1521–1523.

Muhlfeld, C. C., R. P. Kovach, L. A. Jones et al. 2014. Invasive hybridization in a threatened species is accelerated by climate change. Nature Climate Change **4**:620–624.

Navarro, A., and N. H. Barton 2003. Chromosomal speciation and molecular divergence–accelerated evolution in rearranged chromosomes. Science **300**:321–324.

Nelson, M. R., D. Wegmann, M. G. Ehm et al. 2012. An abundance of rare functional variants in 202 drug target genes sequenced in 14,002 people. Science **337**:100–104.

Nice, C. C., Z. Gompert, J. A. Fordyce, M. L. Forister, L. K. Lucas, and C. A. Buerkle 2013. Hybrid speciation and independent evolution in lineages of alpine butterflies. Evolution **67**:1055–1068.

Nolte, A. W., and D. Tautz 2010. Understanding the onset of hybrid speciation. Trends in Genetics **26**:54–58.

Nolte, A. W., Z. Gompert, and C. A. Buerkle 2009. Variable patterns of introgression in two sculpin hybrid zones suggest that genomic isolation differs among populations. Molecular Ecology **18**:2615–2627.

Nosil, P., Z. Gompert, T. E. Farkas et al. 2012. Genomic consequences of multiple speciation processes in a stick insect. Proceedings of the Royal Society of London B: Biological Sciences **279**:5058–5065.

Patterson, N., P. Moorjani, Y. Luo et al. 2012. Ancient admixture in human history. Genetics **192**:1065–1093.

Payseur, B. A. 2010. Using differential introgression in hybrid zones to identify genomic regions involved in speciation. Molecular Ecology Resources **10**:806–820.

Payseur, B. A., J. G. Krenz, and M. W. Nachman 2004. Differential patterns of introgression across the X chromosome in a hybrid zone between two species of house mice. Evolution **58**:2064–2078.

Petit, R. J., C. Bodénès, A. Ducousso, G. Roussel, and A. Kremer 2004. Hybridization as a mechanism of invasion in oaks. The New Phytologist **161**:151–164.

Poelstra, J. W., N. Vijay, C. M. Bossu et al. 2014. The genomic landscape underlying phenotypic integrity in the face of gene flow in crows. Science **344**:1410–1414.

Pritchard, J. K., M. Stephens, and P. Donnelly 2000. Inference of population structure using multilocus genotype data. Genetics **155**:945–959.

R Development Core Team 2015. R: A Language and Environment for Statistical Computing. R Foundation for Statistical Computing, Vienna, Austria, ISBN 3-900051-07-0.

Rhymer, J. M., and D. Simberloff 1996. Extinction by hybridization and introgression. Annual Review of Ecology and Systematics **27**:83–109.

Rieseberg, L. H. 1991. Homoploid reticulate evolution in *Helianthus* (Asteraceae): evidence from ribosomal genes. American Journal of Botany **78**:1218–1237.

Rieseberg, L. H. 2000. Crossing relationships among ancient and experimental sunflower hybrid lineages. Evolution **54**:859–865.

Rieseberg, L. H. 2006. Hybrid speciation in wild sunflowers. Annals of the Missouri Botanical Garden **93**:34–48.

Rieseberg, L. H., and C. A. Buerkle 2002. Genetic mapping in hybrid zones. The American Naturalist **159**:S36–S50.

Rieseberg, L. H., and S. E. Carney 1998. Tansley review no. 102 plant hybridization. The New Phytologist 140:599–624.

Rieseberg, L. H., and J. F. Wendel 1993. Introgression and its consequences in plants. In R. G. Harrison, ed. Hybrid Zones and the Evolutionary Process, pp. 70–109. Oxford University Press, New York, NY.

Rieseberg, L. H., and J. H. Willis 2007. Plant speciation. Science 317:910–914.

Rieseberg, L. H., R. Carter, and S. Zona 1990. Molecular tests of the hypothesized hybrid origin of two diploid *Helianthus* species (Asteraceae). Evolution 44:1498–1511.

Rieseberg, L. H., C. Van Fossen, and A. Desrochers 1995. Hybrid speciation accompanied by genomic reorganization in wild sunflowers. Nature 375:313–316.

Rieseberg, L. H., M. A. Archer, and R. K. Wayne 1999a. Transgressive segregation, adaptation, and speciation. Heredity 83:363–372.

Rieseberg, L. H., J. Whitton, and K. Gardner 1999b. Hybrid zones and the genetic architecture of a barrier to gene flow between two sunflower species. Genetics 152:713–727.

Rieseberg, L. H., O. Raymond, D. M. Rosenthal et al. 2003a. Major ecological transitions in wild sunflowers facilitated by hybridization. Science 301:1211–1216.

Rieseberg, L. H., A. Widmer, A. M. Arntz, and J. M. Burke 2003b. The genetic architecture necessary for transgressive segregation is common in both natural and domesticated populations. Philosophical Transactions of the Royal Society of London B: Biological Sciences 358:1141–1147.

Rockman, M. V. 2012. The QTN program and the alleles that matter for evolution: all that's gold does not glitter. Evolution 66:1–17.

Roe, A. D., C. J. K. MacQuarrie, M. C. Gros-Louis et al. 2014. Fitness dynamics within a poplar hybrid zone: I. Prezygotic and postzygotic barriers impacting a native poplar hybrid stand. Ecology and Evolution 4:1629–1647.

Rutledge, L. Y., S. Devillard, J. Q. Boone, P. A. Hohenlohe, and B. N. White 2015. RAD sequencing and genomic simulations resolve hybrid origins within North American Canis. Biology Letters 11:20150303.

Schwartz, M., and K. McKelvey 2009. Why sampling scheme matters: the effect of sampling scheme on landscape genetic results. Conservation Genetics 10:441–452.

Slatkin, M. 1985. Rare alleles as indicators of gene flow. Evolution 39:53–65.

Soria-Carrasco, V., Z. Gompert, A. A. Comeault et al. 2014. Stick insect genomes reveal natural selection's role in parallel speciation. Science 344:738–742.

Stam, P. 1980. The distribution of the fraction of the genome identical by descent in finite random mating populations. Genetical Research 35:131–155.

Stebbins, G. L. 1950. Variation and Evolution in Plants. Columbia University Press, New York, NY.

Stewart, C. N., M. D. Halfhill, and S. I. Warwick 2003. Transgene introgression from genetically modified crops to their wild relatives. Nature Reviews Genetics 4:806–817.

Sweigart, A. L., A. R. Mason, and J. H. Willis 2007. Natural variation for a hybrid incompatibility between two species of *Mimulus*. Evolution 61:141–151.

Teeter, K. C., L. M. Thibodeau, Z. Gompert, C. A. Buerkle, M. W. Nachman, and P. K. Tucker 2010. The variable genomic architecture of isolation between hybridizing species of house mouse. Evolution 64:472–485.

Vaha, J., and C. Primmer 2006. Efficiency of model-based Bayesian methods for detecting hybrid individuals under different hybridization scenarios and with different numbers of loci. Molecular Ecology 15:63–72.

Vanlerberghe, F., B. Dod, P. Boursot, M. Bellis, and F. Bonhomme 1986. Absence of Y-chromosome introgression across the hybrid zone between *Mus musculus domesticus* and *Mus musculus musculus*. Genetics Research 48:191–197.

Warschefsky, E., R. V. Penmetsa, D. R. Cook, and E. J. von Wettberg 2014. Back to the wilds: tapping evolutionary adaptations for resilient crops through systematic hybridization with crop wild relatives. American Journal of Botany 101:1791–1800.

Weiss, K. M. 2008. Tilting at quixotic trait loci (QTL): an evolutionary perspective on genetic causation. Genetics 179:1741–1756.

Whitham, T., J. Bailey, J. Schweitzer et al. 2006. A framework for community and ecosystem genetics: from genes to ecosystems. Nature Reviews Genetics 7:510–523.

Witherspoon, D., E. Marchani, W. Watkins et al. 2006. Human population genetic structure and diversity inferred from polymorphic L1 (LINE-1) and Alu insertions. Human Heredity 62:30–46.

Wolf, D. E., N. Takebayashi, and L. H. Rieseberg 2001. Predicting the risk of extinction through hybridization. Conservation Biology 15:1039–1053.

Wu, C. I. 2001. The genic view of the process of speciation. Journal of Evolutionary Biology 14:851–865.

Yanchukov, A., S. Hofman, J. M. Szymura, and S. V. Mezhzherin 2006. Hybridization of *Bombina bombina* and *B. variegata* (Anura, Discoglossidae) at a sharp ecotone in Western Ukraine: comparisons across transects and over time. Evolution 60:583–600.

Zalapa, J. E., J. Brunet, and R. P. Guries 2010. The extent of hybridization and its impact on the genetic diversity and population structure of an invasive tree, *Ulmus pumila* (Ulmaceae). Evolutionary Applications 3:157–168.

6

Herbicide cycling has diverse effects on evolution of resistance in *Chlamydomonas reinhardtii*

Mato Lagator,[1*] Tom Vogwill,[1] Nick Colegrave[2] and Paul Neve[1]

[1] School of Life Sciences, University of Warwick Coventry, UK
[2] School of Biological Sciences, Institute of Evolutionary Biology, University of Edinburgh Edinburgh, UK

Keywords

Chlamydomonas reinhardtii, cross-resistance, experimental evolution, fitness costs, herbicide resistance, herbicide rotation

*Correspondence

Mato Lagator, School of Life Sciences, University of Warwick, Coventry CV4 7AL, UK.

e-mail: m.lagator@warwick.ac.uk

Abstract

Cycling pesticides has been proposed as a means of retarding the evolution of resistance, but its efficacy has rarely been empirically tested. We evolved populations of *Chlamydomonas reinhardtii* in the presence of three herbicides: atrazine, glyphosate and carbetamide. Populations were exposed to a weekly, biweekly and triweekly cycling between all three pairwise combinations of herbicides and continuously to each of the three herbicides. We explored the impacts of herbicide cycling on the rate of resistance evolution, the level of resistance selected, the cost of resistance and the degree of generality (cross-resistance) observed. Herbicide cycling resulted in a diversity of outcomes: preventing evolution of resistance for some combinations of herbicides, having no impacts for others and increasing rates of resistance evolution in some instances. Weekly cycling of atrazine and carbetamide resulted in selection of a generalist population. This population had a higher level of resistance, and this generalist resistance was associated with a cost. The level of resistance selected did not vary amongst other regimes. Costs of resistance were generally highest when cycling was more frequent. Our data suggest that the effects of herbicide cycling on the evolution of resistance may be more complex and less favourable than generally assumed.

Introduction

Synthetic herbicides have become the dominant means of controlling weedy plants in agricultural settings (Powles and Shaner 2001), and evolution of resistance to herbicides is widespread (Heap 2011). In general terms, there are two modes of herbicide resistance evolution: target-site resistance and non-target-site resistance (reviewed in Powles and Yu 2010). Target-site resistance confers resistance to a single herbicide mode of action, whereas non-target-site resistance may result in complex patterns of cross-resistance rendering populations resistant to multiple modes of action (Powles and Yu 2010). In evolutionary terms, target-site and non-target-site resistance represent specialist and generalist modes of herbicide resistance, respectively. As both mechanisms can provide resistance to the same herbicide, specialist and generalist phenotypes can coexist.

A key challenge in herbicide, as well as pesticide and antibiotic resistance research, is to design management strategies that effectively deploy a range of modes of action

to retard or prevent evolution of resistance (Georghiou and Taylor 1986; Powles and Yu 2010). A commonly recommended practice is to cycle chemicals with different modes of action (Beckie 2006). Cycling (often referred to as herbicide rotation) introduces temporal environmental heterogeneity so that sequential generations are exposed to different selection pressures. This can potentially affect the rate of resistance evolution in a number of ways. First, over a given time scale, fewer generations are exposed to any single environment, leading to reduced selection for resistance to each component environment (MacArthur 1964; Futuyma and Moreno 1988; Whitlock 1996). Second, if adaptation to one environment incurs a fitness cost in others, cycling may retard or even prevent resistance evolution (Leeper et al. 1986; Gressel and Segel 1990). Additionally, environments in which herbicides are cycled are more complex and may require a greater degree of genetic variation for adaptation to occur. However, ecological and evolutionary theory would predict that environments characterized by a greater degree of temporal heterogeneity would result in the evolution of more generalist phenotypes (Gavrilets and Scheiner 1993; Chesson 2000; Kassen 2002), and hence it may also be the case that cycling exacerbates the spread of generalist resistance phenotypes (Gomulkiewicz and Kirkpatrick 1992; Tufto 2000). This effect is therefore likely to crucially depend on the frequency of cycling between different modes of action, with more rapid rates of switching more strongly favouring generalist types of resistance.

The difficulties associated with performing selection experiments on large weed populations with slow generation times (one generation per year) have limited the testing of these hypotheses mostly to theoretical and simulation models, with only a few experimental studies (Porcher et al. 2004; Roux et al. 2005; Kover et al. 2009; Springate et al. 2011). Models have shown that, in the absence of pleiotropic costs of resistance, cycling may not retard resistance evolution (Diggle et al. 2003; Bergstrom et al. 2004; Roux et al. 2008). It is not possible to generalize on the existence of pleiotropic costs associated with evolved resistance to herbicides, as it seems that fitness costs vary according to the mechanism of resistance (reviewed by Vila-Aiub et al. 2009). A similar lack of understanding of the dynamics of resistance evolution has led to failed attempts to slow the spread of resistance to antibiotics in clinical settings (Bergstrom and Feldgarden 2007).

The techniques of microbial experimental evolution can be applied to a range of fundamental and more applied evolutionary questions (Buckling et al. 2009), including the evolution of resistance to antimicrobials (Perron et al. 2008; Hall et al. 2010; MacLean et al. 2010). Experimental evolution with *Chlamydomonas reinhardtii* offers the potential to better understand the evolution of resistance to herbicides and to experimentally test resistance management strategies. *Chlamydomonas reinhardtii* is a unicellular green chlorophyte, capable of growing as a photoautotroph and a heterotroph. Under laboratory conditions it grows asexually (Harris 2008). It is susceptible to a range of commercial herbicides (Reboud et al. 2007). In the current study, we experimentally evolved populations of *C. reinhardtii* with sequential cycling between pairwise combinations of three herbicides with different modes of action: glyphosate, atrazine and carbetamide. The frequency of cycling between herbicides was varied to explore the impacts of the degree of environmental heterogeneity on the dynamics of resistance evolution. In particular, we were interested in investigating if (i) cycling leads to reduced rates of resistance evolution, (ii) there was a relationship between the frequency of cycling and the rates and outcomes of evolution, (iii) cycling leads to comparable levels of resistance as homogeneous environments and (iv) cycling could result in the selection of more generalist resistance phenotypes.

Methods and materials

Founding population

Chlamydomonas reinhardtii CC-1690, a wild-type positive mating strain obtained from the *Chlamydomonas* Resource Center's core collection, was used in this experiment. Prior to selection experiments, the strain had been adapted to liquid Bold's medium through continuous exposure for over 700 generations. Two weeks before the start of selection procedure, 20 μL of the founding population (approximately 15 000 cells) was spread on an agar plate. After 7 days of growth, a single colony was picked and used to inoculate a Bold's medium liquid culture. This colony was multiplied for 7 days and was used to found all experimentally evolving populations.

Culture conditions

The culture medium used in all experimental conditions is modified Bold's Medium (subsequently BM) (Colegrave et al. 2002). Populations were grown in disposable borosilicate glass 25 × 150 mm tubes, in 20 mL of BM and maintained in an orbital shaker incubator, at 28°C and 180 rpm, under continuous light exposure, provided by six fluorescent tubes mounted in the incubator lid (Osram L30 W/21-840, cool white; light intensity at the location of the tubes was 161 μmol m^{-2}s^{-1}). Cultures were transferred into fresh BM every 7 days (see below), during which time the ancestral population growing in the absence of herbicides would have reached stationary phase (3.1×10^7 cells).

Herbicides

We exposed populations to three herbicides: atrazine, glyphosate and carbetamide. The herbicides have different modes of action (atrazine, photosystem II inhibitor; glyphosate, inhibitor of aromatic amino acid synthesis and carbutamide, mitosis inhibitor). Prior to the selection procedure, we determined the minimum inhibitory concentration (MIC) of each herbicide, this being the minimum concentration that prevented detectable population growth over 4 days.

Cycling regimes

Three experimental conditions involved continuous exposure to a single herbicide (A0 denoting continuous exposure to atrazine, G0 to glyphosate and C0 to carbetamide). A weekly, biweekly and triweekly cycling regime was created for all three possible pairwise combinations of herbicides (AG1 denoting the weekly cycle between atrazine and glyphosate, AG2 the biweekly cycle and so on). Each experimental condition (12 in total) was replicated 6 times, giving rise to 72 independently evolving populations. Six populations were propagated by serial transfer in the absence of herbicides. Throughout the experiment one of these six 'source' populations provided immigration into each of the six replicate treatments as required (see below). These populations also acted as controls.

Approximately 125 000 cells (estimated by absorbance at 750 nm) from the founding population provided the initial population for each of the 78 populations. At each transfer, 200 μL of the evolving culture was transferred into fresh media. If the number of cells in 200 μL of culture medium was estimated to be <125 000, as would happen until resistance was developed, then the appropriate number of cells from one of the source populations was added to make the total cell number at the transfer approximately 125 000. Therefore, the minimum number of cells at the beginning of each cycle was 125 000. For each of the six replicates, the same source population was used for immigration throughout the experiment. According to this protocol, when undergoing sufficient growth (at least 6.64 cell division in 7 days), a population is under soft selection and capable of maintaining itself after the weekly bottleneck event. When growth did not reach this number of cell divisions, weekly bottlenecks would drive the population towards extinction, and these populations were maintained by immigration from the corresponding source population. The experiment was carried out for 12 transfer cycles (12 weeks).

Measuring the rates of evolution

The optical density at 750 nm (OD_{750}) was measured in a Jenway 6315 benchtop-spectrophotometer 4 days after the transfer. OD_{750} was converted into population size using a calibration curve obtained by correlating OD_{750} with cell counts in 70 independent samples. Resistance was considered to have evolved when detectable population growth was consistently measured ($OD_{750} > 0.045$, corresponding to at least three cell divisions). The rate of resistance evolution was quantified by measuring the first week when resistance was observed. The rate of resistance evolution to each component herbicide in cycling regimes was expressed as the number of weeks that the population had been exposed to that herbicide.

Isolation of the evolved populations

In order to ensure that populations used for subsequent resistance and fitness assays contained only herbicide-resistant cells, approximately 20 000 cells of each final population were plated on BM agar plates that contained the MIC of a single herbicide. For cycling regimes, 20 000 cells of

each final population were plated independently onto two plates, one containing each of the herbicides that the population had been exposed to. After 7 days of growth, 200 colonies from each population were randomly selected and used to inoculate a fresh population in liquid BM. If the population had been exposed to two herbicides, 100 colonies were randomly selected from each of the plates containing those herbicides and used to inoculate a fresh population in liquid BM. These populations were grown for 7 days prior to conducting further assays. In addition, for lines evolving under cycling regimes, 10 single colonies from each BM + herbicide plate were picked and multiplied for 7 days in BM. For all 10 populations, 125 000 cells were then transferred into MIC of the second herbicide from that cycling regime. In all cases, populations derived from single cells were resistant to both herbicides in the cycling regime, indicating that evolved populations always consisted of individuals with resistance to both herbicides cycled, rather than to mixtures of individuals with resistance to individual cycle components.

Level of resistance and fitness in the ancestral environment

The growth rate of the evolved populations in the selective environments (hereinafter 'level of resistance') was determined by measuring population growth in liquid culture at the MIC of both herbicides to which that population had been exposed. A total of 125 000 cells were used to inoculate each resistance assay, and population size was determined by measuring optical density at 750 nm after 4 days growth. This was replicated twice for each evolved population, and the mean number of cell divisions completed during 4 days growth was used as an estimate of the level of resistance. In order to assess if adaptation to herbicide environments was associated with a fitness cost, the comparative growth rate of evolved and source populations in the herbicide-free environment was estimated. This measure was replicated twice for each evolved population and the mean number of cell divisions completed during 4 days growth was used as an estimate of fitness in the ancestral environment. Both levels of resistance and fitness in the ancestral environment were expressed as a proportion of the growth of source populations in the ancestral (BM only) environment.

The degree of generality

To test for cross-resistance, we assayed the growth of evolved populations at the MIC of four herbicides to which they had no previous exposure (tembotrione, iodosulfuron-methyl-sodium, isoproturon and S-metolachlor), as well as whichever of atrazine, glyphosate or carbetamide they had not been exposed to (i.e. we also assayed cross-resistance to carbetamide in populations evolved in cycling between atrazine and glyphosate). A total of 125 000 cells of the evolved populations were inoculated into tubes containing one of these herbicides and population growth was measured after 4 days. If growth was significantly different from that of the source populations under the same conditions, the population was deemed to have evolved cross-resistance. Each condition was replicated twice.

Cross-protection assays

To investigate a possible contribution of cross-protection, the phenomenon whereby exposure to one stress provides a degree of physiological acclimation (cross-protection) to subsequent stresses, we grew naïve *C. reinhardtii* populations in the presence of low doses (0.8 MIC for atrazine, 0.7 MIC for glyphosate and carbetamide) of each of our three herbicides. Doses below MIC were used so that detectable population growth was apparent between transfer periods. After 7 days in one herbicide we transferred 125 000 cells into below MIC doses of each of the two other herbicides. We also transferred 125 000 cells without previous herbicide exposure into below MIC doses of all three herbicides as a control. Four days after transfer, growth rates of each population were estimated. Each condition was replicated three times.

Statistical analysis

The rate of resistance evolution (weeks to resistance) was analyzed using a Cox regression. The herbicide regime was fitted as a covariate, with the ancestral immigration source as the strata. For cycling regimes, the number of weeks until resistance evolved to individual herbicide components (weeks exposed to that herbicide) were compared to rates of evolution of resistance when continuously exposed to that herbicide. The Cox regressions were performed in SPSS. The level of resistance and fitness in the ancestral environments of the evolved populations were first analyzed using a General Linear Model with the herbicide cycled and the cycling frequency as fixed factors, and ancestral immigration source as the random factor. We also investigated the interaction between herbicide and cycling frequency. When populations under a cycling regime evolved resistance to only one of the herbicides, we only analyzed the effects of the cycling frequency, making it a fixed factor. The level of resistance to individual herbicide in cycling regimes was subsequently compared to resistance in the continuous exposure treatment using a Dunnett's corrected paired *t*-test, with the herbicide regime fitted as a fixed factor, and the ancestral immigration source as the random factor. When some populations in a regime did

not evolve resistance, we compared them to the continuous exposure treatment using a Dunnett's corrected *t*-test. The level of resistance of the three continuous exposure populations was compared in the same fashion. The fitness in the ancestral environment of all populations was compared to source populations and to populations that underwent continuous exposure using a Dunnett's corrected paired *t*-test, except when some of the populations in a regime did not evolve resistance, in which case Dunnett's corrected *t*-test was used. The fitness in the ancestral environment of the three continuous exposure regimes was compared in the same fashion. Growth rates from the cross-protection assay were compared between the populations that underwent previous exposure to an herbicide and those that did not in a Dunnett's corrected paired *t*-test. The previous herbicide the population was exposed to was fitted as a fixed factor, and the replicate population as the random factor.

Results

Dynamics of herbicide resistance

Evolution of herbicide resistance was observed in many populations, under various continuous exposure and cycling regimes. Resistance evolved in all populations with continuous exposure to atrazine (Fig. 1A) or glyphosate (Fig. 1B), and to both herbicides in all populations that underwent cycling between these two herbicides (Fig. 1A,B). Resistance evolved in two of six populations that underwent continuous carbetamide exposure (Fig. 1C), while resistance to both atrazine and carbetamide evolved in three of six populations that underwent weekly cycling between the two (Fig. 1A,C). Atrazine, but not carbetamide resistance, evolved in all populations under a bi- and triweekly cycle between the two herbicides (Fig. 1A,C). No resistant individuals were observed in the populations cycling between glyphosate and carbetamide (Fig. 1B,C). These results demonstrate that cycling can prevent, accelerate or have no impact on the evolution of resistance to herbicides.

Continuous exposure to glyphosate resulted in significantly more rapid evolution of resistance than continuous exposure to atrazine ($z = 6.096$, $P < 0.05$) or carbetamide ($z = 6.083$, $P < 0.05$). Rates of evolution of atrazine and carbetamide resistance were not significantly different.

The number of weeks until resistance evolved to individual herbicides in cycling regimes was compared for each regime to the rate of evolution in populations that underwent continuous exposure to that herbicide. Resistance to atrazine evolved more rapidly in a weekly cycle between atrazine and glyphosate ($z = 10.169$, $P = 0.001$) (Fig. 1A). Though there was a trend towards more rapid evolution of atrazine resistance in the biweekly ($z = 3.381$, $P = 0.066$) and triweekly cycle with glyphosate ($z = 3.369$, $P = 0.066$),

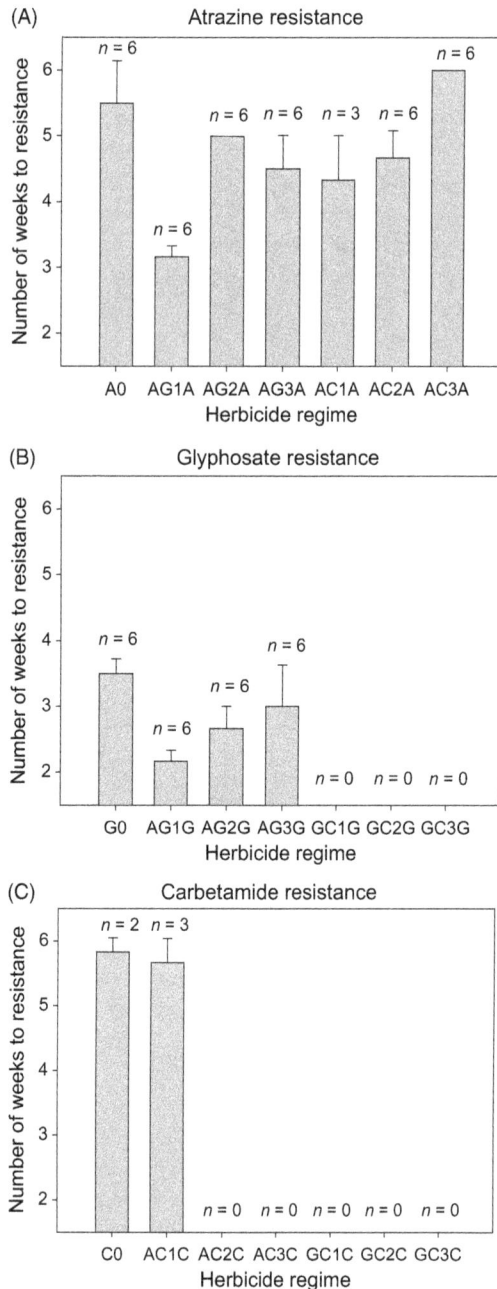

Figure 1 The dynamics of resistance evolution measured as number of weeks until resistance evolved. Bars represent the mean weeks to resistance amongst the replicates where resistance was observed; n is the number of replicate populations that evolved resistance: (A) atrazine resistance (A0 indicates continuous exposure to atrazine, AG1, AG2, AG3 a weekly, biweekly and triweekly rotation between atrazine and glyphosate, respectively. AC1, AC2 and AC3 refer to weekly, biweekly and triweekly rotation between atrazine and carbetamide, respectively); (B) glyphosate resistance (labelling convention as above) and (C) carbetamide resistance. Error bars are standard errors of the mean.

these differences were not significant (Fig. 1A). A weekly cycle between atrazine and glyphosate yielded faster-evolving resistance to glyphosate than continuous exposure to glyphosate ($z = 3.930$, $P = 0.047$) (Fig. 1B). Rates of evolution of carbetamide resistance were not significantly different between any of the regimes in which it evolved.

Level of resistance

We express the level of resistance as the proportion of growth rate retained in populations with evolved resistance in comparison to source populations in herbicide-free environments. In continuous selection regimes, the level of resistance was greater in populations exposed to glyphosate than in atrazine-resistant ($T_{10} = 19.61$, $P < 0.01$) and carbetamide-resistant populations ($T_6 = 5.963$, $P < 0.005$). Carbetamide-resistant populations had a higher level of resistance than atrazine-resistant populations ($T_6 = 4.854$, $P < 0.01$).

Overall, in cycling regimes, the herbicide that atrazine was cycled with had no significant impact on the level of atrazine resistance. However, the frequency of cycling did significantly affect the level of resistance ($F_{2,16} = 8.10$, $P < 0.005$), and there was a significant interaction between the herbicide used and the frequency of cycling ($F_{2,16} = 8.03$, $P < 0.005$). As indicated by Dunnett's corrected t-tests, the levels of atrazine resistance that evolved in the AC1 regime were significantly greater than in continuous atrazine exposure regimes ($T_7 = 5.487$, $P < 0.001$), as well as all other regimes (Fig. 2A). For glyphosate and carbetamide resistance there were no significant differences in the level of evolved resistance in any of the regimes in which resistance evolved (Fig. 2B,C).

Fitness in the ancestral environment

Comparing the fitness in the ancestral environment (defined as the relative growth rate of evolved populations in BM) of evolved populations to the source populations, we found fitness costs (a significant difference between growth rate in BM of the ancestral and evolved populations) to be frequently associated with evolved resistance (Fig. 3). All populations that evolved resistance in continuous exposure to a single herbicide exhibited a significant reduction in fitness in the ancestral environment – exposure to atrazine ($T_{10} = -2.80$, $P < 0.05$), glyphosate ($T_{10} = -9.76$, $P < 0.001$) and carbetamide ($T_6 = -4.711$, $P < 0.05$) (Fig. 3). The fitness in the ancestral environment of populations evolved under continuous exposure to atrazine was significantly higher than in the populations evolved in continuous exposure to glyphosate ($T_{10} = 3.95$,

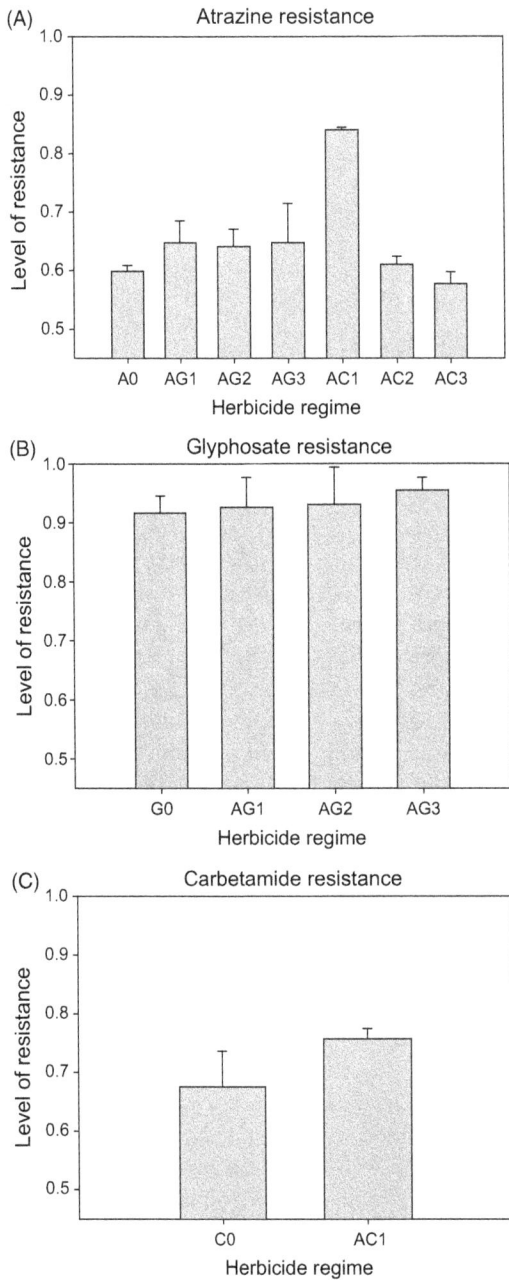

Figure 2 The level of evolved resistance expressed as the proportion of growth retained in herbicide environments in comparison with source populations in herbicide-free environments. Bars are mean values of all the evolved replicates in each condition: (A) atrazine level of resistance, (B) glyphosate level of resistance and (C) carbetamide level of resistance. Error bars are standard errors of the mean.

Figure 3 Growth rates in the absence of herbicides of populations with evolved resistance expressed as the proportion of the source populations' growth rate in herbicide-free environments. Bars are mean values of all the evolved replicates in each condition. Error bars are standard errors of the mean.

triweekly cycle between atrazine and glyphosate or atrazine and carbetamide did not exhibit significant fitness costs.

Cross-resistance

For most selection regimes, no cross-resistance was observed (Fig. 4). Only populations selected under a weekly cycle between atrazine and carbetamide and under continuous exposure to carbetamide exhibited cross-resistance to herbicides to which they had never been exposed (Fig. 4). All of these populations exhibited growth at the MIC of the herbicide tembotrione. All three populations that evolved resistance to both atrazine and carbetamide under a weekly cycle were also resistant to S-metolachlor and iodosulfuron.

Figure 4 Resistance profiles for evolved populations. Hatched shading indicates resistance to herbicides included in corresponding selection regimes. Cross-resistance to herbicides to which populations had no previous exposure is indicated by grey shading. Cross-resistance was only selected in the C0 and AC1 regimes. A = atrazine; G = glyphosate; C = carbetamide; S = S-metolachlor; I = iodosulphuron; Iso = isoproturon; T = tembotrione.

$P < 0.01$) or carbetamide ($T_6 = 3.598$, $P < 0.05$). Reduced fitness in the ancestral environment was also observed in populations under weekly cycle between atrazine and glyphosate ($T_{10} = -5.94$, $P < 0.001$) and weekly cycle between atrazine and carbetamide ($T_7 = -6.034$, $P < 0.001$) (Fig. 3). Populations that evolved in a bi- and

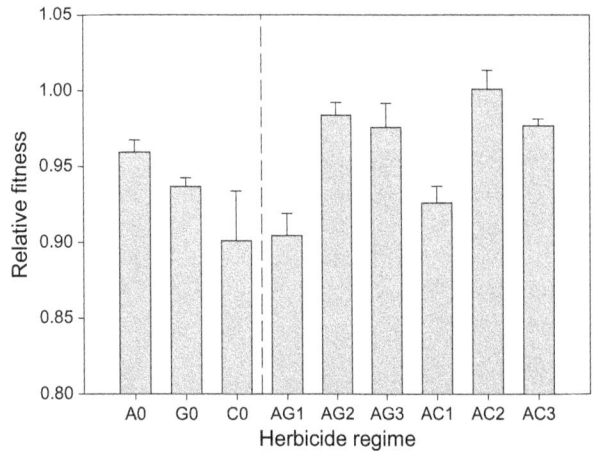

Cross-protection

Seven days of exposure to carbetamide significantly increased the growth rates in 0.8 MIC of atrazine when compared to the populations that had no previous exposure to any herbicides (Fig. 5) ($T_4 = 7.801$, $P < 0.005$). Previous exposure to atrazine significantly increased the growth rates in glyphosate ($T_4 = 7.64$, $P < 0.005$), while the exposure to carbetamide decreased subsequent growth rates in glyphosate ($T_4 = -5.732$, $P < 0.01$).

Discussion

In spite of the lack of evidence for its effectiveness, herbicide cycling has been advocated as a means of slowing or preventing evolution of herbicide resistance (Beckie 2006). A successful cycling strategy must do more than simply extend the chronological time until resistance evolves as this outcome will result simply from the fact that the population is exposed to each component herbicide for less time. A truly effective strategy must increase the time that a population can be exposed (selection time) to, at least one of the cycled herbicides before resistance evolves. In other words, if continuous exposure to herbicide A results in evolution of resistance in selection time x and continuous exposure to herbicide B results in resistance in selection time y, when A and B are cycled, the strategy is successful if either x, y or the sum of x and y is increased. According to these criteria, in this study, we have shown that cycling

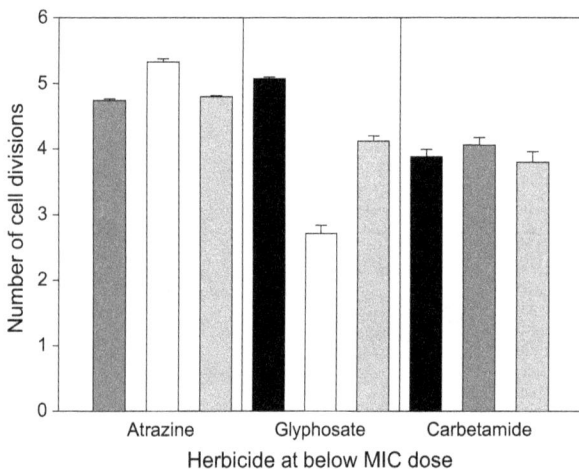

Figure 5 Cross protection. Number of cell divisions the populations underwent after 4 days in below MIC levels of the indicated herbicide. Bars represent mean values. Black bars indicate the populations with previous exposure to atrazine, dark grey bars previous exposure to glyphosate, white bars previous exposure to carbetamide and light grey bars indicate the populations with no previous herbicide exposure. Error bars are standard errors of the mean.

between pairwise combinations of three herbicides can slow, accelerate or have no impact on the dynamics of selection for herbicide resistance. These contrasting outcomes depend on the herbicides being cycled and the frequency of cycling.

Dynamics of resistance under herbicide cycling

Fitness costs associated with resistance are seen as key determinants of the effectiveness of cycling (Leeper et al. 1986; Gressel and Segel 1990; Jasieniuk et al. 1996). In our study, fitness costs (significantly lower growth rates in absence of herbicide) were not universally observed as found in other studies (McCart et al. 2005; Lopes et al. 2008). Models assuming no fitness costs have predicted that cycling will be ineffective in slowing down the evolution of resistance in selection time (Diggle et al. 2003; Neve 2008). Our results support this general trend, as cycling was most effective when occurring between herbicides where evolved resistance yielded the highest cost (glyphosate and carbetamide) and was much less effective when less costly atrazine resistance evolved (Fig. 1).

It seems somewhat counterintuitive that cycling regimes can, in some instances, increase rates of resistance evolution. We offer two explanations: (i) cross-protection, and (ii) population size effects that can account for increased rates of glyphosate and atrazine resistance evolution, respectively, in the AG regimes. Cross-protection gives rise to a temporary increase in growth rates in one stressful environment after exposure to another (Hill et al. 2002), and a variety of sublethal stresses have been shown to alter antibiotic resistance evolution (McMahon et al. 2006). We have found that exposure to atrazine offers positive cross-protection to glyphosate (Fig. 5) and hypothesize that this phenomenon accounts for enhanced rates of glyphosate resistance evolution in the weekly atrazine and glyphosate cycling regime, as it increases the number of non-resistant cells replicating in glyphosate, increasing population size and mutation supply rate. Assuming that cross-protection is a transient effect, this hypothesis is supported by the observation that increased rates of glyphosate resistance evolution are only observed in the weekly cycle. In relation to increased rates of atrazine resistance evolution in the AG1 regime we conclude that increases in population size, driven by the relatively rapid evolution of glyphosate resistance, are resulting in an increased probability of atrazine-resistant mutations arising in the glyphosate-resistant background. Once this occurs, atrazine resistance is selected in both phases of the cycling regime and hence evolution of atrazine resistance (measured in selection-time) is accelerated. We predict that this dynamic is likely to occur when rapid cycling occurs between pesticides where the rate of resistance evolution varies substantially.

Impacts of cycling on the evolution of generalists

The frequency of cycling has the potential to change the trajectory of evolution as evidenced by the evolution of a generalist phenotype in the weekly atrazine and carbetamide cycle and no evolution of resistance in bi- and tri-weekly cycles. Even though this generalist phenotype conferred significantly higher levels of atrazine resistance, it was never selected in the continuous atrazine regime. A number of explanations are possible here. It may be that the generalist phenotype requires fixation of more than one mutation and that the initial mutation confers low levels of resistance to atrazine and carbetamide while carrying a high fitness cost. In a weekly cycle, populations are exposed to carbetamide frequently enough that these mutations are maintained, whereas in other regimes with more frequent or lengthier periods of exposure to atrazine, they are lost due to clonal interference and population bottlenecks. It could also be that the first mutations that get fixed in the population affect the fitness consequences of others, as reported for antibiotic resistance (Trindade et al. 2009; Yeh et al. 2009). Indeed, if the fixation of mutations that confer resistance to atrazine modify the genetic background such that subsequent mutations conferring resistance to carbetamide have a higher selection coefficient (positive epistasis), then generalists are more likely to evolve in a cyclic environment compared to a homogeneous one. In general, it appears that the outcomes of herbicide selection regimes are contingent on complex interactions between the level of resistance, costs of resistance, frequencies of different mutations, cross-resistance phenotypes and the scale of temporal heterogeneity.

In the pesticide and antibiotic literature, generalist resistance usually refers to single mechanisms that confer resistance to multiple toxin modes of action (Delye et al. 2011; for multidrug antibiotic resistance – Alekshun and Levy 2007). The expectation is that generalism will confer lower levels of resistance, often at a higher cost and will therefore only be selected in environments with spatial or temporal variation in selection pressures (Georghiou and Taylor 1986; Futuyma and Moreno 1988; Kassen 2002; Gressel 2009). In this study, broad generalist resistance was selected in the weekly cycle between atrazine and carbetamide, providing some evidence that cycling promotes the evolution of generalist resistance, though in most cycling regimes generalist phenotypes were not observed. Contrary to the major theoretical (Via and Lande 1985; Ravigné et al. 2009), most experimental (Morgan et al. 2009; Legros and Koella 2010; Hall et al. 2010) and the findings in pesticide-resistant organisms (Gressel 2002; Jonsson et al. 2010), we found generalists to have a significantly higher resistance than specialists in both of the selective environments, as well as comparable growth rates in absence of herbicides to the specialists (populations that underwent continuous exposure), a result previously reported for other traits (Turner and Elena 2000; Buckling et al. 2007).

Cycling affects fitness costs

The accumulation of multiple discrete mechanisms of resistance is an alternative means via which a more generalist resistance phenotype may evolve, and it seems likely that this accounts for evolved resistance to atrazine and glyphosate in the atrazine and glyphosate cycling regimes. The evolution of this multiple resistance may be constrained by the accumulation of fitness costs associated with each resistance trait, particularly where these costs are additive, or potentially even synergistic. In populations that evolved resistance in a weekly atrazine and glyphosate cycle, the growth rates in absence of herbicide are significantly lower than in continuous exposure to atrazine and seem to be additive (Fig. 3), suggesting there may be a limit to multiple resistance in the absence of compensations (Andersson and Hughes 2010; Hall et al. 2010). Bi- and triweekly cycles between atrazine and glyphosate resulted in significantly higher growth rates in the absence of herbicides (Fig. 3) than the weekly cycle or continuous exposure to either herbicide. It therefore appears that lower frequencies of cycling favour the compensation of fitness costs as longer periods spent in the non-focal environment will favour selection for reduced costs of resistance. Alternatively, more heterogeneous environments have lower chance of leading to a global optimum (Collins 2011), and as such less rapid rates of cycling could be more effectively selected for mutations with lower fitness cost.

Herbicide cycling: forward with caution

Herbicide cycling has been advocated for resistance management as it introduces environmental complexity and heterogeneity and thus may slow adaptation. Results from this study illustrate that cycling can result in diverse outcomes, though some caution is advisable in translating results to annual weedy plants. Temporal heterogeneity of environments may impact the direction of evolution (Levins 1968; Kassen and Bell 1998; Jasmin and Kassen 2007; Venail et al. 2011), with more fine-grained environments (where environment varies at a rate faster than the generation time) favouring more generalist traits. In our design even the rapid rates of cycling far exceeded the generation time of *C. reinhardtii*, meaning that all the environments were coarse-grained. The herbicide cycling advocated for weed management is fine-grained, generally requiring alternating generations to be exposed to different herbicide modes of action. In addition, the order in which the herbicides are cycled could affect the trajectory of evolution and this was not explored. *Chlamydomonas* is haploid and reproduction in these experiments was asexual. Higher plants have complex and diverse modes of sexual and asexual reproduction. There may also be gene flow between evolving meta-populations of agricultural weeds. Finally, most annual weedy plants have a soil reservoir of dormant seeds that acts as a temporal refuge from herbicide selection. Notwithstanding these important differences, our results clearly demonstrate that herbicide cycling may not always slow the rate of evolution of resistance and may result in the evolution of generalist resistance phenotypes resistant to a broad range of herbicide modes of action.

Acknowledgements

We thank the editor and the anonymous reviewers for helpful and detailed comments. We also thank Carol Evered for her hard work, Andrew Morgan, Charlotte Nellist, Matija Lagator, Ailidh Woodcock and Anthony Carter for helpful comments. This work was funded by Leverhulme Trust.

Literature cited

Alekshun, M. N., and S. B. Levy 2007. Molecular mechanisms of antibacterial multidrug resistance. Cell **128**:1037–1050.

Andersson, D. I., and D. Hughes 2010. Antibiotic resistance and its cost: is it possible to reverse resistance. Nature Reviews Microbiology **8**: 260–271.

Beckie, H. J. 2006. Herbicide-resistant weeds: management tactics and practices. Weed Technology **20**:793–814.

Bergstrom, C. T., and M. Feldgarden 2007. The ecology and evolution of antibiotic-resistant bacteria. In S. Stearns, and J. Koella, eds. Evolution in Health and Disease, pp. 125–137. Oxford University Press, Oxford.

Bergstrom, C. T., M. Lo, and M. Lipsitch 2004. Ecological theory suggests that antimicrobial cycling will not reduce antimicrobial resistance in hospitals. Proceedings of the National Academy of Sciences of the United States of America **101**:13285–13290.

Buckling, A., M. A. Brockhurst, M. Travisano, and P. B. Rainey 2007. Experimental adaptation to high and low quality environments under different scales of temporal variation. Journal of Evolutionary Biology **20**:296–300.

Buckling, A., C. MacLean, M. A. Brockhurst, and N. Colegrave 2009. The beagle in a bottle. Nature **475**:824–829.

Chesson, P. 2000. Mechanisms of maintenance of species diversity. Annual Review of Ecology and Systematics **31**:343–366.

Colegrave, N., O. Kaltz, and G. Bell 2002. The ecology and Genetics of Fitness in *Chlamydomonas*. VIII. The dynamics of adaptation to novel environments after a single episode of sex. Evolution **56**:14–21.

Collins, S. 2011. Many possible worlds: expanding the ecological scenarios in experimental evolution. Evolutionary Biology **38**:3–14.

Delye, C., J. A. C. Gardin, K. Boucansaud, B. Chauvel, and C. Petit 2011. Non-target-site-based resistance should be the centre of attention for herbicide resistance research: *Alopecurus myosuroides* as an illustration. Weed Research **51**:433–437.

Diggle, A. J., P. B. Neve, and F. P. Smith 2003. Herbicides used in combination can reduce the probability of herbicide resistance in finite weed populations. Weed Research **43**:371–382.

Futuyma, D. J., and G. Moreno 1988. The evolution of ecological specialization. Annual Review of Ecology and Systematics **19**:207–233.

Gavrilets, S., and S. M. Scheiner 1993. The genetics of phenotypic plasticity. Theoretical predictions for directional selection. Journal of Evolutionary Biology **6**:49–68.

Georghiou, G. P., and C. E. Taylor 1986. Factor influencing the evolution of resistance. In Pesticide Resistance: Strategies and Tactics for Management, pp. 157–169. National Academy Press, Washington, DC.

Gomulkiewicz, R., and M. Kirkpatrick 1992. Quantitative genetics and the evolution of reaction norms. Evolution **46**:390–411.

Gressel, J. 2002. Molecular Biology of Weed Control. Taylor & Francis, Inc., New York, NY.

Gressel, J. 2009. Evolving understanding of the evolution of herbicide resistance. Pest Management Science **65**:1164–1173.

Gressel, J., and L. A. Segel 1990. Modelling the effectiveness of herbicide rotations and mixtures as strategies to delay or preclude resistance. Weed Technology **4**:186–198.

Hall, A. R., V. F. Griffiths, R. C. MacLean, and N. Colegrave 2010. Mutational neighbourhood and mutation supply rate constrain adaptation in *Pseudomonas aeruginosa*. Proceedings of the Royal Society B: Biological Sciences **277**:643–650.

Harris, E. H. 2008. The *Chlamydomonas* Sourcebook: Introduction to *Chlamydomonas* and its Laboratory Use. Elsevier Inc, New York, NY.

Heap, I. 2011. International Survey of Herbicide Resistant Weeds. Available at: http://www.weedscience.com.

Hill, C., P. D. Cotter, R. D. Sleator, and C. G. M. Gahan 2002. Bacterial stress response in *Listeria monocytogenes*: jumping the hurdles imposed by minimal processing. International Dairy Journal **12**:273–283.

Jasieniuk, M., A. L. Brule-Babel, and I. N. Morrison 1996. The evolution and genetics of herbicide resistance in weeds. Weed Science **44**: 176–193.

Jasmin, J.-N., and R. Kassen 2007. Evolution of a single niche specialist in variable environments. Proceedings of the Royal Society B: Biological Sciences **274**:2761–2767.

Jonsson, N. N., R. J. Miller, D. H. Kemp, A. Knowles, A. E. Ardila, R. G. Verrall, and J. T. Rothwell 2010. Rotation of treatments between spinosad and amitraz for the control of *Rhipicephalus* (*Boophilus*) *microplus* populations with amitraz resistance. Veterinary Parasitology **169**:157–164.

Kassen, R. 2002. The experimental evolution of specialists, generalists, and the maintenance of diversity. Journal of Evolutionary Biology **15**:173–190.

Kassen, B., and G. Bell 1998. Experimental evolution in Chlamydomonas. IV. Selection in environments that vary through time at different scales. Heredity **80**:732–741.

Kover, P. X., J. K. Rowntree, N. Scarcelli, Y. Savriama, T. Eldridge, and B. A. Schaal 2009. Pleiotropic effects of environment-specific adaptation in *Arabidopsis thaliana*. New Phytologist **183**:816–825.

Leeper, J. R., R. T. Roush, and H. T. Reynold 1986. Preventing or managing resistance in arthropods. In Pesticide Resistance: Strategies and Tactics for Management, pp. 335–346. National Academy Press, Washington, DC.

Legros, M., and J. C. Koella 2010. Experimental evolution of specialization by a microsporidian parasite. BMC Evolutionary Biology **10**:159.

Levins, R. 1968. Evolution in Changing Environments. Princeton University Press, Princeton, NJ.

Lopes, P. C., E. Sucena, M. E. Santos, and S. Magalhães 2008. Rapid experimental evolution of pesticide resistance in *C. elegans* entails no costs and affects the mating system. PLoS One 3:e3741.

MacArthur, R. H. 1964. Environmental factors affecting bird species diversity. American Naturalist **98**:387–397.

MacLean, R. C., G. G. Perron, and A. Gardner 2010. Diminishing returns from beneficial mutations and pervasive epistasis shape the fitness landscape for rifampicin resistance in *Pseudomonas aeruginosa*. Genetics **186**:1245–1354.

McCart, C., A. Buckling, and R. H. Constant 2005. DDT resistance in flies carries no cost. Current Biology **15**:R587–R589.

McMahon, M. A. S., J. Xu, J. E. Moore, I. S. Blair, and D. A. McDowell 2006. Environmental stress and antibiotic resistance in food-related pathogens. Applied and Environmental Microbiology **73**:211–217.

Morgan, A. D., R. Maclean, and A. Buckling 2009. Effects of antagonistic coevolution on parasite-mediated host coexistence. Journal of Evolutionary Biology **22**:287–292.

Neve, P. 2008. Simulation modelling to understand the evolution and management of glyphosate resistance in weeds. Pest Management Science **64**:392–401.

Perron, G. G., A. Gonzalez, and A. Buckling 2008. The rate of environmental change drives adaptation to an antibiotic sink. Journal of Evolutionary Biology **21**:1724–1731.

Porcher, E., T. Giraud, I. Goldringer, and C. Lavigne 2004. Experimental demonstration of a causal relationship between heterogeneity of selection and genetic differentiation in quantitative traits. Evolution **58**:1434–1445.

Powles, S., and D. L. Shaner 2001. Herbicide Resistance and World Grains. CRC Press, Boca Raton, FL.

Powles, S. B., and Q. Yu 2010. Evolution in action: plants resistant to herbicides. Annual Review of Plant Biology **61**:317–347.

Ravigné, V., U. Dieckmann, and I. Olivieri 2009. Live where you thrive: joint evolution of habitat choice and local adaptation facilitates specialization and promotes diversity. The American Naturalist **174**: E141–E169.

Reboud, X., N. Majerus, J. Gasquez, and S. Powles 2007. *Chlamydomonas reinhardtii* as a model system for pro-active herbicide resistance evolution research. Bioogical Journal of the Linnean Society **91**:257–266.

Roux, F., C. Camilleri, A. Berard, and X. Reboud 2005. Multigenerational versus single generation studies to estimate herbicide resistance fitness cost in *Arabidopsis thaliana*. Evolution **59**:2264–2269.

Roux, F., M. Paris, and X. Reboud 2008. Delaying weed adaptation to herbicide by environmental heterogeneity: a simulation approach. Pest Management Science **64**:16–29.

Springate, D. A., N. Scarcelli, J. Rowntree, and P. X. Kover 2011. Correlated response in plasticity to selection for early flowering in *Arabidopsis thaliana*. Journal of Evolutionary Biology **24**:2280–2288.

Trindade, S., A. Sousa, K. B. Xavier, F. Dionisio, M. G. Ferreira, and I. Gordo 2009. Positive epistasis drives the acquisition of multidrug resistance. Plos Genetics **5**:e1000578.

Tufto, J. 2000. The evolution of plasticity and nonplastic spatial and temporal adaptations in the presence of imperfect environmental cues. American Naturalist **156**:121–130.

Turner, P. E., and S. F. Elena 2000. Cost of host radiation in an RNA virus. Genetics **156**:1465–1470.

Venail, P. A., O. Kaltz, I. Olivieri, T. Pommier, and N. Mouquet 2011. Diversification in temporally heterogeneous environments: effect of the grain in experimental bacterial populations. Journal of Evolutionary Biology **24**:2485–2495.

Via, S., and R. Lande 1985. Genotype-environment interaction and the evolution of phenotypic plasticity. Evolution **39**:505–522.

Vila-Aiub, M. M., P. Neve, and S. B. Powles 2009. Fitness costs associated with evolved herbicide resistance alleles in plants. New Phytologist **184**:751–767.

Whitlock, M. C. 1996. The red queen beats the jack of all trades: the limitations on the evolution of phenotypic plasticity and niche breadth. American Naturalist **148**:S65–S77.

Yeh, P. J., M. J. Hegreness, A. P. Aiden, and R. Kishony 2009. Drug interactions and the evolution of antibiotic resistance. Nature Reviews Microbiology **7**:460–466.

7

The effects of spatial structure, frequency dependence and resistance evolution on the dynamics of toxin-mediated microbial invasions

Ben Libberton,[1,3] Malcolm J. Horsburgh[1] and Michael A. Brockhurst[2]

1 Department of Integrative Biology, University of Liverpool, Liverpool, UK
2 Department of Biology, University of York, York, UK
3 Karolinska Institute, SE-171 77 Stockholm, Sweden

Keywords
community ecology, experimental evolution, interference competition, invasion, spatial structure, staphylococci, toxin production.

Correspondence
Ben Libberton, Department of Neuroscience, Karolinska Institutet, Retzius väg 8, 17177 Stockholm, Sweden.

e-mail: benjamin.libberton@ki.se

The first author is currently affiliated to the third institution.

Abstract

Recent evidence suggests that interference competition between bacteria shapes the distribution of the opportunistic pathogen *Staphylococcus aureus* in the lower nasal airway of humans, either by preventing colonization or by driving displacement. This competition within the nasal microbial community would add to known host factors that affect colonization. We tested the role of toxin-mediated interference competition in both structured and unstructured environments, by culturing *S. aureus* with toxin-producing or nonproducing *Staphylococcus epidermidis* nasal isolates. Toxin-producing *S. epidermidis* invaded *S. aureus* populations more successfully than nonproducers, and invasion was promoted by spatial structure. Complete displacement of *S. aureus* was prevented by the evolution of toxin resistance. Conversely, toxin-producing *S. epidermidis* restricted *S. aureus* invasion. Invasion of toxin-producing *S. epidermidis* populations by *S. aureus* resulted from the evolution of toxin resistance, which was favoured by high initial frequency and low spatial structure. Enhanced toxin production also evolved in some invading populations of *S. epidermidis*. Toxin production therefore promoted invasion by, and constrained invasion into, populations of producers. Spatial structure enhanced both of these invasion effects. Our findings suggest that manipulation of the nasal microbial community could be used to limit colonization by *S. aureus*, which might limit transmission and infection rates.

Introduction

Staphylococcus aureus colonizes the lower portion of the nasal airway (anterior nares) persistently in around 20% of the human population (Van Belkum et al. 2009). Although persistent nasal colonization by *S. aureus* (carriage) is typically asymptomatic, it is a risk factor for infection in specific patient groups (Von Eiff et al. 2001). These infections can be recurrent and respond poorly to treatment (Kreisel et al. 2006), while the risk of infection is significantly higher for immunocompromised carriers, with increased severity and mortality rates (Yu et al. 1986; Hoen et al. 1995; Senthilkumar et al. 2001).

Studies have revealed many diverse host, bacterial and environmental factors that influence *S. aureus* carriage. Host factors include genetic variation of the immune response (Van den Akker et al. 2006; Ruimy et al. 2010) and being part of certain patient groups give higher rates of carriage (Atela et al. 1997; Lederer et al. 2007). *S. aureus* determinants that affect carriage include secreted components associated with immune system interaction (De Haas et al. 2004; Genestier et al. 2005; Rooijakkers et al. 2005) or components of the bacterial cell surface (Kreikemeyer et al. 2002; Clarke et al. 2004; Heilmann et al. 2004).

The nasal microbial community is mainly comprised of *Corynebacterium*, *Propionibacterium* and *Staphylococcus*, with the latter genus constituting between 15% and 60% of the nasal microbial community and mainly comprising the species *S. aureus* and *Staphylococcus epidermidis* (Wos-Oxley et al. 2010). There is increasing evidence that the nasal microbial community may contribute to determining *S. aureus* carriage (Peacock et al. 2001; Frank et al. 2010;

Wos-Oxley et al. 2010; Yan et al. 2013; Libberton et al. 2014). One well-described staphylococcal mechanism is via competition arising from allelic variation within *agr*-dependent signal transduction (Regassa et al. 1992; Yarwood et al. 2002; Weinrick et al. 2004; Schlievert et al. 2007; Horswilll and Nauseef 2008; Peterson et al. 2008). Several studies report negatively associated distributions of *S. epidermidis* and *S. aureus* across nasal communities, suggesting that these species engage in one-way or mutual exclusion (Lina et al. 2003; Frank et al. 2010; Wos-Oxley et al. 2010; Libberton et al. 2014). Several potential biochemical mechanisms for these observed patterns have been suggested. Iwase et al. (2010) identified that *S. epidermidis* can displace *S. aureus* from the nasal niche by serine protease-mediated biofilm disruption; Lina et al. (2003) showed that quorum sensing interference could contribute to competition whereby different agr types of *S. aureus* and *S. epidermidis* could not inhabit the same community. In addition, *S. aureus* and *S. epidermidis* both secrete a variety of toxins, which can kill interspecific competitors (Nascimento et al. 2012; Sandiford and Upton 2012; Peschel and Otto 2013)

Here, we constructed simple *in vitro* communities of *S. epidermidis* and *S. aureus* to explore the hypothesis that toxin-mediated killing of competitor species (interference competition) could contribute to the observed negatively associated distributions of these species in nasal communities. Theory predicts that interference competition can both promote and prevent invasion of resident communities. Invasion is promoted when invading populations produce toxin(s) that can kill the resident. However, the cost of producing toxins must be lower than the benefits gained from producing them, and the benefits must not be shared between invader and resident populations. If these criteria are not met, then the interference competition will reduce the chance of invasion (Chao and Levin 1981). Resident populations that produce toxins have been shown to restrict invasion by toxin-sensitive populations (Adams et al. 1979; Chao and Levin 1981; Durrett and Levin 1994; Frank 1994; Duyck et al. 2006; Allstadt et al. 2012). We explored two scenarios in which toxin production by *S. epidermidis* could drive exclusion of *S. aureus*: first, where resident toxin-producing *S. epidermidis* prevent invasion by susceptible *S. aureus,* and second, where invading toxin-producing *S. epidermidis* displace a resident susceptible *S. aureus* population. In addition, we manipulated two ecological parameters that influence the success of toxin-mediated interference competition, specifically, the spatial structure of the environment and the starting frequency of invaders.

In bacteria, interference competition is typically mediated by environmentally secreted toxins, and therefore, it is likely to be affected by environmental spatial structure.

Experiments with *Escherichia coli* have demonstrated that in spatially structured environments (agar plates), bacteriocin producers invaded from very low starting frequency (0.001) into bacteriocin-sensitive populations. By contrast, in the absence of spatial structure (shaken liquid broth), much higher initial frequencies of producers (0.1) were required for successful invasion (Chao and Levin 1981). Spatially structured environments were proposed to promote invasion of toxin producers because clustering of producers enables toxins to reach higher local concentrations (Majeed et al. 2011). As such, the benefits of costly toxin production can accrue to small founding populations. By contrast, in spatially unstructured environments, rapid diffusion of the bacteriocin and quorum sensing molecules away from producing cells of *E. coli* required bacteriocin producers to exceed a higher threshold frequency before the benefits of bacteriocin production could be realized (Chao and Levin 1981; Tait and Sutherland 2002; Greig and Travisano 2004). Similar frequency-dependent invasion effects of toxin producers were demonstrated in spatially structured populations of the yeast *Saccharomyces cerevisiae* (Greig and Travisano 2004). We predicted therefore that toxin-producing *S. epidermidis* strains would be better able to invade-from-rare than nonproducing strains and would do so from lower starting frequencies in more highly spatially structured populations.

Ecological theory proposes that interference competition by a resident species should prevent invasion by a susceptible species irrespective of spatial structure (Adams and Traniello 1981; Doyle et al. 2003). When a toxin kills susceptible immigrants, invaders are unable to sustain a viable population; in population ecology, such hostile environmental patches are often termed black hole sinks (Holt and Gaines 1992). Evolutionary theory also proposes that there is potential for a susceptible invading population to evolve resistance to a toxin and that the probability of this will depend upon the frequency of invaders and the spatial structure of the environment (Chao and Levin 1981; Holt et al. 2003). Several theoretical models predict that the likelihood of adaptation to a black hole sink environment increases with the frequency of immigrants from the source population (Gomulkiewicz et al. 1999; Holt et al. 2003). Higher immigration rates will increase the probability that immigrants carry beneficial mutations that are pre-adapted to survive the conditions of the black hole sink (Holt and Gaines 1992; Perron et al. 2008). Therefore, invading *S. aureus* populations are more likely to contain mutants resistant to *S. epidermidis* toxins when invading from higher starting frequencies. However, the spread of these beneficial resistance mutations is likely to be impeded in more highly spatially structured environments. This is because competition of the beneficial mutant can only occur at the edge of a colony, and as the colony grows, a smaller proportion of the mutant population will be competing with the ancestral genotype (Habets et al. 2007). Taken together, we predict therefore that nonproducing residents will be more easily invaded, that resistance of the invader to inhibitory toxins is more likely to evolve when invaders are at a high starting frequency and that resistant mutants that evolve will be more likely to invade in unstructured environments.

To test these predictions, we performed competition experiments whereby toxin-producing and nonproducing nasal isolates of *S. epidermidis* were invaded from three starting frequencies (0.1, 0.01 and 0.001) into resident populations of toxin-sensitive *S. aureus*. Conversely, to test whether *S. aureus* invasion could be restricted by *S. epidermidis* toxin production, we performed the reciprocal invasion of *S. aureus* from three starting frequencies (0.1, 0.01 and 0.001) into resident populations of toxin-producing and nonproducing *S. epidermidis*. All competitions were propagated for 7 days on solid agar with daily transfer of communities to fresh medium; in half of the replicates, population structure was maintained at each transfer, whereas in the other half of the replicates, the population structure was homogenized at each transfer.

Materials and methods

Culture conditions

All bacterial strains used in this study were cultured at 37°C in 10 mL BHI broth shaken at 200 rpm and on agar-solidified BHI medium (brain–heart infusion solids (porcine), 17.5 g/L; tryptose, 10.0 g/L; glucose, 2.0 g/L; sodium chloride, 5.0 g/L; disodium hydrogen phosphate, 2.5 g/L) (Lab M, Heywood, UK). Chemicals were obtained from Sigma-Aldrich Co., UK.

Selection of nasal isolates

Four independent *S. epidermidis* isolates were selected from a previous study that sampled the anterior nares of 60 healthy volunteers (Libberton et al. 2014): two isolates were toxin producers as revealed in a deferred inhibition assay by their killing of *S. aureus* [zone of clearing when a lawn of *S. aureus* strain SH1000 was sprayed over them (Nascimento et al. 2012)]; two isolates were toxin nonproducers based on not reducing viability of strain SH1000. SH1000 displayed no growth inhibition activity against any of the selected *S. epidermidis* strains in the deferred inhibition assay (Nascimento et al. 2012). Of the two toxin-producing *S. epidermidis* strains, B180 produced an inhibition area that was around ten times greater than that of B155. We first established that the *S. epidermidis* strains had comparable growth rates to SH1000. An overnight culture of each strain (Table 1) was inoculated (1% inoculum) into

Table 1. Strains used in this study.

Species	Strain identification	Reference
S. aureus	SH1000	Horsburgh et al. (2002)
S. epidermidis	B155 (inhibitor producing)	Libberton et al. (2014)
S. epidermidis	B180 (inhibitor producing)	Libberton et al. (2014)
S. epidermidis	B035 (noninhibitor producing)	Libberton et al. (2014)
S. epidermidis	B115 (noninhibitor producing)	Libberton et al. (2014)

Table 2. Doubling times of strains used in this study. The doubling times in minutes were compared to SH1000 (*S. aureus*) as a control using a *post hoc* Dunnett's test. There is no significant difference between any of the *S. epidermidis* strains tested and the *S. aureus* strain SH1000 used in this study.

	Doubling time (min)	*T*-value	*P*-value
SH1000	116.45	NA	NA
B180	116.06	−0.074	1.0000
B155	110.11	−1.203	0.5689
B115	120.49	0.767	0.8579
B035	123.34	1.307	0.4970

200 μL of BHI broth in a 96-well plate. The 96-well plates were incubated at 37°C for 8 h, and OD_{600} readings were taken at 20-min intervals. The doubling time (min) was then calculated (Table 2) using the following formula where T_d is the doubling time; t_1 and t_2 are two consecutive time points throughout the bacterial growth; and d_1 and d_2 are the corresponding OD_{600} readings at t_1 and t_2.

$$T_d = (t_2 - t_1) * \frac{\log(2)}{\log(d_2/d_1)}.$$

Competition experiments

All strains were cultured on BHI agar plates prior to competition experiments. Bacteria were cultured for 18 h on 50-mm-diameter BHI agar plates, and the lawns of *S. aureus* (SH1000) and *S. epidermidis* strains (resident and invader – Table 1) were then scraped off the agar plates and suspended in 10 mL of PBS by vortexing thoroughly. The cfu/mL in each tube was equalized by diluting the cell suspensions in PBS and comparing the OD_{600} of each suspension (approximately 5×10^8 cfu/mL for *S. aureus* and *S. epidermidis*, determined by viable count). Both species were then mixed together in a final volume of 10 mL PBS, with the invader at different frequencies (ratios) to the resident (0.1:1, 0.01:1, 0.001:1). For brevity, these ratios are referred to in this manuscript as frequencies, and only the first number in the ratio pair is used to define each frequency. The mixtures were vortexed thoroughly before 50 μL (containing approximately 2.5×10^6 cells) was plated onto 25 mL BHI agar and incubated at 37°C. Six replicate communities (structured and unstructured, in triplicate) were established at each starting frequency. The communities were transferred to a new agar plate every day for 7 days. Half of the replicates underwent a regime whereby the transfers were made by replica plating with velvet (Lederberg and Lederberg 1952) to maintain spatial structure. While the other half of replicates underwent a mixed regime whereby the spatial structure was destroyed every 24 h transfer by scraping the entire bacterial lawn off the plate and transferring to 10 mL of sterile PBS, before thoroughly vortexing and pipetting 50 μL onto a new plate to complete the transfer. Each set was performed in triplicate. Viable counts for each isolate were calculated every second day. On the structured plates, this was achieved after replica plating from viable counts of the remaining lawn; colonies were differentiated by colony morphology and pigmentation. *S. aureus* SH1000 possesses a distinct yellow carotenoid pigment which was stable over the course of these experiments. Raw data for the experiments are presented in appendices (Figs A1 and B1).

Deferred inhibition spray assay

A deferred inhibition spray assay was performed to determine whether *S. aureus* clones had developed resistance to the toxin-producing *S. epidermidis* strains. The assay was performed on 10 clones from each experiment. A 25-μL spot (approximately 10^8 cells) of an overnight bacterial culture was pipetted onto the centre of an agar plate containing 15 mL of BHI agar (Lab M). The plates were incubated for 18 h at 37°C before 250 μL of a 10-fold diluted overnight culture of a different strain (10^6 cfu) was sprayed over the plate. The plates were incubated for a further 18 h after when the size of the inhibition zones produced by the central spot on the overlaid strain was assessed. The clarity of the inhibition zone was scored based on a simple scoring system of 1–4, 4 being completely clear and 1 being no

detectable zone. The areas of any detectable zones were also recorded by measuring the diameter of the inhibition zone and the central colony.

Data analysis

To quantify the success of the invasion, we calculated the selection rate constant for each invader using relative

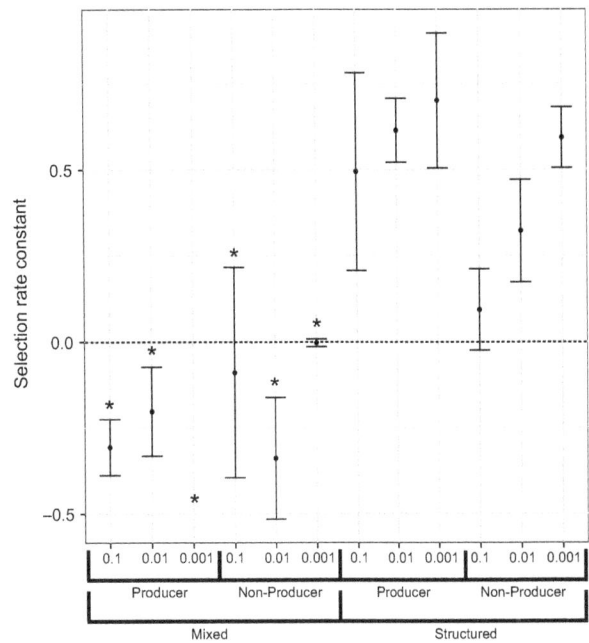

Figure 1 Selection rate coefficients for *Staphylococcus epidermidis* invading populations of *S. aureus* (SH1000). Toxin-producing *S. epidermidis* isolates (155 and 180) and nonproducing isolates (035 and 115) were invaded into populations of *S. aureus* (SH1000) at relative frequencies of 10, 100 and 1000. Each of the invasions was also carried out under a spatially structured treatment and a mixed treatment. Asterisks mark negative selection rate coefficients where invasion did not occur. Error bars represent the standard error of the mean.

Table 3. Analysis of variance testing the main effects of successful invasion of *S. epidermidis* into populations of *S. aureus*. The table shows the results of a multifactorial ANOVA. Both main effects and interactions are shown.

	df	Sum sq	Mean sq	F value	P value
Frequency	1	0.1382	0.1382	3.4920	0.0662444
Structure	1	12.7710	12.7710	322.7662	<2.2e-16***
Inhibition	1	0.1452	0.1452	3.6690	0.0599020
Frequency × Structure	1	0.5359	0.5359	13.5452	0.0004798***
Frequency × Inhibition	1	0.0243	0.0243	0.6141	0.4361327
Structure × Inhibition	1	0.5652	0.5652	14.2852	0.0003474***
Frequency × Structure × Inhibition	1	0.0746	0.0746	1.8858	0.1744717

df, Degrees of freedom; Sum sq, sum of squares; Mean sq, Mean of squares; F value, F statistic for terms in the row; P value, significance. Asterisks indicate the significance levels at different thresholds. ***P < 0.001.

bacterial frequencies from day 0 and day 7 with the following equation.

$$C_{ir} = \frac{\ln[N_i(1)/N_i(0) - \ln[N_r(1)/N_r(0)]}{1 \text{ day}},$$

where $N_i(0)$ and $N_r(0)$ represent the initial densities of the competing populations i (invader) and r (resident), and N_i

(1) and $N_r(1)$ represent their densities after 1 day (Travisano & Lenski, 1996).

Negative values indicated that invasion was not possible, whereas positive values indicated invasion was possible. The invasion time-course data were visualized using plots of the natural log of the invader to resident ratio over time; selection rate constants were analysed in a three-way ANOVA.

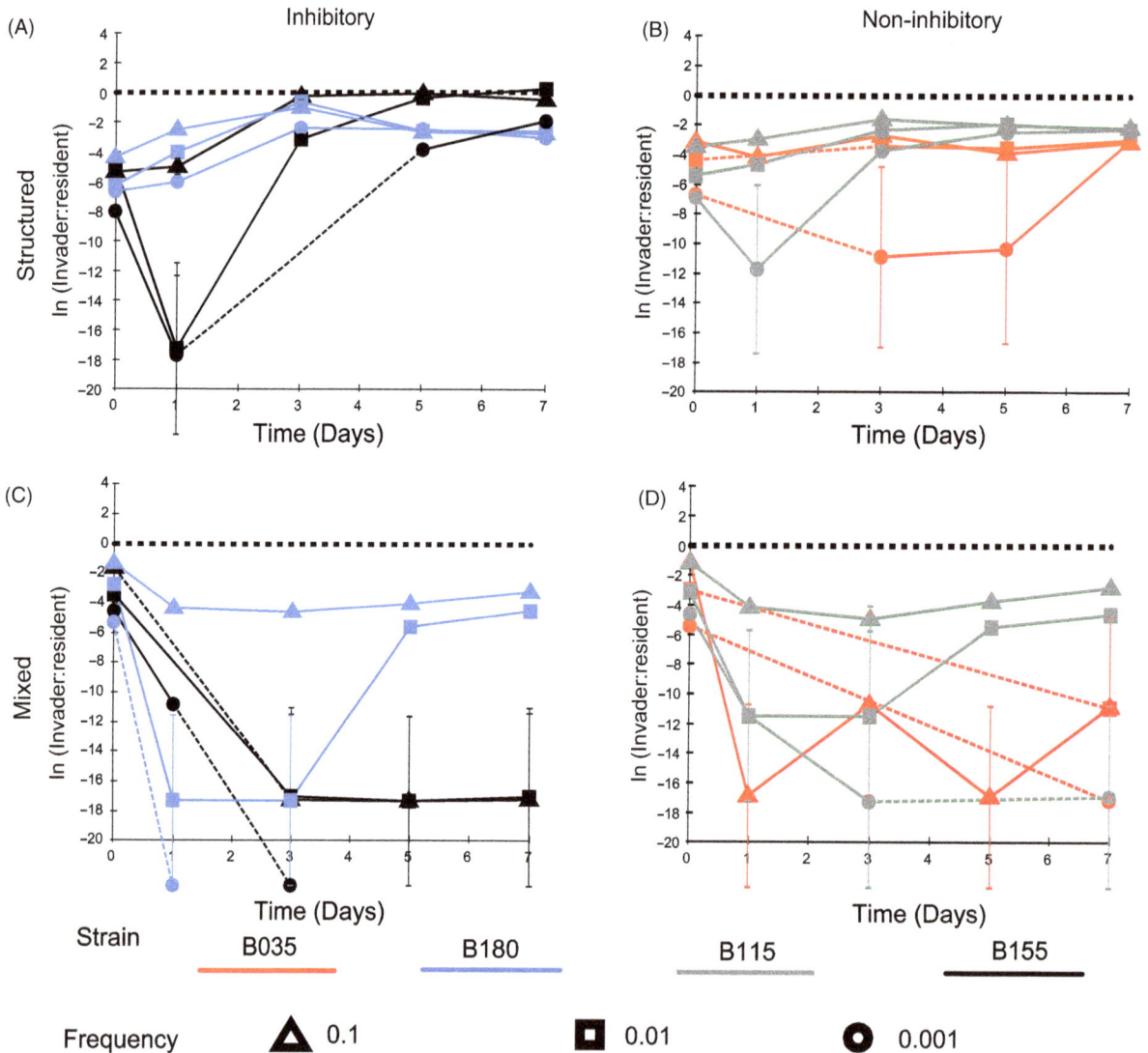

Figure 2 Toxin-producing (blue and black) and nonproducing (red and grey) isolates of *Staphylococcus epidermidis* invading populations of *S. aureus* (SH1000) at frequencies of 0.1 (triangle), 0.01 (square) and 0.001 (circle). Toxin-producing *S. epidermidis* isolates (155 and 180) and nonproducing *S. epidermidis* isolates (035 and 115) were introduced into a population of *S. aureus* (SH1000) at three different frequencies. This was carried under a spatially structured regime (A and B) and under a mixed regimen (C and D). The x-axis is the time in days, and the y-axis is the natural log of the invader to resident ratio. A dotted line in the time course shows when the population dipped below the experiment detection threshold (for clarity, these lines also cross the x-axis if the population went to extinction). There is a heavy dotted line at 0 on the y-axis to indicate an equal invader to resident ratio. The line crossing the x-axis symbolizes that the population went to extinction. Error bars represent the standard error of the mean ($n = 3$).

Results

Spatial structure promotes invasion by inhibitor-producing S. epidermidis

Environmental structure promoted *S. epidermidis* invasion (structure, $F_{1,64} = 322.77$, $P < 0.001$) (Fig. 1 and Table 3), and this effect was stronger for *S. epidermidis* toxin producers than for nonproducers (structure × inhibition, $F_{1,64} = 14.29$, $P < 0.001$) (Table 3). *S. epidermidis* was never able to successfully invade under mixed conditions (Fig. 1 and Table 3). However, *S. epidermidis* was more likely to persist at low frequencies and avoid extinction in mixed environments when initiated at a higher starting frequency (frequency × structure, $F_{1,64} = 13.55$, $P < 0.001$).

Invasion was impeded by evolution of resistance

Under structured conditions, the two invading, toxin-producing strains of *S. epidermidis* show different dynamics over time (Fig. 2A). All starting frequencies of strain B155 increase after day 1 and approach a 1:1 invader to resident ratio, whereas strain B180 (starting frequencies 0.1 and 0.01) increases until day 3, after which they decrease. Spray assays were performed to test whether the decline in frequency of strain B180 populations (of starting frequency 0.1 and 0.01) was caused by resistance evolution in the resident *S. aureus* population. Ancestral and evolved resident *S. aureus* clones were sprayed over ancestral and evolved *S. epidermidis* strain B180 (Fig. 3). These assays show that after 7 days, the resident *S. aureus* had evolved resistance to the invading *S. epidermidis* under structured conditions at starting frequencies of 0.1 and 0.01 (Fig. 3) (Fisher's exact test, $P = 0.0022$). Resistance was not seen in the *S. aureus* resident population when invaded with strain B180 at a starting frequency of 0.001 (Fig. 3) (Fisher's exact test, $P = 1$). Of note, evolved *S. epidermidis* strains (Fig. 3B) produced larger inhibition zones against susceptible *S. aureus* than the ancestral *S. epidermidis* strains (Fig. 3B) (paired *t*-test: $T = 2.69$, $P = 0.03$).

Toxin-producing S. epidermidis strains resist invasion, especially in structured environments

Toxin-producing *S. epidermidis* strains were more resistant to invasion than nonproducing strains (inhibition, $F_{1,64} = 124.95$, $P < 0.0001$, Table 4) and restricted invasion more effectively under structured environmental conditions (Figs 4 and 5A) (structure × inhibition, $F_{1,64} = 6.14$, $P < 0.05$, Table 4). Invasion of *S. aureus* into a toxin-producing *S. epidermidis* resident was positively frequency-dependent with highest initial frequencies invading the fastest and lower initial frequencies going to extinc-

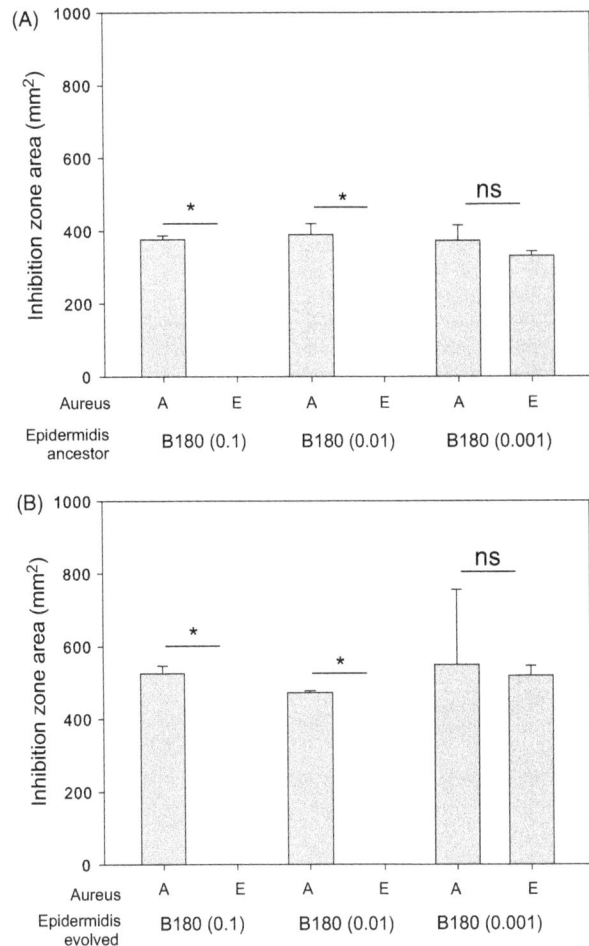

Figure 3 Resistance of evolved SH1000 resident after *Staphylococcus epidermidis* (B180) invasion. Panel A shows inhibition zone produced by the ancestral *S. epidermidis* strains, and panel B shows the inhibition zones produced by the evolved *S. epidermidis* strains. Both panels A and B show the inhibition zone area (mm²) produced by the toxin-producing *S. epidermidis* strains against the ancestral SH1000 (A) and the evolved SH1000 (E). Asterisks represent a significant difference between the inhibition zone areas of ancestral (A) and evolved (E) *S. aureus* strains as determined by a Fisher's exact test. Each significance star represents a *P* value of 0.0022 which is significant when Bonferroni corrected for multiple comparisons with an alpha value of 0.1. Error bars represent the standard error of the mean.

tion (Fig. 5A,C) (frequency × inhibition, $F_{1,64} = 46.5$, $P < 0.001$).

Evolved resistance promotes S. aureus invasion

Staphylococcus aureus was only able to invade toxin-producing *S. epidermidis* under mixed conditions (Fig. 5C). To test whether the evolution of inhibitory toxin resistance by *S. aureus* was responsible for the invasion in a mixed environment (Figs 4 and 5C), ancestral and evolved

Table 4. Analysis of variance testing the main effects of successful invasion of *S. aureus* into populations of *S. epidermidis*. The table shows the results of a multifactorial ANOVA. Both main effects and interactions are shown.

	df	Sum sq	Mean sq	F value	P value
Frequency	1	2.721	2.721	8.1457	0.005810**
Structure	1	2.949	2.949	8.8266	0.004177**
Inhibition	1	41.744	41.744	124.9525	<2.2e-16***
Frequency × Structure	1	0.007	0.007	0.0198	0.88449
Frequency × Inhibition	1	15.554	15.554	46.5589	3.794e-09***
Structure × Inhibition	1	2.051	2.051	6.1382	0.015880*
Frequency × Structure × inhibition	1	0.398	0.398	1.1900	0.279418

df, Degrees of freedom; Sum sq, sum of squares; Mean sq, Mean of squares; *F* value, *F* statistic for terms in the row; *P* value, significance. Asterisks indicate the significance levels at different thresholds. *$P < 0.05$; **$P < 0.01$, ***$P < 0.001$.

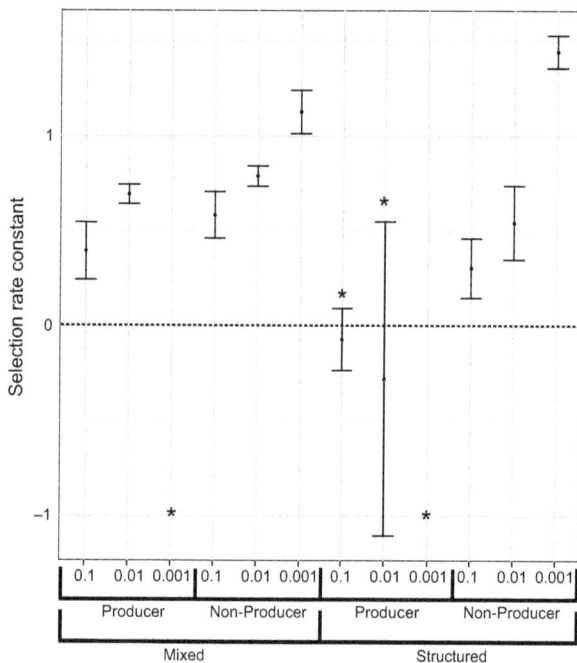

Figure 4 Selection rate coefficients for *Staphylococcus aureus* (SH1000) invading populations of *S. epidermidis*. *S. aureus* was introduced into populations of toxin-producing *S. epidermidis* isolates (155 and 180) and nonproducing isolates (035 and 115) at relative frequencies of 10, 100 and 1000. Each of the invasions was also carried out under a spatially structured treatment and a mixed treatment. Asterisks mark negative selection rate coefficients where invasion did not occur. Error bars represent the standard error of the mean.

S. aureus strains were sprayed over ancestral and evolved *S. epidermidis* toxin-producing residents. In all cases, evolved *S. aureus* were resistant to the *S. epidermidis* toxin (Fig. 6) (Fisher's exact test, $P = 0.0022$).

Discussion

We show that antimicrobial toxin production by *S. epidermidis* nasal isolates can have important effects on competi-

tion with *S. aureus*: interference competition acts both to promote invasion by, and to prevent invasion into, *S. epidermidis* populations. This supports the growing body of evidence that species interactions can play an important role determining species distributions in nasal microbial communities (Lina et al. 2003; Frank et al. 2010; Wos-Oxley et al. 2010; Yan et al. 2013), and more specifically that trait variation, in this case in the toxins mediating interference competition, could act to prevent *S. aureus* nasal carriage (Libberton et al. 2014).

Our findings highlight the critical role for spatial structure in determining the outcome of interference competition. If spatial structure is maintained, then inhibitor-producing bacteria can better prevent the invasion-from-rare of *S. aureus*, whereas unstructured environments generally do not favour the production of inhibitory toxins. Spatial structure is likely to be an important component of life in the anterior nares. While nutrient agar is clearly not equivalent to this environment and transfers using velvet may select those bacteria growing nearest to the colony surface, our manipulation of spatial structure is arguably more relevant to the nasal environment than comparing agar plates to liquid culture (Chao and Levin 1981). The macro-topography of the nares is irregular, with ridges and recesses providing spatially discrete surfaces. The base layer of the nares is comprised of a squamous epithelium that microbes colonize and form spatially discrete groups (Uraih and Maronpot 1990; Yuki et al. 2000; Dongari-Bagtzoglou et al. 2009). Spatial population structure is therefore expected to be present, but is likely to be disrupted by changes to the squamous epithelium, the flow of air (Churchill et al. 2004) and mucus (Proctor et al. 1973) through the nasal passages and mechanical disruptions (e.g. nose picking). Factors that reduce the spatial structuring in nasal communities could weaken the ability of inhibitory resident species to prevent invasion by *S. aureus*.

Further, our data demonstrate that rapid evolution of resistance to antimicrobial toxins can determine the

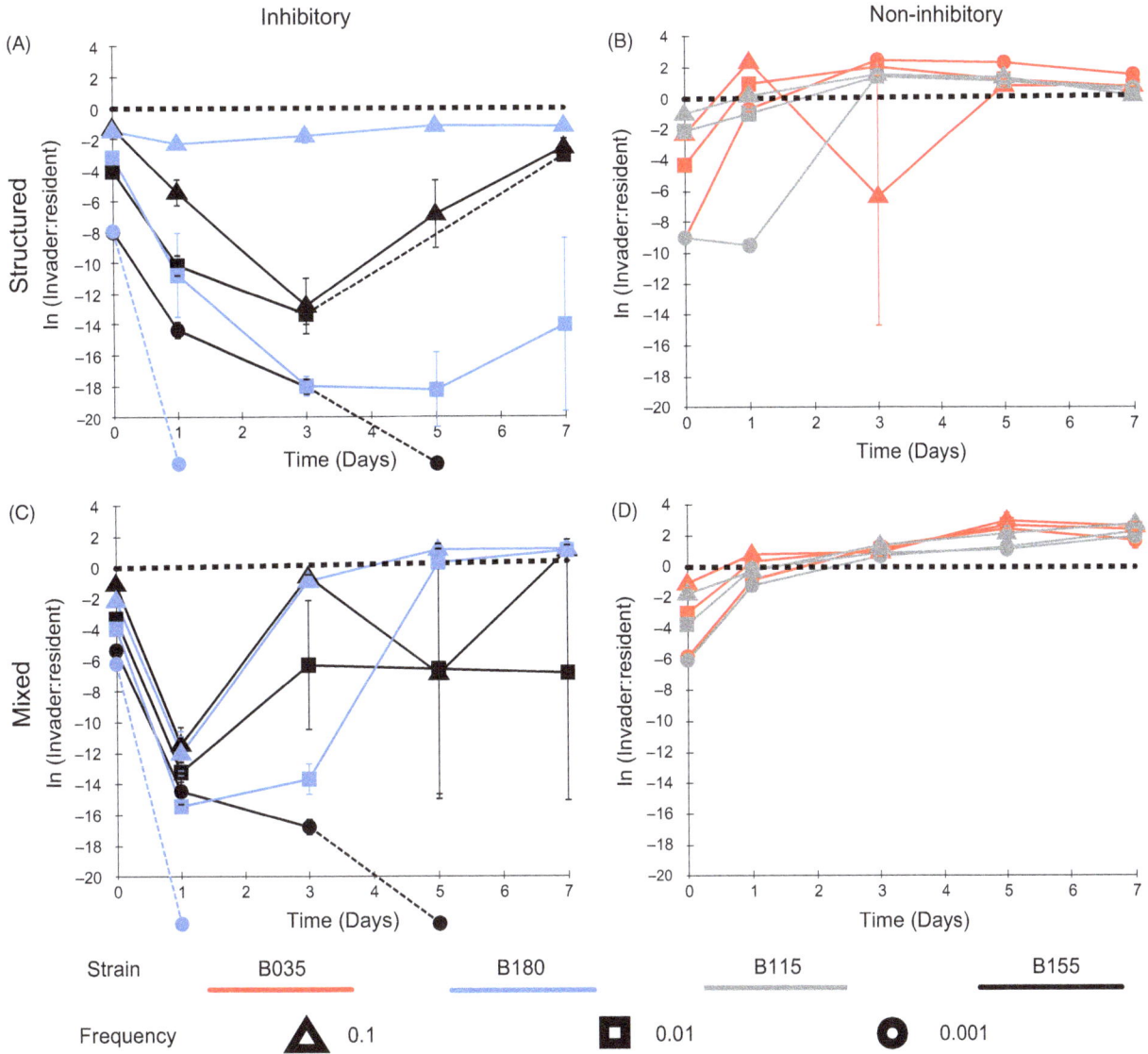

Figure 5 *Staphylococcus aureus* invading populations of toxin-producing (blue and black) and nonproducing (red and grey) *S. epidermidis* at frequencies of 0.1 (triangle), 0.01 (square) and 0.001 (circle). *S. aureus* strain (SH1000) was introduced into two different toxin-producing *S. epidermidis* populations (155 and 180), and two different nonproducing populations (035 and 115) at three different frequencies. This was carried under a spatially structured regime (A and B) and under a mixed regime (C and D). The *x*-axis is the time in days, and the *y*-axis is the natural log of the invader to resident ratio. A dotted line in the time course shows when the population dipped below the experiment detection threshold (for clarity, these lines also cross the *x*-axis if the population went to extinction). There is a heavy dotted line at 0 on the *y*-axis to indicate an equal invader to resident ratio. The line crossing the *x*-axis symbolizes that the population went to extinction. Error bars represent the standard error of the mean (*n* = 3).

outcome of interference competition. Positive frequency-dependent fitness was observed for *S. aureus* invading inhibitory residents. This may have occurred due to a protective effect from a lager inoculum neutralizing the effect of the toxin. This is unlikely, however, as protection caused by large bacterial densities is typically a result of quenching and lowering the local concentration of available toxin. In this experimental set-up, toxin would be produced during growth phases every time the population

is transferred, which would overcome any possible quenching effect. The more likely explanation for resistance evolution is that the higher inoculation frequencies increased the chance of these invading populations containing beneficial resistance mutations. This outcome is similar to theory predicting conditions for adaptation to black hole sink environments where an increased immigration rate (i.e., increased frequency of the invading population) increases the probability of adaptation (Holt and Gaines 1992;

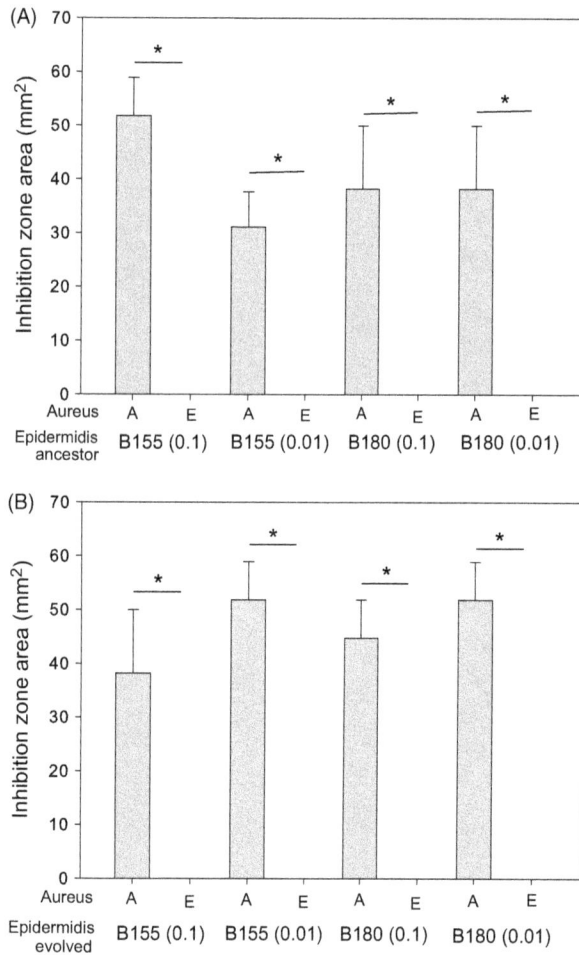

Figure 6 Resistance of evolved SH1000 after successful invasion. Panel A shows inhibition zone produced by the ancestral *Staphylococcus epidermidis* strains, and panel B shows the inhibition zones produced by the evolved *S. epidermidis* strains. Both panels A and B show the inhibition zone area (mm²) produced by the inhibitory *S. epidermidis* strains against the ancestral SH1000 (A) and the evolved SH1000 (E). The * represents a significant difference between the inhibition zone areas of ancestral (A) and evolved (E) *S. aureus* strains as determined by a Fisher's exact test. Each significance star represents a *P* value of 0.0022, which is significant when Bonferroni corrected for multiple comparisons with an alpha value of 0.1. Error bars represent the standard error of the mean.

Perron et al. 2008). Moreover, this scenario suggests that an inhibitor-producing community could resist invasion from rare by *S. aureus*, because of a low probability of resistance evolution. When toxin-producing *S. epidermidis* were invading resident *S. aureus*, evolution of resistance in *S. aureus* resident populations was most likely when the toxin-producing invaders were relatively common (Fig. 3). Higher frequencies of invading toxin producers would have produced more of the toxin, generating stronger selection for resistance; additionally, a larger fraction of the *S. aureus* populations would have

been exposed to these toxins. This relationship suggests that resistance evolution by residents may frequently impede invasion by toxin-producing strains, because resident populations are unlikely to be mutation-limited and selection for resistance progressively strengthens as an invasion proceeds. Intriguingly, in spatially structured environments, the evolved invading *S. epidermidis* (B180) showed greater inhibitory activity on ancestral SH1000 than the ancestral B180 genotype (Fig. 3). This suggests that the toxin producer coevolved to meet the survival challenges posed by increasingly resistant *S. aureus* populations. The evolved *S. epidermidis* may have upregulated production of the inhibitory toxin, or alternatively initiated production of alternative toxins. However, in the absence of knowledge of the mechanism of inhibition, this remains unclear.

One strength of this study is the use of toxin-producing strains isolated from the nares of healthy volunteers and not isogenic toxin-producing and nonproducing laboratory strains. Although nasal isolates are more difficult to compare with well-characterized laboratory strains, they have greater relevance to future development of therapeutic strategies and provide added realism to laboratory models of colonization. There are also other limitations, for example the zones of inhibition produced from B180 and B155 had different areas, which implies differential expression of the same toxin or discrete toxins. Resource competition and adaptation to the growth medium could affect the outcomes described and contribute to the interactions between the pairs of bacteria, but these aspects would require further study to describe.

It is stated that preventing *S. aureus* carriage significantly reduces the risk of infection (Von Eiff et al. 2001). Our findings support the possibility that manipulation of the microbial community in the human nose to increase the frequency of inhibitor-producing residents could reduce *S. aureus* colonization. The human gut has been a model system for therapeutic manipulation of the microbial flora for many years (Borody et al. 1989; Landy et al. 2011). If the models of colonization and their outcomes can be replicated in the nasal environment, it would represent a novel way to limit *S. aureus* carriage that is correlated with associated life-threatening infections.

Acknowledgements

This study was supported by a BBSRC studentship (BB/D526529/1) to B.L. awarded to M.A.B. and M.J.H.

Literature cited

Adams, E. S., and J. F. A. Traniello 1981. Chemical interference competition by *Monomorium minimum* (Hymenoptera: Formicidae). Oecologia **51**:265–270.

Adams, J., T. Kinney, S. Thompson, L. Rubin, and R. B. Helling 1979. Frequency dependent selection for plasmid containing cells of *Escherichia coli*. Genetics **91**:627–637.

Allstadt, A., T. Caraco, F. Molnár, and G. Korniss 2012. Interference competition and invasion: spatial structure, novel weapons and resistance zones. Journal of Theoretical Biology **306**:46–60.

Atela, I., P. Coll, J. Rello, E. Quintana, J. Barrio, F. March, F. Sanchez et al. 1997. Serial surveillance cultures of skin and catheter hub specimens from critically ill patients with central venous catheters: molecular epidemiology of infection and implications for clinical management and research. Journal of Clinical Microbiology **35**:1784–1790.

Borody, T. J., L. George, P. Andrews, S. Brandl, S. Noonan, P. Cole, L. Hyland et al. 1989. Bowel-flora alteration: a potential cure for inflammatory bowel disease and irritable bowel syndrome. Medical Journal of Australia **150**:604.

Chao, L., and B. R. Levin 1981. Structured habitats and the evolution of anticompetitor toxins in bacteria. Proceedings of the National Academy of Sciences of the United States of America **78**:6324–6328.

Churchill, S. E., L. L. Shackelford, J. N. Georgi, and M. T. Black 2004. Morphological variation and airflow dynamics in the human nose. American Journal of Human Biology **16**:625–638.

Clarke, S. R., M. D. Wiltshire, and S. J. Foster 2004. IsdA of *Staphylococcus aureus* is a broad spectrum, iron-regulated adhesin. Molecular Microbiology **51**:1509–1519.

De Haas, C. J. C., K. E. Veldkamp, A. Peschel, F. Weerkamp, W. J. B. Van Wamel, E. C. J. M. Heezius, M. J. Poppelier et al. 2004. Chemotaxis inhibitory protein of *Staphylococcus aureus*, a bacterial antiinflammatory agent. Journal of Experimental Medicine **199**:687–695.

Dongari-Bagtzoglou, A., H. Kashleva, P. Dwivedi, P. Diaz, and J. Vasilakos 2009. Characterization of mucosal *Candida albicans* biofilms. PLoS One **4**:e7967.

Doyle, R. D., M. D. Francis, and R. M. Smart 2003. Interference competition between *Ludwigia repens* and *Hygrophila polysperma*: two morphologically similar aquatic plant species. Aquatic Botany **77**:223–234.

Durrett, R., and S. Levin 1994. The importance of being discrete (and spatial). Theoretical Population Biology **46**:363–394.

Duyck, P. F., P. David, G. Junod, C. Brunel, R. Dupont, and S. Quilici 2006. Importance of competition mechanisms in successive invasions by polyphagous tephritids in La Reunion. Ecology **87**:1770–1780.

Frank, S. A. 1994. Spatial polymorphism of bacteriocins and other allelopathic traits. Evolutionary Ecology **8**:369–386.

Frank, D. N., L. M. Feazel, M. T. Bessesen, C. S. Price, E. N. Janoff, and N. R. Pace 2010. The human nasal microbiota and *Staphylococcus aureus* carriage. PLoS One **5**:e10598.

Genestier, A. L., M. C. Michallet, G. Prevost, G. Bellot, L. Chalabreysse, S. Peyrol, F. Thivolet et al. 2005. *Staphylococcus aureus* Panton-Valentine leukocidin directly targets mitochondria and induces Bax-independent apoptosis of human neutrophils. Journal of Clinical Investigation **115**:3117–3127.

Gomulkiewicz, R., R. D. Holt, and M. Barfield 1999. The effects of density dependence and immigration on local adaptation and niche evolution in a black-hole sink environment. Theoretical Population Biology **55**:283–296.

Greig, D., and M. Travisano. 2004. The Prisoner's Dilemma and polymorphism in yeast SUC genes. Proceedings of the Royal Society of London. Series B: Biological Sciences **271**:S25–S26.

Habets, M. G. J. L., T. Czaran, R. F. Hoekstra, and J. de Visser. 2007. Spatial structure inhibits the rate of invasion of beneficial mutations in asexual populations. Proceedings of the Royal Society of London. Series B: Biological Sciences **274**:2139–2143.

Heilmann, C., S. Niemann, B. Sinha, M. Herrmann, B. E. Kehrel, and G. Peters 2004. *Staphylococcus aureus* fibronectin-binding protein (FnBP)-mediated adherence to platelets, and aggregation of platelets induced by FnBPA but not by FnBPB. Journal of Infectious Diseases **190**:321–329.

Hoen, B., M. Kessler, D. Hestin, and D. Mayeux 1995. Risk-factors for bacterial-infection in chronic heamodialysis adult patients – a multicenter prospective survey. Nephrology Dialysis Transplant **10**:377–381.

Holt, R. D., and M. S. Gaines 1992. Analysis of adaptation in heterogeneous landscapes – implications for the evolution of fundamental niches. Evolutionary Ecology **6**:433–447.

Holt, R. D., R. Gomulkiewicz, andM. Barfield. 2003. The phenomenology of niche evolution via quantitative traits in a "black-hole" sink. Proceedings of the Royal Society of London. Series B: Biological Sciences **270**:215–224.

Horsburgh, M. J., J. L. Aish, I. J. White, L. Shaw, J. K. Lithgow, and S. J. Foster 2002. SigmaB modulates virulence determinant expression and stress resistance: characterization of a functional rsbU strain derived from *Staphylococcus aureus* 8325–4. Journal of Bacteriology **184**:5457–5467.

Horswilll, A. R., and W. M. Nauseef 2008. Host interception of bacterial communication signals. Cell Host & Microbe **4**:507–509.

Iwase, T., Y. Uehara, H. Shinji, A. Tajima, H. Seo, K. Takada, T. Agata et al. 2010. *Staphylococcus epidermidis* Esp inhibits *Staphylococcus aureus* biofilm formation and nasal colonization. Nature **465**:346–349.

Kreikemeyer, B., D. McDevitt, and A. Podbielski 2002. The role of the Map protein in *Staphylococcus aureus* matrix protein and eukaryotic cell adherence. International Journal of Medical Microbiology **292**:283–295.

Kreisel, K., K. Boyd, P. Langenberg, and M. C. Roghmann 2006. Risk factors for recurrence in patients with *Staphylococcus aureus* infections complicated by bacteremia. Diagnostic Microbiology Infectious Disease **55**:179–184.

Landy, J., H. O. Al-Hassi, S. D. McLaughlin, A. W. Walker, P. J. Ciclitira, R. J. Nicholls, S. K. Clark et al. 2011. Review article: faecal transplantation therapy for gastrointestinal disease. Alimentary Pharmacology and Therapeutics **34**:409–415.

Lederberg, J., and E. M. Lederberg 1952. Replica plating and indirect selection of bacterial mutants. Journal of Bacteriology **63**:399–406.

Lederer, S. R., G. Riedelsdorf, and H. Schiffl 2007. Nasal carriage of methicillin resistant *Staphylococcus aureus*: the prevalence, patients at risk and the effect of elimination on outcomes among outclinic haemodialysis patients. European Journal of Medical Research **12**:284–288.

Libberton, B., R. E. Coates, M. A. Brockhurst, and M. J. Horsburgh 2014. Evidence that intraspecific trait variation among nasal bacteria can shape the distribution of *Staphylococcus aureus*. Infection and Immunity **82**:3811–3815.

Lina, G., F. Boutite, A. Tristan, M. Bes, J. Etienne, and F. Vandenesch 2003. Bacterial competition for human nasal cavity colonization: role of Staphylococcal agr alleles. Applied and Environmental Microbiology **69**:18–23.

Majeed, H., O. Gillor, B. Kerr, and M. A. Riley 2011. Competitive interactions in *Escherichia coli* populations: the role of bacteriocins. The ISME Journal **5**:71–81.

Nascimento, J. D., M. L. V. Coelho, H. Ceotto, A. Potter, L. R. Fleming, Z. Salehian, I. F. Nes et al. 2012. Genes involved in immunity to and secretion of aureocin A53, an atypical class II bacteriocin produced by *Staphylococcus aureus* A53. Journal of Bacteriology **194**:875–883.

Peacock, S. J., I. de Silva, and F. D. Lowy 2001. What determines nasal carriage of *Staphylococcus aureus*? Trends in Microbiology **9**:605–610.

Perron, G. G., A. Gonzalez, and A. Buckling 2008. The rate of environmental change drives adaptation to an antibiotic sink. Journal of Evolutionary Biology **21**:1724–1731.

Peschel, A., and M. Otto 2013. Phenol-soluble modulins and staphylococcal infection. Nature Reviews Microbiology **11**:667–673.

Peterson, M. M., J. L. Mack, P. R. Hall, A. A. Alsup, S. M. Alexander, E. K. Sully, Y. S. Sawires et al. 2008. Apolipoprotein B is an innate barrier against invasive *Staphylococcus aureus* infection. Cell Host & Microbe **4**:555–566.

Proctor, D., I. Andersen, and G. Lundqvist 1973. Clearance of inhaled particles from the human nose. Archives of Internal Medicine **131**:132–139.

Regassa, L. B., R. P. Novick, and M. J. Betley 1992. Glucose and nonmaintained pH decrease expression of the accessory gene regulator (agr) in *Staphylococcus aureus*. Infection and Immunity **60**:3381–3388.

Rooijakkers, S. H. M., M. Ruyken, A. Roos, M. R. Daha, J. S. Presanis, R. B. Sim, W. J. van Wamel et al. 2005. Immune evasion by a staphylococcal complement inhibitor that acts on C3 convertases. Nature Immunology **6**:920–927.

Ruimy, R., C. Angebault, F. Djossou, C. Dupont, L. Epelboin, S. Jarraud, L. A. Lefevre et al. 2010. Are host genetics the predominant determinant of persistent nasal *Staphylococcus aureus* carriage in humans? Journal of Infectious Diseases **202**:924–934.

Sandiford, S., and M. Upton 2012. Identification, characterization, and recombinant expression of epidermicin NI01, a Novel unmodified bacteriocin produced by *Staphylococcus epidermidis* that displays potent activity against Staphylococci. Antimicrobial Agents and Chemotherapy **56**:1539–1547.

Schlievert, P. M., L. C. Case, K. A. Nemeth, C. C. Davis, Y. P. Sun, W. Qin, F. Wang et al. 2007. Alpha and beta chains of hemoglobin inhibit production of *Staphylococcus aureus* exotoxins. Biochemistry **46**:14349–14358.

Senthilkumar, A., S. Kumar, and J. N. Sheagren 2001. Increased incidence of *Staphylococcus aureus* bacteremia in hospitalized patients with acquired immunodeficiency syndrome. Clinical Infectious Diseases **33**:1412–1416.

Tait, K., and I. W. Sutherland 2002. Antagonistic interactions amongst bacteriocin-producing enteric bacteria in dual species biofilms. Journal of Applied Microbiology **93**:345–352.

Travisano, M., and R. E. Lenski 1996. Long-term experimental evolution in *Escherichia coli*. IV. Targets of selection and the specificity of adaptation. Genetics. **143**:15–26.

Uraih, L. C., and R. R. Maronpot 1990. Normal histology of the nasal cavity and application of special techniques. Environmental Health Perspectives **85**:187–208.

Van Belkum, A., N. J. Verkaik, C. P. de Vogel, H. A. Boelens, J. Verveer, J. L. Nouwen, J. L. Nouwen et al. 2009. Reclassification of *Staphylococcus aureus* nasal carriage types. Journal of Infectious Diseases **199**:1820–1826.

Van den Akker, E. L. T., J. L. Nouwen, D. C. Melles, E. F. C. van Rossum, J. W. Koper, A. G. Uitterlinden, A. Hofman et al. 2006. *Staphylococcus aureus* nasal carriage is associated with glucocorticoid receptor gene polymorphisms. Journal of Infectious Diseases **194**:814–818.

Von Eiff, C., K. Becker, K. Machka, H. Stammer, and G. Peters 2001. Nasal carriage as a source of *Staphylococcus aureus* bacteremia. Study Group. New England Journal of Medicine **344**:11–16.

Weinrick, B., P. M. Dunman, F. McAleese, E. Murphy, S. J. Projan, Y. Fang, and R. P. Novick 2004. Effect of mild acid on gene expression in *Staphylococcus aureus*. Journal of Bacteriology **186**:8407–8423.

Wos-Oxley, M. L., I. Plumeier, C. von Eiff, S. Taudien, M. Platzer, R. Vilchez-Vargas, K. Becker et al. 2010. A poke into the diversity and associations within human anterior nare microbial communities. The ISME Journal **4**:839–851.

Yan, M., S. J. Pamp, J. Fukuyama, P. H. Hwang, D. Y. Cho, S. Holmes, and D. A. Relman 2013. Nasal microenvironments and interspecific interactions influence nasal microbiota complexity and *S. aureus* carriage. Cell Host & Microbe **14**:631–640.

Yarwood, J. M., J. K. McCormick, M. L. Paustian, V. Kapur, and P. M. Schlievert 2002. Repression of the *Staphylococcus aureus* accessory gene regulator in serum and in vivo. Journal of Bacteriology **184**:1095–1101.

Yu, V. L., A. Goetz, M. Wagener, P. B. Smith, J. D. Rihs, J. Hanchett, and J. J. Zuravleff 1986. *Staphylococcus aureus* nasal carriage and infection in patients on hemodialysis – efficacy of antibiotic prophylaxis. New England Journal of Medicine **315**:91–96.

Yuki, N., T. Shimazaki, A. Kushiro, K. Watanabe, K. Uchida, T. Yuyama, and M. Morotomi 2000. Colonization of the stratified squamous epithelium of the nonsecreting area of horse stomach by lactobacilli. Applied and Environmental Microbiology **66**:5030–5034.

Appendix A

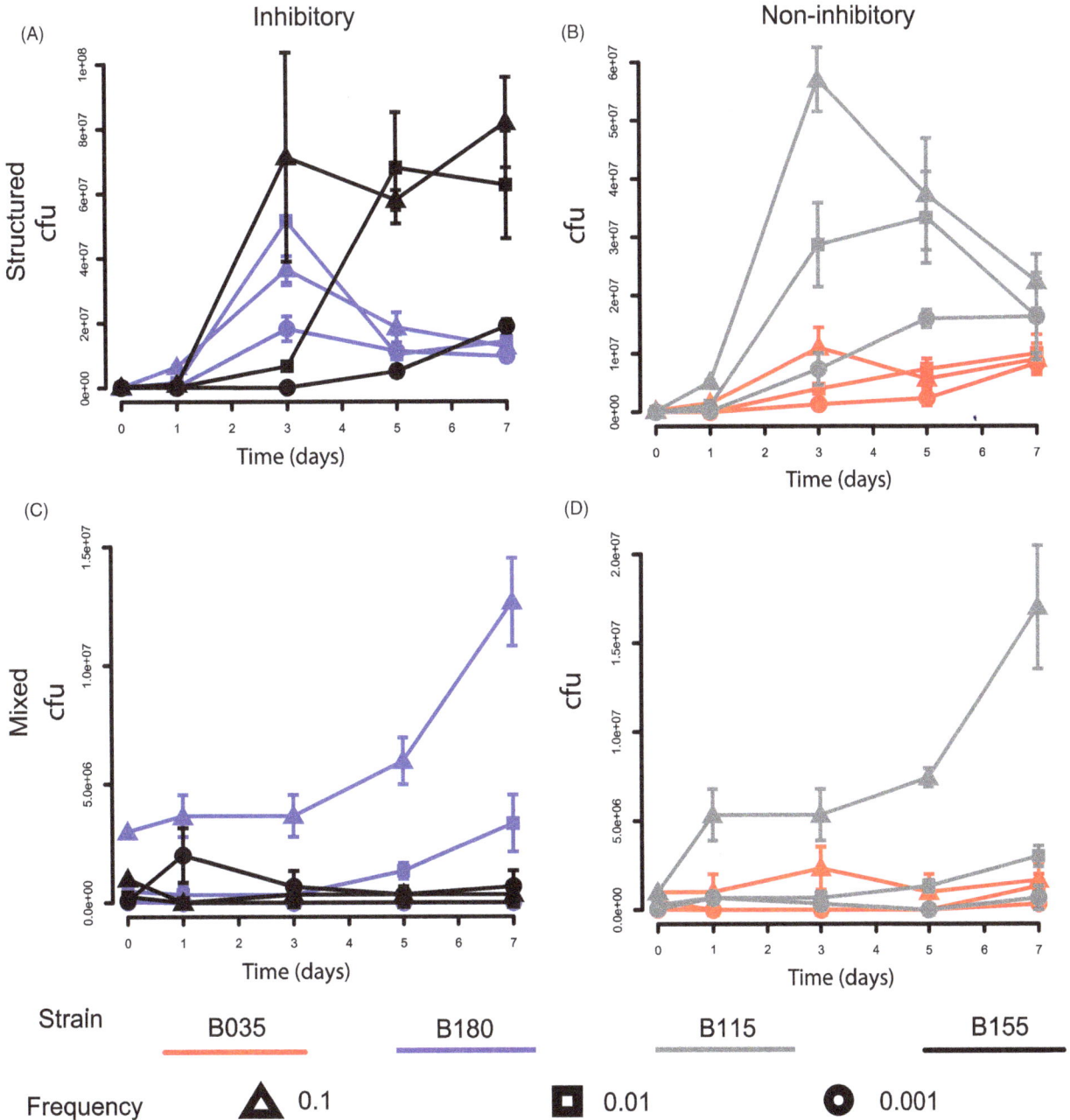

Figure A1 Toxin-producing (—— and ——) and nonproducing (——and ——) isolates of *Staphylococcus epidermidis* invading populations of *S. aureus* (SH1000) at frequencies of 0.1 (▲), 0.01 (■) and 0.001 (●). Toxin-producing *S. epidermidis* isolates (155 and 180) and nonproducing *S. epidermidis* isolates (035 and 115) were introduced into a population of *S. aureus* (SH1000) at three different frequencies. This was carried under a spatially structured regime (A and B) and under a mixed regimen (C and D). The *x*-axis is the time in days, and the *y*-axis is the colony-forming units (cfu) per plate. Error bars represent the standard error of the mean (*n* = 3).

Appendix B

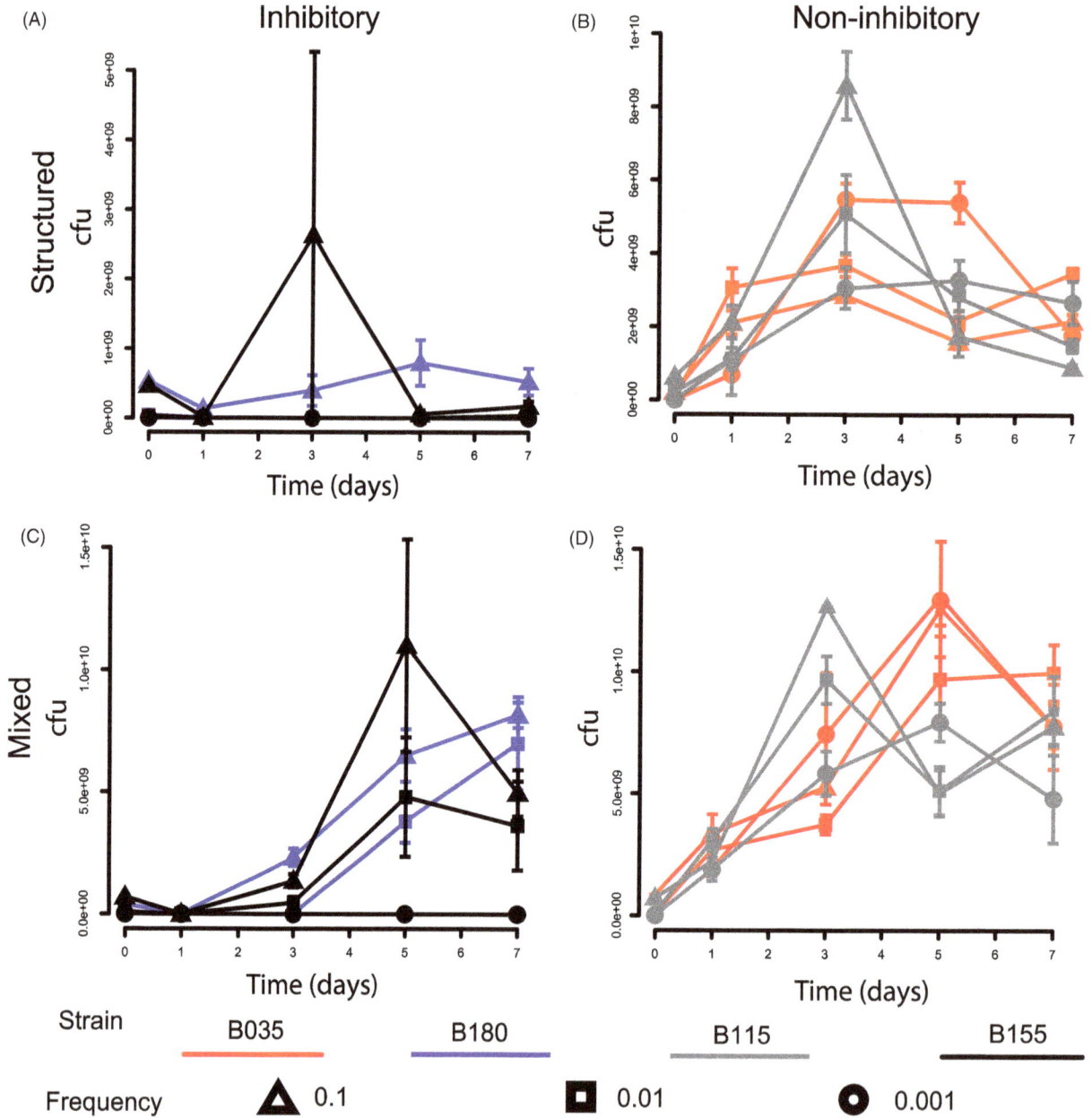

Figure B1 *Staphylococcus aureus* invading populations of toxin-producing (—— and ——) and nonproducing (—— and ——) *S. epidermidis* at frequencies of 0.1 (▲), 0.01 (▣) and 0.001 (●). *S. aureus* strain (SH1000) was introduced into two different toxin-producing *S. epidermidis* populations (155 and 180), and two different nonproducing populations (035 and 115) at three different frequencies. This was carried under a spatially structured regime (A and B) and under a mixed regime (C and D). The *x*-axis is the time in days, and the *y*-axis is the colony-forming units (cfu) per plate. Error bars represent the standard error of the mean (*n* = 3).

8

Swift thermal reaction norm evolution in a key marine phytoplankton species

Luisa Listmann,[1],* Maxime LeRoch,[1],* Lothar Schlüter,[1] Mridul K. Thomas[2] and Thorsten B. H. Reusch[1]

1 Evolutionary Ecology of Marine Fishes, GEOMAR Helmholtz-Centre for Ocean Research Kiel, Kiel, Germany
2 Department of Aquatic Ecology, –Eawag, Swiss Federal Institute of Aquatic Science and Technology, Dübendorf, Switzerland

Keywords

adaptation, coccolithophore, experimental evolution, global warming, phytoplankton, reaction norm, temperature.

Correspondence

Thorsten B. H. Reusch, Evolutionary Ecology of Marine Fishes, GEOMAR Helmholtz-Centre for Ocean Research Kiel, Düsternbrooker Weg 20, 24105 Kiel, Germany

e-mail: treusch@geomar.de

*These authors contributed equally.

Abstract

Temperature has a profound effect on the species composition and physiology of marine phytoplankton, a polyphyletic group of microbes responsible for half of global primary production. Here, we ask whether and how thermal reaction norms in a key calcifying species, the coccolithophore *Emiliania huxleyi*, change as a result of 2.5 years of experimental evolution to a temperature ≈2°C below its upper thermal limit. Replicate experimental populations derived from a single genotype isolated from Norwegian coastal waters were grown at two temperatures for 2.5 years before assessing thermal responses at 6 temperatures ranging from 15 to 26°C, with pCO_2 (400/1100/2200 μatm) as a fully factorial additional factor. The two selection temperatures (15°/26.3°C) led to a marked divergence of thermal reaction norms. Optimal growth temperatures were 0.7°C higher in experimental populations selected at 26.3°C than those selected at 15.0°C. An additional negative effect of high pCO_2 on maximal growth rate (8% decrease relative to lowest level) was observed. Finally, the maximum persistence temperature (T_{max}) differed by 1–3°C between experimental treatments, as a result of an interaction between pCO_2 and the temperature selection. Taken together, we demonstrate that several attributes of thermal reaction norms in phytoplankton may change faster than the predicted progression of ocean warming.

Introduction

Temperature has an overriding effect on species composition, photosynthetic performance and growth rates of marine phytoplankton (Eppley 1972; Raven and Geider 1988; Thomas et al. 2012; Boyd et al. 2013). The thermal physiology of phytoplankton species broadly corresponds to mean temperature values within their climate zone. For example, in tropical species, optimal temperatures and upper thermal limits are higher compared to temperate or polar species (Thomas et al. 2012, 2016; Boyd et al. 2013). There is also preliminary evidence that temperature niche width is correlated with the annual temperature variation, with temperate species displaying a wider temperature range than both tropical/subtropical and polar species (Boyd et al. 2013) although this pattern was not observed in a recent literature compilation (Thomas et al. 2016).

At the within-species level, there are also appreciable differences in thermal responses on growth and photosynthesis rates (Brand 1982; Wood and Leatham 1992) that likely have a heritable basis (Zhang et al. 2014). In light of global change impacting all marine and terrestrial ecosystems today, such within-population diversity may provide essential standing genetic variation for populations to track climate change via genotypic selection and hence adaptive evolution at the population level (Reusch and Boyd 2013; Collins et al. 2014).

Although there have been some recent evolution experiments in the phytoplankton to ocean acidification and warming (e.g., Lohbeck et al. 2012; Schlüter et al. 2014; Hutchins et al. 2015), we currently do not know the time scales over which thermal reaction norms evolve, the response in growth and other traits to a range of environmental temperatures. The question of how rapidly biologically meaningful differences in these reaction norms can arise via adaptive evolution is highly important to under-standing future ocean biogeochemical cycles, as the environment for phytoplankton is changing on the timescale of decades (e.g., Boyd et al. 2010). As a tool relatively new to marine science, long-term experiments can directly address whether and how organismal responses to global change may evolve at the population level (Collins et al. 2014; Sunday et al. 2014). Here, fast-dividing marine microbes (approx. 1 cell division day^{-1}) are prime examples for observing evolution in action over timescales of several months to years (Reusch and Boyd 2013; Collins et al. 2014; Hutchins et al. 2015).

The response of a given phytoplankton species or genotype to temperature is described by a thermal reaction norm, which is typically unimodal and left (negatively) skewed (Huey and Kingsolver 1989) as above the optimal temperature T_{opt}, growth rates decline more rapidly than below it. The maximum of the curve depicting the optimal growth temperature is correlated with (and generally higher than) the mean environmental temperature at the locale from which a population or genotype has been isolated for laboratory cultivation, reflecting adaptation to local temperature conditions (Thomas et al. 2012, 2016). As in other ectothermic species (Huey and Kingsolver 1989), phytoplankton thermal reaction norms are evolutionarily constrained by thermal trade-offs that prevent any one species from dominating across all temperatures found in the world's oceans (Boyd et al. 2013). For example, species can be categorized as having a narrow or a wide thermal niche (specialist vs. generalist), with maximal growth rates that are traded off against generalist growth performance and vice versa (Angilletta et al. 2003; Izem and Kingsolver 2005). Other conceivable trade-offs could be envisaged between maximal persistence temperature, thus stress tolerance, and maximal growth rates, but experimental data from phytoplankton demonstrating such trade-offs are lacking.

Our model species is the world's most abundant calcifying microalgae, the coccolithophore *Emiliania huxleyi* (Paasche 2002), that is one of the most intensely studied eukaryotic phytoplankton species with a near-worldwide distribution. Recently, this species has also become a model for combining experimental evolution and phytoplankton ecology (Reusch and Boyd 2013). Selection experiments subjecting this important phytoplankton species have shown that rapid adaptation to ocean acidification is possible within the time frame of 1 year either through genotypic sorting or via the occurrence of novel mutations within asexually dividing replicate populations (Lohbeck et al. 2012). In terms of temperature adaptation, previous experiments have shown that this species can adapt to a temperature only 1–2°C below the maximal growth temperature by rapid adaptive evolution within a timeframe of 1 year (corresponding to $\approx$500 asexual generations). Inter-

estingly, increases in fitness relative to control populations were amplified by simultaneous exposure to ocean acidification levels of 1100 and 2200 µatm (Schlüter et al. 2014). The work by Schlüter et al. (2014) only tested two temperatures in the assay experiment, a control temperature (15.0°C) and one a few degrees below the lethal threshold (26.3°C). It is thus currently unknown how the entire thermal reaction norm may have changed upon thermal adaptation in this species, as well as in any other phytoplankton species.

Here, we address whether or not selection for a single temperature close to the upper thermal limit (26.3°C) resulted in a reconfiguration of the entire reaction norm shape relative to populations evolving at 15°C, the approximate isolation temperature of the coccolithophore ecotypes at Bergen, Norway (Lohbeck et al. 2012). Note that all genotypes/replicate populations had been grown previously for 4 years at 15°C such that the warm temperature is a novel environment, while we do not deny that there is also some long-term adaptation to 15°C still ongoing during the experimental phase of this study. This was studied in full factorial combination with two levels of ocean acidification (1100 and 2200 µatm pCO$_2$) along with ambient controls (400 µatm pCO$_2$). We were particularly interested if adaptation to high temperature also changed the optimal growth temperature T_{opt}, maximum persistence temperature T_{max} (i.e., the temperature above which growth rate becomes negative), and maximal growth rates μ_{max} (Boyd et al. 2013; Thomas et al. 2016). Moreover, we studied possible trade-offs, for example with respect to T_{max} T_{opt}, and μ_{max}. Evolution experiments are particularly suited to address trade-offs because trait correlations, among the above three attributes of thermal reaction norms, have to evolve within an identical genetic background (Fry 2003).

Material and methods

Study species, culturing, and experimental design
The coccolithophore *Emiliania huxleyi* is the most abundant calcifying organisms in the world oceans, distributed from subpolar to subtropical waters (Paasche 2002). When forming blooms, their areal extent can be seen from outer space owing to the calcite platelets that reflect a proportion of the incoming solar radiation. Previous studies in *Emiliania huxleyi* have demonstrated swift evolutionary adaptation to ocean acidification and warming in asexual populations within the time frame of 1 year (approx. 500 asexual divisions) (Schlüter et al. 2014). Here, we built upon a previous CO$_2$ (Lohbeck et al. 2012) and temperature selection experiment (Schlüter et al. 2014) and ask whether and how the entire thermal reaction norm differs in two sets of asexual experimental populations that evolved for 2.5 years under a control and one high temper-

ature close to the upper thermal limit. Note that maximal water temperatures off Bergen, Norway, are at most 19°C (see August maxima at http://www.seatemperature.org/europe/norway/bergen-august.htm), thus while the control temperature is within the conditions encountered by the culture genotypes, this was not the case for the high temperature.

The temperature evolution experiment started in February 2013 when *Emiliania huxleyi* semi-continual batch cultures at three CO_2-levels ($N = 5$) were subdivided into a 'cold' (15.0°C) and a 'warm' (26.3°C) treatment, resulting in six fully factorial 'temperature by CO_2 treatment' combinations (Fig. 1). Phenotypic changes after 1 year (approx. 500 asexual generations) of temperature selection, tested at only two assay conditions, that is, the two selection regimes, have already been published elsewhere (Schlüter et al. 2014). The exact level of the 'high' temperature was determined in pilot experiments because initially, daily specific growth rates were approximately similar at both temperatures, thus the elapsed number of generations would also be similar across any occurring evolutionary adaptation (Schlüter et al. 2014). The temperature treatment was run for 1200 asexual generations or 2.5 years, thus ≈700 generations longer than the results reported in Schlüter et al. (2014). The original selection lines were founded in 2009 from a single cell isolated from a natural phytoplankton assemblage in the coastal waters off Bergen (Lohbeck et al. 2012). *Emiliania huxleyi* cultures were unialgal but not axenic as checked by monthly light microscopy and flow cytometry.

Three CO_2 levels were combined with the two temperatures in a full factorial way: along with a control treatment at 400 μatm, we subjected *E. huxleyi* to 1100 μatm as simulation of an end-of-the-century levels and 2200 μatm as the highest possible future level of ocean acidification in the year 2300 (Caldeira and Wickett 2003). CO_2 levels were reached by bubbling the medium with CO_2-enriched air before inoculating the experimental populations at the respective temperature treatments for 24 h, see (Schlüter et al. 2014) for details on the seawater chemistry and CO_2/ carbonate measurements.

For long-term experiments and temperature assays, we used artificial seawater ASW according to Kester et al. (1967) with the following nutrient additions: 64 μmol kg^{-1} of nitrate, 6 μmol kg^{-1} phosphate, trace metals, and vitamins according to f/8 composition, 10 nmol kg^{-1} of selenium, and 2 mL kg^{-1} filtered North Sea water to avoid limitation by micronutrients.

Cultures were kept under continuous rotation (0.5 min^{-1} in two Sanyo MLR-351 light cabinets) at 150 μmol m^{-2} s^{-1} photon flux density under a 16:8 light:-dark conditions during each 5-day cycle. To start the next batch cycle, and renew the medium, every 5 days 10^5 cells

were transferred from cultures into fresh medium. The cell abundance and diameter were measured in triplicate before each transfer, using a Beckman Coulter Z2 Particle and Size Analyzer. Daily specific growth rates (μ) were calculated from cell abundance as $\mu = (\ln N_d - \ln N_0)/d$. All culture work, including the ASW preparation, was performed under sterile conditions (laminar flow) (Schlüter et al. 2014).

Assay experiment to determine the thermal reaction norm

The reaction norm of all 30 replicates was determined at six assay temperatures after one full cultivation cycle of transfer of 100 000 cells to a new culture flask for acclimation (5 days =5-8 cell divisions), namely at 15, 18, 22, 24, 26, and 27°C. While the two lower temperature values are attained at Bergen (Norway) coastal waters, the isolation location of the tested *E. huxleyi* genotype, we were particularly interested in the possible changes of upper range of temperature tolerance. This also implies that our design inappropriate to address correlated responses at low temperatures (i.e., <15°C). Batch cycles to assess growth rates differed in length because we needed to maintain approximately similar maximal cell numbers in the assays. To accommodate for the different growth rates, batch cycles lasted 6 days in the 15°C treatment, 5 days in the 18°C treatment, 4 days in the 22°C, 5 days for 24°C, 7 days (15°C) and 5 days (26.3°C) for 26°C, and 10 days (26.3°C) in the 27°C treatment. Preparation of ASW was similar to the long-term culturing phase, except that during CO_2 manipulation via bubbling the medium, the seawater was kept at the planned assay temperature (15°C, 18°C, 22°C, 24°C, 26°C, and 27°C ($\pm$3°C)) for 24 h, and not at the respective selection temperature (15.0 and 26.3°C). Nutrients were never limiting given the cell abundance achieved at the end of the batch cycles. At the end of each growth cycle, the cell densities and cell diameter were determined triplicate by a Z2 Coulter Particle Count and Size Analyzer (Beckman® Coulter Counter; Krefeld, Germany), and specific daily growth rates (μ) were calculated from cell abundances as above.

Statistical analyses

Where strains exhibited no growth at either 26 or 27°C, we removed the 27°C measurement from our calculations. The absence of growth can indicate either zero or negative growth rate, but we were unable to measure negative growth rates. Therefore, zero is almost certainly an overestimate at 27°C in these cases.

Thermal reaction norm parameter estimation
Thermal reaction norms are typically left-skewed and have been described using a variety of functions. We used the

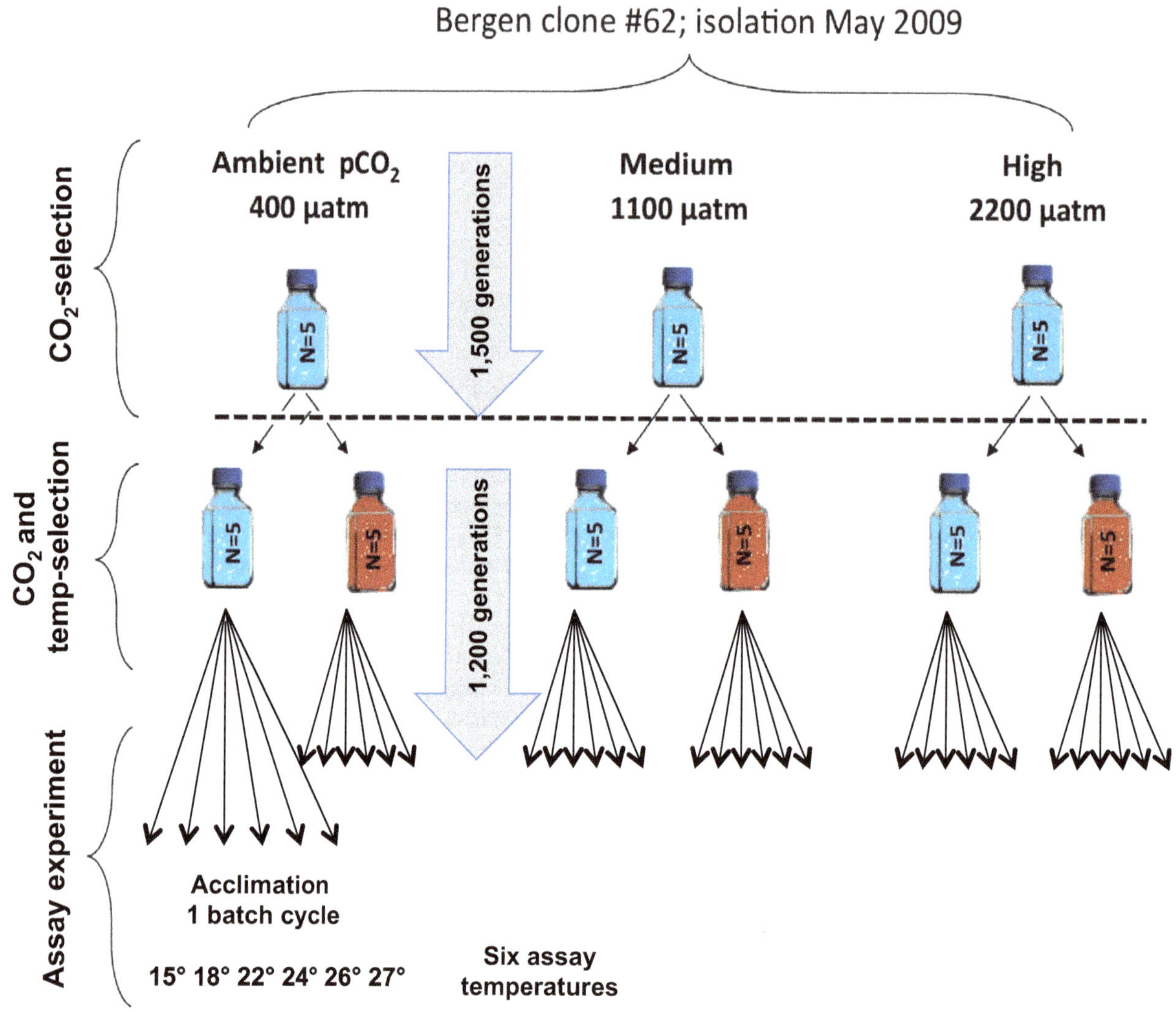

Figure 1 Five years of experimental evolution in *Emiliania huxleyi* in semi-continuous batch cultures. Schematic representation of the experimental design and the selection history.

equation and parameter and bootstrap-based uncertainty estimation procedures described in Thomas et al. (2012) and Boyd et al. (2013).

$$f(T) = ae^{bT}\left[1 - \left(\frac{T - z}{w/2}\right)^2\right] \qquad (1)$$

where specific growth rate f depends on temperature, T, as well as parameters z, w, a, and b. w is the temperature niche width (the range of temperatures over which growth rate is positive), while the other three (z,a,b) possess no explicit biological meaning but interact to influence the rate of increase in growth rate with temperature, the maximum growth rate and the optimum temperature for growth. We fit (1) to the growth data for each strain using maximum

likelihood to obtain estimates for parameters z, w, a, and b. In addition, we estimated the optimum temperature for growth (T_{opt}) and maximum persistence temperature (T_{max}) through numerical estimation. For maximum growth rate μ_{max}, we instead used the highest growth rate measured in our growth assays, as we are less confident in these estimates from our fitted reaction norms.

Estimating the influence of temperature selection and pCO₂ levels on traits

We tested whether the three thermal traits changed as a result of selection at different temperatures and pCO₂ levels while accounting for uncertainty in our estimates of these traits using a parametric bootstrapping approach. For each replicate, we fitted the thermal reaction norm function to the growth rate measurements and extracted the residuals from this fit. We then performed 1000 residual bootstraps, a procedure in which the residuals are randomly 'reassigned' to the predicted values (each of which corresponds to a growth rate measurement) and added to them, thereby generating a slightly different thermal reaction norm. For each iteration, we refitted the function and estimated the parameters (z, w, a, b) and also two of the derived traits, T_{opt} and T_{max}.

Examining the distribution of the traits (T_{opt} and T_{max}) over the 1000 bootstraps allowed us to quantify the uncertainty in our trait estimates, which we then incorporated in models seeking to explain how selection temperature and pCO_2 influenced them. For each set of bootstraps of all 30 experimental units, we fitted a linear model explaining each trait as a function of selection temperature (coded as a categorical variable), pCO_2 level and their interaction. pCO_2 level was standardized by subtracting the mean and dividing by two standard deviations to improve fitting procedures and generate readily comparable model parameter values (Gelman 2008; Gelman and Su 2015. We then examined the distribution of linear model parameter estimates over the 1000 bootstraps to determine whether temperature selection, pCO_2 or their interaction significantly influenced the traits. If the 95% confidence interval of a linear model parameter did not overlap zero, we concluded that the model parameter had a significant influence on the trait. If

the interaction between selection temperature and pCO_2 was an important predictor, we did not draw conclusions about significance of the main effects.

In the case of μ_{max}, we fitted the linear model only the highest measured growth rate of each strain (no bootstraps), using selection temperature (coded as a categorical variable), pCO_2 level and their interaction as explanatory variables. As with the other two traits, we standardized pCO_2 level by subtracting the mean and dividing by two standard deviations. After this, we estimated the 95% confidence intervals on the fitted model parameters.

Parameter estimation and bootstrapping was performed in the R Statistical Environment 3.2.2. (R Core Team 2015).

Results

Fitted reaction norms of all five replicates within the two selection treatments are depicted in Fig. 2, grouped by selection treatment and CO_2 condition. 2.5 years of selection at 15 and 26.3°C and three CO_2 levels resulted in marked differences in reaction norm shape, which we capture in important temperature traits (Fig. 3). The effect sizes of the selection conditions (i.e., the regression coefficients of the parameters) on T_{opt}, T_{max}, and μ_{max} are presented in Fig. 4. The variance explained (adjusted R^2) by the models for these three traits was 0.37, 0.74, and 0.65,

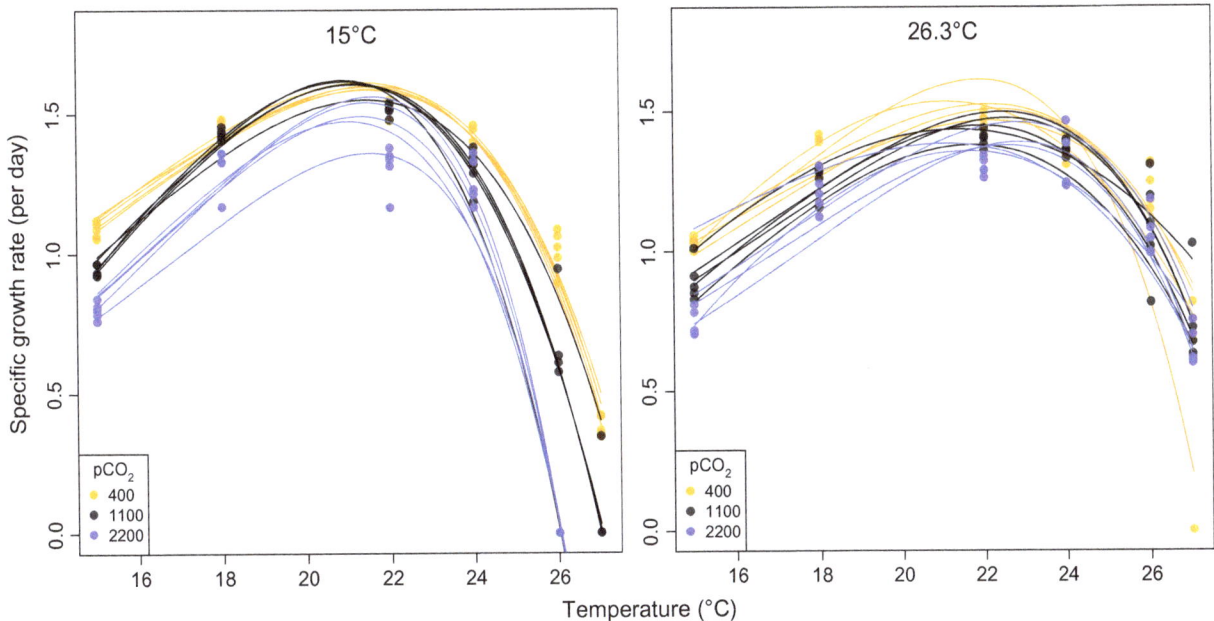

Figure 2 Fitted temperature reaction norms for individual replicate asexual populations according to eqn (1) in *Emiliania huxleyi*. Curves are grouped according to two different selection temperatures, 15.0 (left) and 26.3°C (right panel) fully crossed with three CO_2 environments, depicted by color (orange = 400 μatm, black = 1100 μatm, blue = 2200 μatm). Note that the CO_2 environment was similar during selection and temperature assay.

respectively. Temperature had a strong effect on T_{opt} and T_{max}, while CO_2 was important for determining the maximal growth rate G_{max} and for T_{max}. Both μ_{max} and T_{max} were influenced by interactions between CO_2 and selection temperature. We discuss the changes in the three traits below. We do not interpret our estimates of thermal niche width as we lack critical values at the lower temperature threshold to characterize them accurately.

Optimum temperature for growth (T_{opt})

T_{opt} differed by 0.7°C on average between the low- and high-temperature selection treatments (Fig. 3), being 21–21.5°C in the former and 22°C in the latter (Fig. 4). pCO_2 did not have a detectable influence on T_{opt} either directly or as part of an interaction with temperature selection (Figs. 3 and 4).

Maximum persistence temperature (T_{max})

T_{max} varied substantially between treatments and was influenced by an interaction between temperature selection and pCO_2 level (Fig. 4). In the high-temperature selection treatments, T_{max} reached 28.5–29°C regardless of pCO_2 level (Fig. 3). However, in the low-temperature selection

treatments, T_{max} varied by 2°C depending on pCO_2 level (Fig. 4). At 400 µatm (control), T_{max} was nearly 28°C, but this decreased by 1°C at 1100 µatm and 2°C at 2200 µatm (Fig. 3).

Maximum growth rate (μ_{max})

μ_{max} was reduced strongly at higher pCO_2 levels, but the extent of the reduction was influenced by an interaction with temperature (Figs. 3 and 4). μ_{max} decreased more strongly with increasing pCO_2 at 15°C (by nearly 0.2 per day at 2200 µatm relative to 400 µatm) than at 26.3°C (by approximately 0.09 per day).

Discussion

As a response to climate forcing via emission of greenhouse gasses, in particular CO_2, atmospheric warming has already produced pronounced ocean warming during the past few decades in the world's oceans, and even down to depths of 2000 m (Roemmich et al. 2015). As a result, among diverse plankton species, observed range shifts have been attributed to mean ocean warming (or increased variability) and concomitant poleward shifts of distributional ranges (Poloczanska et al. 2013). Adaptive evolution is a nonmutually exclusive process to range shifts that allows populations to 'stay' within their geographical range and has

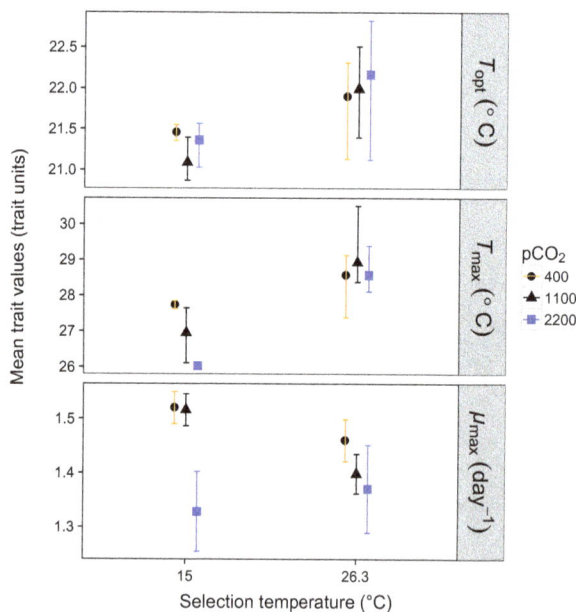

Figure 3 Mean temperature traits (T_{opt}, T_{max}, and μ_{max}) in the different treatments. T_{opt} and T_{max} were estimated from reaction norm fits depicted in Fig. 2, while μ_{max} was calculated from the measured growth rates. Confidence intervals (±95%) are based on residual bootstraps of the fitted reaction norms for the first two traits; for μ_{max}, calculations were based on the empirically measured growth rates and assumed normality of the data.

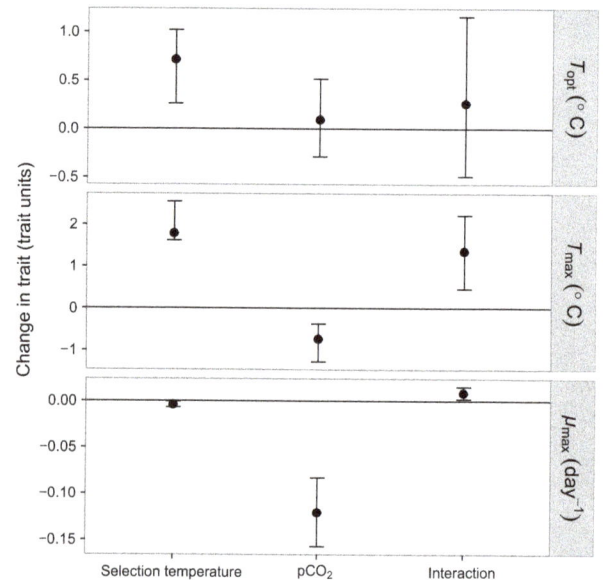

Figure 4 Changes in *Emiliania huxleyi* temperature traits T_{opt}, T_{max} [based on the nonlinear curve fit according to eqn (1)] and μ_{max} (based on measured growth rates) as a function of selection temperature, pCO_2, and their interaction. Standardized trait coefficients and their interaction are depicted; if confidence intervals do not overlap with zero, effects are statistically significant.

recently come into the research focus (Hoffmann and Sgro 2011).

Here, we have addressed how the rate at which adaptive evolution will produce different thermal reaction norms in a genetically near-uniform base population. We started our experimental evolution with a single isolate from Bergen, Norway. Thus, due to the lack of standing genetic variation, evolutionary adaptation most likely occurred via novel mutations. Mutant genotypes with under positive selection displayed higher Darwinian fitness and rose to (near) fixation in the asexually dividing population. Our scenario thus simulates that sexual reproduction and recombination is rare or absent (Reusch and Boyd 2013), which may actually be the case in some *E. huxleyi* populations (von Dassow et al. 2015). For our isolates, population-level data indicate that they are composed of many genotypes (unpublished data), which mirror results from the English channel that even in bloom situations genotypes are in Hardy–Weinberg equilibrium (Krueger-Hadfield et al. 2014) suggesting frequent sexual reproduction.

Although we started with one single genotype, we nevertheless found marked divergence among the two temperature selection treatments in several reaction norm parameters such as T_{opt}, T_{max}, and μ_{max}. These effects of temperature selection interacted with the CO_2 environment. Specifically, effects on T_{max}, in particular, were markedly negatively affected by medium and high ocean acidification levels but only in high-temperature selection treatments. This highlights how important interactions of major projected ocean perturbations are not only to understand future physiology, but also the evolutionary trajectories in plankton populations (Collins 2011; Schlüter et al. 2014). Note that our design cannot strictly test for temporal changes in reaction norms, because we cannot retest our initial base population or genotype as is possible in model systems such as *E.coli* or yeast with the help of cryopreservation (Elena and Lenski 2003). The salient experimental test is therefore the synchronic test of different evolution treatments at the same time, which also provides the necessary control for laboratory adaptation.

We cannot exclude that the genotype #62, originally isolated in 2009 off Bergen, Norwegian coastal waters, is special in terms of adaptation rates or magnitudes toward ocean warming and acidification. In other words, it is highly desirable to compare different genetic backgrounds with respect to their rates and magnitudes of adaptation, including associated phenotypic traits.

Our data support results of a shorter (1 year = 500 generations) evolution experiment reported in Schlüter et al. (2014). He tested thermal adaptation only at the two selection temperatures, 15.0°C and 26.3° over 1 year and found pronounced adaptation of growth rates to the long-term selection temperature in reciprocal exposure assays. The relative fitness of high-temperature-adapted replicates compared to low-temperature-adapted ones (as quotient of exponential growth rates) was 1.08–1.25, when tested at 26.3°C. The higher the CO_2 level in the experiment of Schlüter et al. (2014), the greater were the fitness gains of adapted populations, relative to the control populations evolving at today's pCO_2 and at a temperature of 15°C. When focusing on the rather similar assay temperature 26°C assessed here, the present data reveal an even stronger thermal adaptation after about 700 additional generations of evolution. Depending on CO_2, we find a relative fitness of 1.22 and 2.05 in ambient and medium (1100 μatm) pCO_2. Note that at the highest CO_2 level (2200 μatm), no growth was observed in the 15°C adapted populations at an assay temperature of 26°C. We also found a pronounced correlated response at 15°C – low-temperature-adapted populations grew faster at 15°C than high-temperature-adapted ones (Fig. 2).

Contrary to Schlüter et al. (2014), at an assay temperature of 26°C, we now find zero growth in some 15°C adapted replicate populations, which may be due to further adaptive evolution toward low-temperature specialization. However, this may have been influenced by differences in assay methods. In the present study, the initiation of the assay experiment (which started with a 1-week acclimation phase in all cases) was abrupt (i.e., within one single day), while the assay temperature was reached at a rate of 1°C per day in the earlier experiment by Schlüter et al. (2014). The concomitant carry-over effects may have negatively affected growth rates in the 2nd assay cycle where the growth rates presented in this paper were measured.

In the ideal case, we would have started the thermal evolution experiment with identical genetic material at the onset. However, replicate lines had already 3 years time (about 1500 generations) to accumulate slightly favorable and concomitant hitchhiking mutations (Lang et al. 2013) under long-term CO_2 selection before thermal selection had started. Thus, the starting genetic diversity was likely higher than a pure uni-clonal inoculum derived immediately from a single cell directly before the thermal selection started. This may explain the relatively fast pace and extent of reaction norm evolution compared to previous assessments of adaptation to ocean acidification (Lohbeck et al. 2012). Note, however, that also the CO_2-ambient control lines (at 400 μatm) always evolving under ambient CO_2 levels had also changed their thermal reaction norm relative to the 26.3°C selected replicates, thus previous high-CO_2 selection was apparently not prerequisite for adaptive responses.

In any case, the standing genetic variation in natural phytoplankton populations is much higher: ample standing genetic variation in coccolithophores has often been observed in genetic marker studies (Iglesias-Rodriguez et al. 2006; Krueger-Hadfield et al. 2014) or physiological assessments (Brand 1982; Wood and Leatham 1992; Kremp et al. 2012; Zhang et al. 2014), which provide abundant possibilities for selection to operate.

The major result of evolution at different temperatures was that the optimal growth temperature T_{opt} shifted upwards, which was expected. As growth rates increased only little during the 5 years of selection at 15°C (Schlüter and Reusch, unpublished data), we attribute most of the divergence in reaction norm shape to changes at high temperature. This may reflect a trade-off with between low- and high-temperature performance. A second major finding is that T_{max} shifted upwards very strongly relative to the low-temperature-adapted populations, indicating that adaptation to lethal temperatures is possible over monthly-to-yearly timescales even in very small populations (relative to natural populations). While this upward shift was present in all high-temperature selection replicates, pCO_2 had a negative effect on T_{max} in the low-temperature selection treatments. Currently, we have no explanation for this interaction, but it clearly deserves further testing in this and other phytoplankton species. In contrast to these two traits, μ_{max} was only weakly affected by temperature selection but strongly decreased with increasing pCO_2, especially at 15°C. Unfortunately, since we could not assess the niche width, our findings cannot be interpreted within a generalist–specialist trade-off scenario (Angilletta et al. 2003; Izem and Kingsolver 2005), However, it suggests that a general 'flattening' and broadening of the reaction norm may have occurred, superimposed onto a right-hand shift of the entire reaction norm curve.

Although our upper thermal selection temperature was very unrealistic with respect to the temperatures *E. huxleyi* may experience throughout the North Atlantic (maximal temperatures at Bergen, the isolation site 19°C in August), our results nevertheless provide a proof-of-principle of swift evolution of reaction norms and provide first insights into trade-offs of important traits associated with phytoplankton temperature reaction norms. Biogeochemical models of future ocean productivity contain thermal sensitivities of major phytoplankton groups as key parameters (Taucher and Oschlies 2011). Our results show that these thermal traits are parameters that can change by evolution and may to some extent track the expected increases in sea surface temperatures. A big open question is whether in the ocean, thermally sensitive species will be replaced by taxa possessing higher optimal growth temperatures and upper tolerances (i.e., ecological compositional change), or whether *in situ* evolution of thermal reaction norms will occur as a nonexclusive additional response, favouring the persistence of existing species and communities (Collins and Gardner 2009).

Acknowledgements

We thank Renate Ebbinghaus and Katrin Beining for laboratory assistance and Kai T. Lohbeck for his advice. This work was partly funded through the Kiel Excellence Cluster The Future Ocean. Maxime LeRoch was funded through a stipend of the University du Bretagne-Sud. Comments by P.W. Boyd and one anonymous referee greatly improved the manuscript.

References

Angilletta, M. J. Jr, R. S. Wilson, C. A. Navas, and R. S. James 2003. Tradeoffs and the evolution of thermal reaction norms. Trends Ecology Evolution 18:234–240.

Boyd, P. W., R. Strzepek, F. U. Feixue, and D. A. Hutchins 2010. Environmental control of open-ocean phytoplankton groups: Now and in the future. Limnology & Oceanography 55:1353–1376.

Boyd, P. W., T. A. Rynearson, E. A. Armstrong, F. Fu, K. Hayashi, Z. Hu, D. A. Hutchins et al. 2013. Marine phytoplankton temperature versus growth responses from polar to tropical waters – outcome of a scientific community-wide study. PLoS One 8:e63091.

Brand, L. E. 1982. Genetic variability and spatial patterns of genetic differentiation in the reproductive rates of the marine coccolithophores *Emiliania huxleyi* and *Gephyrocapsa oceanica*. Limnology & Oceanography 27:236–245.

Caldeira, K., and M. E. Wickett 2003. Oceanography: anthropogenic carbon and ocean pH. Nature 425:365–365.

Collins, S. 2011. Many possible worlds: expanding the ecological scenarios in experimental evolution. Evolutionary Biology 38:3–14.

Collins, S., and A. Gardner 2009. Integrating physiological, ecological and evolutionary change: a Price equation approach. Ecology Letters 12:744–757.

Collins, S., B. Rost, and T. A. Rynearson 2014. Evolutionary potential of marine phytoplankton under ocean acidification. Evolutionary Applications 7:140–155.

von Dassow, P., U. John, H. Ogata, I. Probert, E. M. Bendif, J. U. Kegel, S. Audic et al. 2015. Life-cycle modification in open oceans accounts for genome variability in a cosmopolitan phytoplankton. ISME Journal 9:1365–1377.

Elena, S. F., and R. E. Lenski 2003. Evolution experiments with microorganisms: the dynamics and genetic bases of adaptation. Nat Rev Genet 4:457–469.

Eppley, R. W. 1972. Temperature and phytoplankton growth in the sea. Fisheries Bulletin 70:1063–1085.

Fry, J. D. 2003. Detecting ecological trade-offs using selection experiments. Ecology 84:1672–1678.

Gelman, A. 2008. Scaling regression inputs by dividing by two standard deviations. Statistics in Medicine 27:2865–2873.

Gelman, A., and Y.-S. Su (2015). arm: Data Analysis Using Regression and Multilevel/Hierarchical Models. R package version 1.8-6. http://CRAN.R-project.org/package=arm (accessed on 1 November 2015).

Hoffmann, A. A., and C. M. Sgro 2011. Climate change and evolutionary adaptation. Nature 470:479–485.

Huey, R. B., and J. G. Kingsolver 1989. Evolution of thermal sensitivity of ectotherm performance. Trends in Ecology & Evolution 4:131–135.

Hutchins, D. A., N. G. Walworth, E. A. Webb, M. A. Saito, D. Moran,

M. R. McIlvin, J. Gale et al. 2015. Irreversibly increased nitrogen fixation in *Trichodesmium* experimentally adapted to elevated carbon dioxide. Nature Communications 6:8155.

Iglesias-Rodriguez, M. D., O. M. Schofield, J. Batley, L. K. Medlin, and P. K. Hayes 2006. Intraspecific genetic diversity in the marine coccolithophore *Emiliania huxleyi* (Prymnesiophyceae): the use of microsatellite analysis in marine phytoplankton population studies. Journal of Phycology 42:526–536.

Izem, R., and J. G. Kingsolver 2005. Variation in continuous reaction norms: quantifying directions of biological interest. The American Naturalist 166:277–289.

Kester, D. R., I. W. Duedall, D. N. Connors, and R. M. Pytkowicz 1967. Preparation of artificial seawater. Limnology & Oceanography 12:176–179.

Kremp, A., A. Godhe, J. Egardt, S. Dupont, S. Suikkanen, S. Casabianca, and A. Penna 2012. Intraspecific variability in the response of bloomforming marine microalgae to changed climate conditions. Ecology & Evolution 2:1195–1207.

Krueger-Hadfield, S. A., C. Balestreri, J. Schroeder, A. Highfield, K. T. Lohbeck, U. Riebesell, T. B. H. Reusch et al. 2014. Genotyping an *Emiliania huxleyi* (Prymnesiophyceae) bloom event in the North Sea reveals evidence of asexual reproduction. Biogeosciences 11:5215–5234.

Lang, G. I., D. P. Rice, M. J. Hickman, E. Sodergren, G. M. Weinstock, D. Botstein, and M. M. Desai 2013. Pervasive genetic hitchhiking and clonal interference in forty evolving yeast populations. Nature 500:571–574.

Lohbeck, K. T., U. Riebesell, and T. B. H. Reusch 2012. Adaptive evolution of a key phytoplankton species to ocean acidification. Nature Geoscience 5:346–351.

Paasche, E. 2002. A review of the coccolithophorid *Emiliania huxleyi* (Prymnesiophyceae), with particular reference to growth, coccolith formation, and calcification-photosynthesis interactions. Phycologia 40:503–529.

Poloczanska, E. S., C. J. Brown, W. J. Sydeman, W. Kiessling, D. S.

Schoeman, P. J. Moore, K. Brander et al. 2013. Global imprint of climate change on marine life. Nature Climate Change 3:919–925.

Raven, J., and R. Geider 1988. Temperature and algal growth. New Phytologist 110:441–461.

R Core Team. 2015. R: A language and environment for statistical computing. R Foundation for Statistical Computing, Vienna, Austria. Available from: https://www.R-project.org/.

Reusch, T. B. H., and P. W. Boyd 2013. Experimental evolution meets marine phytoplankton. Evolution 67:1849–1859.

Roemmich, D., J. Church, J. Gilson, D. Monselesan, P. Sutton, and S. Wijffels 2015. Unabated planetary warming and its ocean structure since 2006. Nature climate change 5:240–245.

Schlüter, L., K. T. Lohbeck, M. A. Gutowska, J. P. Gröger, U. Riebesell, and T. B. H. Reusch 2014. Adaptation of a globally important coccolithophore to ocean warming and acidification. Nature Climate Change 4:1024–1030.

Sunday, J. M., P. Calosi, S. Dupont, P. L. Munday, J. H. Stillman, and T. B. H. Reusch 2014. Evolution in an acidifying ocean. Trends in Ecology & Evolution 29:117–125.

Taucher, J., and A. Oschlies 2011. Can we predict the direction of marine primary production change under global warming? Geophysical Research Letters 38:L02603.

Thomas, M. K., C. T. Kremer, C. A. Klausmeier, and E. Litchman 2012. A global pattern of thermal adaptation in marine phytoplankton. Science 338:1085–1088.

Thomas, M. K., C. T. Kremer, and E. Litchman 2016. Environment and evolutionary history determine the global biogeography of phytoplankton temperature traits. Global Ecology and Biogeography. 25:75–86.

Wood, A. M., and T. Leatham 1992. The species concept in phytoplankton ecology. Journal of Phycology 28:723–729.

Zhang, Y., R. Klapper, K. T. Lohbeck, L. T. Bach, K. G. Schulz, T. B. H. Reusch, and U. Riebesell 2014. Between- and within-population variations in thermal reaction norms of the coccolithophore *Emiliania huxleyi*. Limnology & Oceanography 59:1570–1580.

9

Fight evolution with evolution: plasmid-dependent phages with a wide host range prevent the spread of antibiotic resistance

Ville Ojala,[1] Jarkko Laitalainen[1] and Matti Jalasvuori[1,2]

1 Centre of Excellence in Biological Interactions, Department of Biological and Environmental Science and Nanoscience Center University of Jyväskylä, Jyväskylä, Finland
2 Division of Ecology, Evolution and Genetics, Research School of Biology, Australian National University, Canberra, ACT, Australia

Keywords

evolution of antibiotic resistance, conjugation, conjugative plasmid-dependent phages, phage therapy

Correspondence

Matti Jalasvuori, Department of Biological and Environmental Science, University of Jyväskylä, Ambiotica Building, Survontie 9, P.O.Box 35, 40014 Jyväskylä, Finland.

e-mail: matti.jalasvuori@jyu.fi

Abstract

The emergence of pathogenic bacteria resistant to multiple antibiotics is a serious worldwide public health concern. Whenever antibiotics are applied, the genes encoding for antibiotic resistance are selected for within bacterial populations. This has led to the prevalence of conjugative plasmids that carry resistance genes and can transfer themselves between diverse bacterial groups. In this study, we investigated whether it is feasible to attempt to prevent the spread of antibiotic resistances with a lytic bacteriophage, which can replicate in a wide range of gram-negative bacteria harbouring conjugative drug resistance–conferring plasmids. The counter-selection against the plasmid was shown to be effective, reducing the frequency of multidrug-resistant bacteria that formed via horizontal transfer by several orders of magnitude. This was true also in the presence of an antibiotic against which the plasmid provided resistance. Majority of the multiresistant bacteria subjected to phage selection also lost their conjugation capability. Overall this study suggests that, while we are obligated to maintain the selection for the spread of the drug resistances, the 'fight evolution with evolution' approach could help us even out the outcome to our favour.

Introduction

The rapidly increasing number of antibiotic-resistant bacterial infections is of a major concern to modern health care worldwide, causing both substantial financial loss and numerous deaths (Taubes 2008; Bush et al. 2011). From an evolutionary standpoint, the problem is hardly a surprising one, because the constant application (both appropriate and inappropriate) of antibiotics has exerted a strong pressure on bacteria to develop resistance (Cohen 1992; Levin et al. 1997; Austin et al. 1999; Levy and Marshall 2007). Nevertheless, the predictability of the issue has not made it any easier to deal with, and in the case of many serious bacterial pathogens, such as methicillin-resistant *Staphylococcus aureus* (MRSA) or certain strains of *Pseudomonas aeruginosa* and *Klebsiella pneumoniae*, the number of viable treatment options is dangerously declining (MacKenzie et al. 1997; Livermore 2002; Taubes 2008; Wise et al. 2011). The situation is further complicated by the fact that only a few novel classes of antibiotics have been introduced during the past 50 years (Walsh 2003; Coates et al. 2011).

Resistant bacteria commonly harbour mobile genetic elements, such as conjugative plasmids that contain genes conferring resistance to several classes of antibiotics (Bennett 2008). Conjugative plasmids replicate independently of the host genome and they can facilitate their own transfer from one bacterial strain or species to another by coding for a channel through which a copy of the plasmid is transferred from donor to recipient cell (Brinton 1965). This horizontal gene transfer (HGT) allows for a highly efficient spread of resistances in bacterial communities (Davies 1994; Grohmann et al. 2003). Autonomous replication and the ability to move between (sometimes distantly related) bacteria mean that conjugative plasmids are independently evolving genetic elements (Norman et al. 2009). Consequently, their presence and horizontal movement can, depending on the circumstances, be an advantage, a disadvantage or neutral in terms of fitness of both the host they reside in and the other bacteria in the microbial community (Eberhard 1990; Kado 1998; Dionisio et al. 2005; Slater et al. 2008; Norman et al. 2009). For example, the transfer of a conjugative plasmid from a bacterial donor to a (unrelated) recipient could potentially lower the fitness of the donor and increase the fitness of the recipient if the two bacteria compete over resources, and the possession of the plasmid provides some competitive advantage, such as resistance to the antibiotics present in the system (Jalasvuori 2012).

Interfering with the process of bacterial conjugation has been proposed as one potential way of combating the spread of plasmid-mediated antibiotic resistances (Smith and Romesberg 2007; Williams and Hergenrother 2008). Certain bacteriophages (phages) specifically infect and kill conjugative plasmid–harbouring bacteria (Caro and Schnös 1966). These phages use conjugative plasmid–encoded proteins as their receptor to gain entrance to a host cell. The host range of a given conjugative plasmid–dependent (or male-specific) phage is therefore mainly determined by the host range of suitable conjugative plasmids (Olsen et al. 1974). In practice, conjugative plasmid–dependent phages are natural enemies of both the conjugative plasmids and the bacteria that harbour them. A previous study suggests that in the absence of antibiotic selection and other bacteria, the presence of a lytic conjugative plasmid–dependent phage can efficiently select for bacteria that either have lost their conjugative plasmids or harbour a conjugation-deficient version of the plasmid (Jalasvuori et al. 2011). However, the capability of these phages to limit the rate of horizontal transfer of plasmids between bacteria was not investigated. Other studies have shown that nonlytic filamentous phages are capable of preventing the spread of conjugative plasmids by physically inhibiting conjugation (Novotny et al. 1968; Lin et al. 2011).

Elaborating from these previous studies, we here investigated whether a lytic conjugative plasmid–dependent phage can prevent the emergence of new multiresistant strains by selecting against the plasmid or, more specifically, the plasmid-encoded sex apparatus facilitating the transfer of the plasmid to other bacteria. Moreover, we measure how much the presence of nonlethal antibiotic selection favouring different plasmid and bacterium combinations alters the counter-selective effect of phages (Fig. 1). In our experiments, two antibiotic-resistant bacterial strains of *Escherichia coli* K-12 were cultivated in daily replenished cultures together for 3 days. One of the used strains, JE2571(RP4), contains a conjugative plasmid RP4 conferring resistance to several antibiotics of different classes (ampicillin, kanamycin and tetracycline), whereas the other strain HMS174 is plasmid free but resistant to rifampicin due to a chromosomal mutation. In this experimental setup, the potential

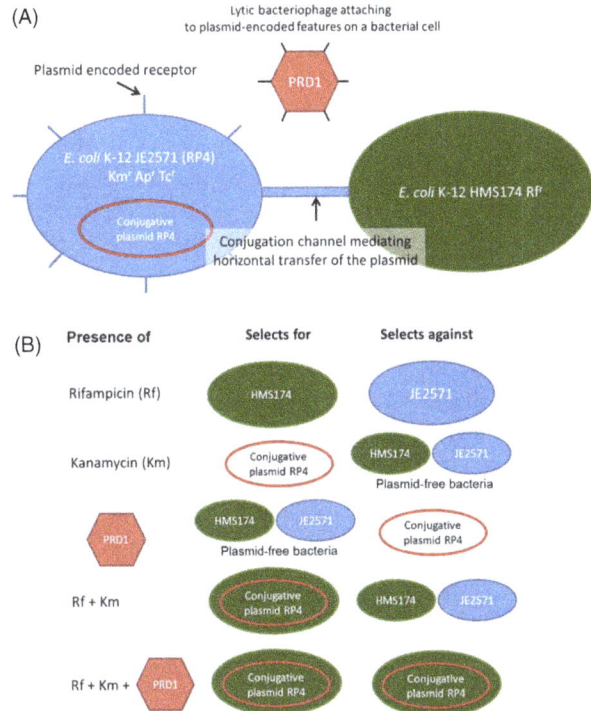

Figure 1 Schematic presentation of the experimental setup and the selection pressures.

conjugative transfer of the RP4 plasmid from JE2571(RP4) to HMS174 would create a new multiresistant strain HMS174(RP4). The presence of the conjugative plasmid–dependent phage PRD1 selects against all bacteria representing plasmid-encoded receptors on the cell surface. Bacteria are resistant to phage infections if they are free of the plasmid or they harbour a conjugation-defective mutant (Jalasvuori et al. 2011).

Altogether, we here demonstrate that conjugative plasmid–dependent phage PRD1 effectively restricts the emergence of the multiresistant HMS174(RP4) strain even in the presence of nonlethal antibiotic selection. While growth-reducing antibiotic concentrations may play an important role in the evolution of bacterial antibiotic resistance (Andersson and Hughes 2012), these results suggest that is possible to combat this evolution with counter-selective attempts.

Materials and methods

Bacterial strains, bacteriophages and culture conditions

Escherichia coli K-12 strains JE2571(RP4) (Bradley 1980), HMS174 (Campbell et al. 1978) and JM109(pSU19) were used in this study. JE2571 harbours a conjugative incompatibility group P plasmid RP4 (Datta et al. 1971), which induces antibiotic resistance to kanamycin, ampicillin and tetracycline. HMS174 contains chromosomal rifampicin resistance. JM109(pSU19) contains a nonconjugative plasmid pSU19 (Bartolomé et al. 1991) that induces chloramphenicol resistance. All strains were cultivated in Luria–Bertani (LB) medium (Sambrook et al. 1989) at 37°C. Shaking at 200 revolutions per minute (rpm) was used, with the exception of the evolution experiments where the cultures were unshaken. For general antibiotic selection, kanamycin, rifampicin and chloramphenicol were used in final concentrations of 32 μg/mL, 55 μg/mL and 25 μg/mL, respectively. The bacteriophage used in this study was PRD1; a lytic conjugative plasmid–dependent phage infecting a wide range of gram-negative bacteria that contain conjugative plasmids belonging to incompatibility groups P, N and W (Olsen et al. 1974).

Evolution experiments

5 μL of JE2571(RP4) and HMS174 overnight cultures were inoculated into the same tube containing 5 mL of fresh LB medium. The mixed cultures were treated with (i) no antibiotics, (ii) kanamycin, (iii) rifampicin or (iv) kanamycin and rifampicin. When appropriate, kanamycin and rifampicin were added in nonlethal but growth-reducing concentrations of 3.2 μg/mL and 3.7 μg/mL, respectively (Fig S1A,B). Each antibiotic treatment was performed both in the presence and in the absence of conjugative plasmid–dependent phage PRD1. Immediately after the transfer of the bacteria, 5 μL of phage stock containing approximately 10^{11} plaque-forming units per millilitre (pfu/mL) was added to the appropriate treatments. Cultures were grown at 37°C without shaking. The length of the experiment was approximately 72 hours, and the cultures were renewed at 24- and 48-hour time points by transferring 5 μL of culture to 5 ml of fresh LB medium (containing the appropriate antibiotics; no new phage was added during the refreshments). Each treatment was sampled during the culture renewals and at the end of the experiment. These samples were diluted and plated on either regular or antibiotic-containing (kanamycin and rifampicin) 1% LB agar plates to obtain the total bacterial densities and the number of bacteria resistant to both antibiotics. Also, from all treatments, a random sample of clones ($n_{total} = 210$) growing on kanamycin- and rifampicin-containing plates were transferred to kanamycin-, tetracycline-, ampicillin- and rifampicin-containing plates to further confirm that the formed multiresistant clones harbour a plasmid (i.e. controlling the frequency of spontaneous antibiotic-resistant mutants). In addition, final phage densities were determined at the end of the experiment by plating diluted samples (on 1% LB agar plates with a 0.7% soft agar overlay) from phage-containing treatments with the ancestral form of JE2571(RP4) bacteria.

Conjugation assay

To study the ability of evolved multiresistant bacteria to further transfer their resistance-conferring plasmid through conjugation, random individual bacterial clones (both from phage-containing and from phage-free treatments) were transferred from the kanamycin–rifampicin plates to 5 mL of fresh LB medium and then grown overnight at 37°C and 200 rpm. Similar culture was made of strain JM109(pSU19). Next day, the cultures of the multiresistant clones were mixed in 1:1 ratio with JM109(pSU19), and fresh LB medium was added (12.5% of the combined volume of the two bacteria). These cultures were then grown for 24 hours at 37°C without shaking. A sample from each culture was plated on 1% LB agar plate containing chloramphenicol, kanamycin, ampicillin and tetracycline to see whether the RP4 plasmid had transferred itself to JM109(pSU19) and again formed a new multiresistant strain: JM109(pSU19)(RP4). Clones were scored conjugation defective if no colonies formed on chloramphenicol–kanamycin–tetracycline–ampicillin plates. Five randomly selected clones that turned out to be conjugation deficient were further grown in LB medium with JM109(pSU19), now in the presence of chloramphenicol and kanamycin in nonlethal but growth-reducing concentrations (in final concentrations of 0.625 μg/mL and 1.25 μg/mL, respectively (Fig S1C–D)), to see whether the selective pressure posed by the antibiotics would revert the conjugation ability. Five clones that had been capable of conjugation in the first experiment were used as a control. This experiment lasted for 72 hours with the initiation, culture renewing, sampling and plating carried out similarly to the main experiment with the exception of using different antibiotics. The number of potential JM109(pSU19)(RP4) bacteria was measured every day.

Data analysis

The frequencies of multiresistant bacteria in different treatments were calculated by dividing the density of multiresistant bacteria by the total bacterial density. For statistical tests, arcsine transformation was performed on the obtained frequencies, and the transformed frequencies were compared between phage-free and phage-containing treatments using one-way ANOVA. The level of statistical significance was adjusted with Bonferroni correction to control the effects of multiple comparisons.

Results

The presence of the conjugative plasmid–dependent phage PRD1 significantly reduced the formation of multidrug-resistant E. coli HMS714(RP4) bacteria by infecting all bac-

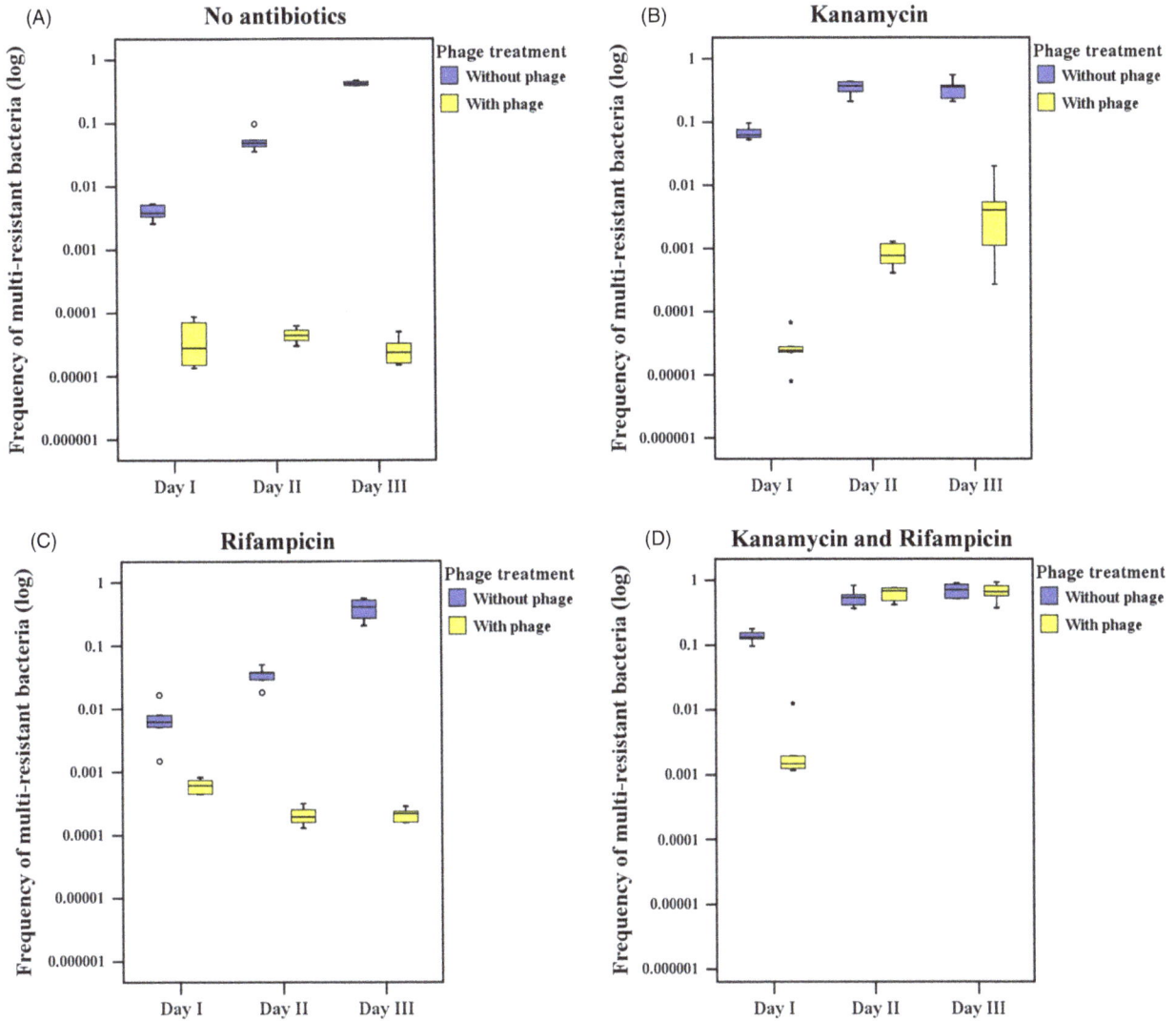

Figure 2 Frequencies of multiresistant (kanamycin + rifampicin) bacteria in the presence and absence of phages. In antibiotic treatments (A) 'no antibiotics', (B) 'kanamycin' and (C) 'rifampicin', there were significantly less multiresistant bacteria in phage-containing treatments throughout the experiment; in treatment (D) 'kanamycin and rifampicin', the difference was significant only after the first experimental day (Table S1).

teria harbouring actively conjugating resistance–conferring plasmids (Fig. 2). In the absence of phages, multiresistant bacteria quickly became common in both antibiotic-free and all antibiotic-containing treatments. The addition of phages resulted in several orders of magnitude lower levels of multiresistant bacteria. In treatments where neither or only one of the two antibiotics (kanamycin or rifampicin) was present in nonlethal but growth-reducing concentration, the phages reduced the prevalence of multiresistance for the entire length of the three-day experiment (Fig. 2A–C). With the double-antibiotic selection, there were still significantly fewer multiresistant bacteria in the phage-containing treatment after the first experimental day, but by the second day, the difference to the phage-free treatment had mostly disappeared (Fig. 2D). Descriptive statistics for

the frequencies of multiresistant bacteria in different treatments and the statistical comparisons thereof (one-way ANOVA) are given in Table S1. The addition of phages did not considerably affect the total number of bacteria (Table S2) in any treatment (all cultivations grew to a saturated density of approximately 10^8 colony-forming units per millilitre; cfu/mL), but rather had an effect on the relative numbers of different bacterial types in a population, often selecting against the multiresistant ones. Furthermore, infective PRD1 particles were still abundant at the end of the experiment in all phage-containing treatments (Table S3).

Given that multiresistant bacteria were capable of taking over the double-antibiotic system (Fig. 2D) despite the presence of phages, we decided to further investigate the properties of these particular bacteria. They may be harbouring RP4 plasmids with mutations in the cell surface complex that PRD1 uses as a receptor and which is also needed for successful conjugation (Kornstein et al. 1992; Kotilainen et al. 1993). The kanamycin- and rifampicin-resistant clones in all experiments were resistant also to ampicillin and tetracycline, confirming that it was the plasmid and not spontaneous resistance that produced the observed resistance pattern. Nevertheless, such mutated plasmids would provide the host bacteria with simultaneous antibiotic and phage resistance but, due to disturbed conjugation machinery, be unable to transfer the plasmid. Following this line of reasoning, we tested whether the multiresistant bacteria emerged under the simultaneous double-antibiotic and phage selection were still capable of conjugation and, indeed, found that only 35% of randomly selected kanamycin–rifampicin-resistant clones ($n = 72$) were capable of transferring the RP4 plasmid to a third chloramphenicol-resistant *E. coli* strain, JM109(pSU19). Moreover, in three of the total five independent selection experiments, the lost conjugation capability did not revert even after 3 days of subsequent cultivation under antibiotic selection that would have favoured the reversion. In two selection experiments, few multiresistant clones of total ~5×10^8 bacteria appeared, but they remained at very low quantities (~10^2) throughout the three-day experiment. This suggests that after phage exposure, a prolonged selective condition would be required for the potential conjugative multiresistant strains to become abundant in the population. In contrast, all multiresistant bacterial clones isolated from the phage-free treatment retained the conjugation ability.

Discussion

Our results demonstrate that conjugative plasmid–dependent bacteriophage PRD1 can significantly reduce the horizontal spread of antibiotic resistance genes in a bacterial community even when the bacteria are exposed to antibiotic selection that should favour the evolution of multidrug-resistant strains via conjugation. The addition of conjugative plasmid–dependent phages to any of the antibiotic treatments acted as counter-selection against the spread of multiresistance commonly reducing it by several orders of magnitude. Only the selection specifically for the formation of HMS174(RP4) transconjugants coupled with 48 hours of evolution was a strong enough selective pressure to cancel the differences between the phage-containing and phage-free treatments. However, most bacteria in this phage-containing treatment had also lost their conjugation ability, whereas all bacteria in the phage-free treatment were still capable of conjugation.

It is known that conjugative plasmids can regulate their rate of transfer in several ways (Gasson and Willetts 1975). More specifically to this study, previous empirical work has shown that the presence of PRD1 can select for plasmid-harbouring bacteria that are phage resistant but conjugation deficient (Kotilainen et al. 1993; Jalasvuori et al. 2011). Theoretical models have suggested that heterogeneity in the rate of transfer is essential for the stable maintenance of conjugative plasmids in bacterial communities when conjugative plasmid–dependent phages are present (Dionisio 2005). From this heterogeneity, it follows that phages may be unlikely to be able to completely eradicate conjugative plasmids from a bacterial community but they can, nevertheless, potentially hinder the further spread of plasmid-mediated antibiotic resistances to other bacterial species that may already possess some other resistances (thus being candidates for new multiresistant agents). In our experiments, the lost conjugation ability of a given multiresistant bacterial clone did not seem to revert easily even when the phage selection was lifted and a three-day antibiotic selection favouring the reversion was added.

Recently Zhang and Buckling (2012) demonstrated that the combined bacteriosidic effect of antibiotic kanamycin and a lytic bacteriophage significantly decreased the rate at which bacteria developed resistance against the antibiotic. Therefore, these studies, along with the present results, suggest that it is reasonable to presume the combination of both plasmid-dependent phages with other lytic phages will induce significant constraints for bacteria to maintain resistances, acquire them horizontally or develop resistances *in situ*. However, Escobar-Páramo et al. (2012) showed that application of antibiotic rifampicin against the host bacteria of a phage decreased the survival of phages in the system and would therefore potentially hinder the efficacy of combined phage and antibiotic treatments. We noticed similar effects when rifampicin alone was used in the system. In these experiments, the phage densities at the end of the three-day serial culture were more than 10-fold lower in comparison with other selection pressures (Table S3). This, nevertheless, is what was expected given that rifampicin selects against the initial plasmid-harbouring bacterium JE2571. Yet, the frequency of multiresistant bacteria in the end of the three-day experiment was relatively high in the presence of rifampicin, suggesting that the lower number of phages eased the selection pressure on the formed HMS174(RP4) transconjugants. However, and in contrast to rifampicin experiments, presence of kanamycin alone or both kanamycin and rifampicin elevated the phage densities above those of antibiotic-free experiments. This was also as predicted as kanamycin selects for the plasmid and thus the hosts of phage PRD1.

The concept of preventing the horizontal transfer of antibiotic resistance genes has been explored by a handful of earlier *in vitro* studies that have successfully used different nonphage molecules, phage coat proteins or replicative nonlytic and lytic conjugative plasmid–dependent phages to interfere with the bacterial conjugation (Novotny et al. 1968; Ou 1973; Fernandez-Lopez et al. 2005; Garcillán-Barcia et al. 2007; Lujan et al. 2007; Jalasvuori et al. 2011; Lin et al. 2011). Our study, to our knowledge, is the first one to demonstrate that lytic conjugative plasmid–dependent phages can, in principle, be effective selective agents against conjugative elements and thus the spread of drug resistances even when the bacteria are under sublethal antibiotic selection favouring the horizontal spread of resistance. Such growth-reducing concentrations of antibiotics have been thought to generate new multiresistant strains (Andersson and Hughes 2012). There are, however, important caveats to keep in mind when assessing these promising results. For example, it is unclear whether the evolutionary trajectories observed in this one particular experimental system are also common in other systems with different sets of bacteria, antibiotics, conjugative plasmids and conjugative plasmid–dependent phages. Also more generally, the relevance of results of *in vitro* experiments to the situation in natural environments is always uncertain.

As it currently seems inevitable that the development of new antibiotics will not be able to keep up with the worldwide emergence of resistance in pathogenic bacteria, it is increasingly important that we come up with alternative and complementary methods of treatment. Phage therapy has traditionally been overlooked by the Western medicine, whereas in Eastern Europe and Soviet Union, it was extensively studied and applied, although not always accordingly to the standards and rigour expected in Western science (Alisky et al. 1998; Chanishvili 2012). The reluctance in the West has largely been due to various technical, financial and safety challenges associated with developing and applying phage therapy. However, the worsening resistance epidemic has led to a revived interest in looking into phages as potential antibacterial agents (Lu and Koeris 2011). We suggest that, along with direct attempts to eliminate pathogenic bacteria via phages, the use of conjugative plasmid–dependent viruses could be one interesting avenue to explore. Characteristics of PRD1-like viruses are particularly promising for the development of phage applications. While phages are usually very host specific infecting only some strains of a given species, PRD1 has an extremely wide host range for a phage and it can exploit plasmids from various incompatibility groups (Olsen et al. 1974). PRD1 can also be produced easily in sufficient quantities (Mesquita et al. 2010) and stored stably over long times (Ackermann et al. 2004). Therefore, it may be possible to develop a wide host range cocktail of phages recognizing a wide variety of conjugation apparatuses and be thus usable in different contexts where antibiotic resistances cause problems. For instance, in hospitals, antibiotics are often administrated both before and after a surgical operation to reduce the risk of complications caused by bacterial infections. The number of postoperative hospitalization days under antibiotic treatment correlates positively with the probability of the emergence of life-threatening multiresistant infections (Schentag et al. 1998). Given that antibiotic resistances rise via horizontal gene transfer in various bacterial groups, including both opportunistic pathogens such as *Actinobacter baumannii* (Joshi et al. 2003) and common nosocomial pathogens like *Escherichia coli*, *Klebsiella pneumoniae* (Harajly et al. 2010) and *Staphylococcus aureus* (Lyon and Skurray 1987; Chang et al. 2003; Weigel et al. 2003), the presence of plasmid-dependent phages could hypothetically give antibiotics and the immune system more time to clear the infection before the emergence of highly resistant strains and also restrict the spread of resistances within the hospital in general. Yet, while this concept appears promising, future research in actual *in vivo* systems that are inevitably much more complex in all respects is essential to evaluate the real potential of plasmid-dependent phages.

Acknowledgements

This work was supported by the Academy of Finland Centre of Excellence in Biological Interactions and by Academy of Finland personal grant to MJ.

Literature cited

Ackermann, H. W., D. Tremblay, and S. Moineau 2004. Long-term bacteriophage preservation. WFCC Newsletter **38**:35–40.

Alisky, J., K. Iczkowski, A. Rapoport, and N. Troitsky 1998. Bacteriophages show promise as antimicrobial agents. Journal of Infection **36**:5–15.

Andersson, D. I., and D. Hughes 2012. Evolution of antibiotic resistance at non-lethal drug concentrations. Drug Resistance Updates **15**:162–172.

Austin, D. J., K. G. Kristinsson, and R. M. Anderson 1999. The relationship between the volume of antimicrobial consumption in human communities and the frequency of resistance. Transactions of New York Academy of Sciences USA **96**:1152–1156.

Bartolomé, B., J. Jubete, E. Martinez, and F. de la Cruz 1991. Construction and properties of a family of pACYC184-derived cloning vectors compatible with pBR322 and its derivatives. Gene **102**:75–78.

Bennett, P. M. 2008. Plasmid encoded antibiotic resistance: acquisition and transfer of antibiotic resistance genes in bacteria. British Journal of Pharmacology **153**(Suppl 1):S347–S357.

Bradley, D. E. 1980. Morphological and serological relationships of conjugative pili. Plasmid **4**:155–169.

Brinton, C. C. Jr 1965. The structure, function, synthesis and genetic control of bacterial pili and a molecular model for DNA and RNA transport in gram negative bacteria. Transactions of the New York Academy of Sciences **27**:1003–1054.

Bush, K., P. Courvalin, G. Dantas, J. Davies, B. Eisenstein, P. Huovinen, G. A. Jacoby et al. 2011. Tackling antibiotic resistance. Nature Reviews Microbiology **9**:894–896.

Campbell, J. L., C. C. Richardson, and F. W. Studier 1978. Genetic recombination and complementation between bacteriophage T7 and cloned fragments of T7 DNA. Proceedings of National Academy of Sciences USA **75**:2276–2280.

Caro, L. G., and M. Schnös 1966. The attachment of the male-specific bacteriophage F1 to sensitive strains of *Escherichia coli*. Proceedings of National Academy of Sciences USA **56**:126–132.

Chang, S., D. M. Sievert, J. C. Hageman, M. L. Boulton, F. C. Tenover, F. P. Downes, S. Shah et al. 2003. Infection with vancomycin-resistant *Staphylococcus aureus* containing the vanA resistance gene. New England Journal of Medicine **348**:1342–1347.

Chanishvili, N. 2012. Phage therapy–history from Twort and d'Herelle through Soviet experience to current approaches. Advances in Virus Research **83**:3–40.

Coates, A. R., G. Halls, and Y. Hu 2011. Novel classes of antibiotics or more of the same? British Journal of Pharmacology **163**:184–194.

Cohen, M. L. 1992. Epidemiology of drug resistance: implications for a post-antimicrobial era. Science **257**:1050–1055.

Datta, N., R. W. Hedges, E. J. Shaw, R. B. Sykes, and M. H. Richmond 1971. Properties of an R Factor from *Pseudomonas aeruginosa*. Journal of Bacteriology **108**:1244–1249.

Davies, J. 1994. Inactivation of antibiotics and the dissemination of resistance genes. Science **264**:375–382.

Dionisio, F. 2005. Plasmids survive despite their cost and male-specific phages due to heterogeneity of bacterial populations. Evolutionary Ecology Research **7**:1089–1107.

Dionisio, F., I. C. Conceição, A. C. Marques, L. Fernandes, and I. Gordo 2005. The evolution of a conjugative plasmid and its ability to increase bacterial fitness. Biology Letters **1**:250–252.

Eberhard, W. G. 1990. Evolution in bacterial plasmids and levels of selection. The Quarterly Review of Biology **65**:3–22.

Escobar-Páramo, P., C. Gougat-Barbera, and M. E. Hochberg 2012. Evolutionary dynamics of separate and combined exposure of *Pseudomonas fluorescens* SBW25 to antibiotics and bacteriophage. Evolutionary Applications **5**:583–592.

Fernandez-Lopez, R., C. Machón, C. M. Longshaw, S. Martin, S. Molin, E. L. Zechner, M. Espinosa et al. 2005. Unsaturated fatty acids are inhibitors of bacterial conjugation. Microbiology **151**:3517–3526.

Garcillán-Barcia, M. P., P. Jurado, B. González-Pérez, G. Moncalián, L. A. Fernández, and F. de la Cruz 2007. Conjugative transfer can be inhibited by blocking relaxase activity within recipient cells with intrabodies. Molecular Microbiology **63**:404–416.

Gasson, M. J., and N. S. Willetts 1975. Five control systems preventing transfer of *Escherichia coli* K-12 Sex Factor F. Journal of Bacteriology **122**:518–525.

Grohmann, E., G. Muth, and M. Espinosa 2003. Conjugative plasmid transfer in gram-positive bacteria. Microbiology and Molecular Biology Reviews. **67**:277–301.

Harajly, M., M. T. Khairallah, J. E. Corkill, G. F. Araj, and G. M. Matar 2010. Frequency of conjugative transfer of plasmid-encoded ISEcp1 - blaCTX-M-15 and aac(6')-lb-cr genes in Enterobacteriaceae at a tertiary care center in Lebanon - role of transferases. Annals of Clinical Microbiology and Antimicrobials **9**:19.

Jalasvuori, M. 2012. Vehicles, replicators, and intercellular movement of genetic information: evolutionary dissection of a bacterial cell. International Journal of Evolutionary Biology **2012**:874153.

Jalasvuori, M., V. P. Friman, A. Nieminen, J. K. Bamford, and A. Buckling 2011. Bacteriophage selection against a plasmid-encoded sex apparatus leads to the loss of antibiotic-resistance plasmids. Biology Letters **7**:902–905.

Joshi, S. G., G. M. Litake, V. S. Ghole, and K. B. Niphadkar 2003. Plasmid-borne extended-spectrum beta-lactamase in a clinical isolate of *Acinetobacter baumannii*. Journal of Medical Microbiology **52**:1125–1127.

Kado, C. I. 1998. Origin and evolution of plasmids. Antonie van Leeuwenhoek **73**:117–126.

Kornstein, L. B., V. L. Waters, and R. C. Cooper 1992. A natural mutant of plasmid RP4 that confers phage resistance and reduced conjugative transfer. FEMS Microbiology Letters **70**:97–100.

Kotilainen, M. M., A. M. Grahn, J. K. Bamford, and D. H. Bamford 1993. Binding of an *Escherichia coli* double-stranded DNA virus PRD1 to a receptor coded by an IncP-type plasmid. Journal of Bacteriology **175**:3089–3095.

Levin, B. R., M. Lipsitch, V. Perrot, S. Schrag, and R. Antia 1997. The population genetics of antibiotic resistance. Clinical Infectious Diseases **24**:S9–S16.

Levy, S. B., and B. Marshall 2007. Antibacterial resistance worldwide: causes, challenges and responses. Nature Medicine **10**:S122–S129.

Lin, A., J. Jimenez, J. Derr, P. Vera, M. L. Manapat, K. M. Esvelt, L. Villanueva et al. 2011. Inhibition of bacterial conjugation by phage M13 and its protein g3p: quantitative analysis and model. PLoS ONE **6**: e19991.

Livermore, D. M. 2002. Multiple mechanisms of antimicrobial resistance in *Pseudomonas aeruginosa*: our worst nightmare? Clinical Infectious Diseases **34**:634–640.

Lu, T. K., and M. S. Koeris 2011. The next generation of bacteriophage therapy. Current Opinions in Microbiology **14**:524–531.

Lujan, S. A., L. M. Guogas, H. Ragonese, S. W. Matson, and M. R. Redinbo 2007. Disrupting antibiotic resistance propagation by inhibiting the conjugative DNA relaxase. Proceedings of National Academy of Sciences USA **104**:12282–12287.

Lyon, B. R., and R. Skurray 1987. Antimicrobial resistance of *Staphylococcus aureus*: genetic basis. Microbiology Reviews **51**:88–134.

MacKenzie, F. M., K. J. Forbes, T. Dorai-John, S. G. Amyes, and I. M. Gould 1997. Emergence of a carbapenem-resistant *Klebsiella pneumonia*. Lancet **350**:783.

Mesquita, M. M., J. Stimson, G. T. Chae, N. Tufenkji, and C. J. Ptacek 2010. Optimal preparation and purification of PRD1-like bacteriophages for use in environmental fate and transport studies. Water Research **44**:1114–1125.

Norman, A., L. H. Hansen, and S. J. Sørensen 2009. Conjugative plasmids: vessels of the communal gene pool. Philosophical Transactions of the Royal Society of London B: Biological Sciences **364**:2275–2289.

Novotny, C., W. S. Knight, and C. C. Jr Brinton 1968. Inhibition of bacterial conjugation by ribonucleic acid and deoxyribonucleic acid male-specific bacteriophages. Journal of Bacteriology **95**:314–326.

Olsen, R. H., J. S. Siak, and R. H. Gray 1974. Characteristics of PRD1, a plasmid-dependent broad host range DNA bacteriophage. Journal of Virology **14**:689–699.

Ou, J. T. 1973. Inhibition of formation of *Escherichia coli* mating pairs by f1 and MS2 bacteriophages as determined with a coulter counter. Journal of Bacteriology **114**:1108–1115.

Sambrook, J., E. F. Fritsch, and T. Maniatis 1989. Molecular Cloning: A Laboratory Manual. 2nd edn. Cold Spring Harbor Laboratory Press, Cold Spring Harbor, NY.

Schentag, J. J., J. M. Hyatt, J. R. Carr, J. A. Paladino, M. C. Birmingham, G. S. Zimmer, and R. J. Cumbo 1998. Genesis of methicillin-resistant *Staphylococcus aureus* (MRSA), how treatment of MRSA Infections has selected for vancomycin-resistant *Enterococcus faecium*, and the importance of antibiotic management and infection control. Clinical Infectious Diseases **26**:1204–1214.

Slater, F. R., M. J. Bailey, A. J. Tett, and S. L. Turner 2008. Progress towards understanding the fate of plasmids in bacterial communities. FEMS Microbiology Ecology **66**:3–13.

Smith, P. A., and F. E. Romesberg 2007. Combating bacteria and drug resistance by inhibiting mechanisms of persistence and adaptation. Nature Chemical Biology **3**:549–556.

Taubes, G. 2008. The bacteria fight back. Science **321**:356–361.

Walsh, C. 2003. Where will new antibiotics come from? Nature Reviews Microbiology **1**:65–70.

Weigel, L. M., F. B. Clewell, S. R. Gill, N. C. Clark, L. K. McDougal, S. E. Flannagan, J. F. Kolonay et al. 2003. Genetic analysis of a high-level vancomycin-resistant isolate of *Staphylococcus aureus*. Science **302**:1569–1571.

Williams, J. J., and P. J. Hergenrother 2008. Exposing plasmids as the Achilles' heel of drug-resistant bacteria. Current Opinions Chemical Biology **12**:389–399.

Wise, R., R. Bax, F. Burke, I. Chopra, L. Czaplewski, R. Finch, D. Livermore et al. 2011. The urgent need for new antibacterial agents. Journal of Antimicrobial Chemotherapy **66**:1939–1940.

Zhang, G.-C., and A. Buckling 2012. Phages limit the evolution of bacterial antibiotic resistance in experimental microcosms. Evolutionary Applications **5**:575–582.

10

Crop pathogen emergence and evolution in agro-ecological landscapes

Julien Papaïx,[1,2,3,4] Jeremy J. Burdon,[4] Jiasui Zhan[5] and Peter H. Thrall[4]

1 UMR 1290 BIOGER, INRA, Thiverval-Grignon, France
2 UR 341 MIA, INRA, Jouy-en-Josas, France
3 UR 546 BioSP, INRA, Avignon, France
4 CSIRO Agriculture Flagship, Canberra, ACT, Australia
5 Fujian Key Lab of Plant Virology, Institute of Plant Virology, Fujian Agriculture and Forestry University, Fuzhou, China

Keywords
agro-ecological interface, landscape epidemiology, pathogen evolution, plant disease management.

Correspondence
Julien Papaïx, Inra - Unité BioSp, Domaine Saint-Paul Site Agroparc, 84914 Avignon Cedex 9, France.

e-mail: julien.papaix@paca.inra.fr

Abstract

Remnant areas hosting natural vegetation in agricultural landscapes can impact the disease epidemiology and evolutionary dynamics of crop pathogens. However, the potential consequences for crop diseases of the composition, the spatial configuration and the persistence time of the agro-ecological interface – the area where crops and remnant vegetation are in contact – have been poorly studied. Here, we develop a demographic–genetic simulation model to study how the spatial and temporal distribution of remnant wild vegetation patches embedded in an agricultural landscape can drive the emergence of a crop pathogen and its subsequent specialization on the crop host. We found that landscape structures that promoted larger pathogen populations on the wild host facilitated the emergence of a crop pathogen, but such landscape structures also reduced the potential for the pathogen population to adapt to the crop. In addition, the evolutionary trajectory of the pathogen population was determined by interactions between the factors describing the landscape structure and those describing the pathogen life histories. Our study contributes to a better understanding of how the shift of land-use patterns in agricultural landscapes might influence crop diseases to provide predictive tools to evaluate management practices.

Introduction

Integrating ecosystem processes occurring at large spatial scales into the design of agricultural landscapes with the aim of improving productivity while decreasing the negative impact of agricultural practices on the environment is increasingly recognized as key to addressing global food security concerns (Bianchi et al. 2006; Tscharntke et al. 2012; Bommarco et al. 2013). At the landscape scale, remnant areas hosting wild vegetation (weeds, exotic or native plant communities) have the potential to promote desired ecosystem services because of their influence on the community ecology of crop pests and beneficial organisms such as pollinators and predators (Bianchi et al. 2006; Chaplin-Kramer et al. 2011), and through their impact on the disease epidemiology and evolutionary dynamics of crop pathogens (Wisler and Norris 2005; Burdon and Thrall 2008; Alexander et al. 2014). However, management plans

to hinder the evolution of crop pests are rarely designed at the landscape scale and if so, they do not consider remnant patches of wild vegetation. A classic example in plant–pathogen interactions was the campaign to eradicate barberry (*Berberis vulgaris*) growing along wheat field margins in the United States (Roelfs 1982; Kolmer et al. 2007). Barberry is the sexual host for wheat stem rust (*Puccinia graminis tritici*) and when present, provides early inoculum and new infectivity combinations for the pathogen. In this study, we quantified the impacts of the area, spatial configuration and average persistence time of wild vegetation fragments in agricultural landscapes on the emergence of crop pathogen over time.

In plant epidemiology, the breakdown of qualitative resistance in crop cultivars can lead to spectacular disease outbreaks (McDonald and Linde 2002) and the erosion of quantitative resistance causes partly resistant hosts to become increasingly susceptible (Lannou 2012). The role of wild hosts in influencing the outcome of such evolutionary processes has been reported for various plant–pathogen systems with different possible scenarios (Jones 2009; Alexander et al. 2014) including: emergence of pathogens from native flora (Wang et al. 2010; van der Merwe et al. 2013), use of sources of resistance from wild hosts (Garry et al. 2005; Lebeda et al. 2008; Leroy et al. 2014) and reciprocal influence of native flora and cultivated plants (Webster et al. 2007; Lê Van et al. 2011). The importance of considering interactions across the wild and cultivated compartments in agricultural landscapes is clearly seen in the emergence of Fusarium wilt disease in Australian cotton-growing regions in the early 1990s (*Fusarium oxysporum* f. sp. *vasinfectum* on *Gossypium hirsutum* L.). Several *Gossypium* species are native to Australia, and at least two of these species have distributions that overlap cultivated cotton-growing regions. Isolates of *F. oxysporum* from these wild hosts were found to cause mild symptoms of Fusarium wilt in cultivated cotton (Wang et al. 2004). A detailed study of the genetic structure of *Fusarium oxysporum* f. sp. *vasinfectum* and *F. oxysporum* populations indicated that Fusarium wilt in cultivated cotton evolved locally through specialization of native *F. oxysporum* strains to the crop (Wang et al. 2010).

Changes in agricultural practices imply changes in the potential for contact between crops and remnant vegetation in agricultural landscapes. The area where such contacts occur is termed the agro-ecological (AE) interface (Burdon and Thrall 2008). Variation in AE interfaces among farming systems implies highly diverse situations for crops and wild plants to interact (Fig. 1). Wild and cultivated elements can be either almost undistinguishable and highly intricately intermeshed such as in tropical agro-forestry landscapes (Tscharntke et al. 2011) or strongly separated with markedly different species diversity such as in intensive monoculture landscapes. The area and spatial distribution of remnant wildlands can vary greatly: for example, in some European regions, landscapes are composed of small fields separated by hedges with significant areas of woodland and the AE interface is highly developed. In contrast, this interface is extremely reduced in agricultural regions dominated by intensive monoculture of broadacre crops (Fig. 1). Seasonality can also differ between the agricultural and wild elements of these landscapes, with grasses and herbs in hedgerows providing year-long green bridges when annual crop cycles finish and fields are fallowed.

Variation in the temporal (through seasonal changes) and spatial (through landscape composition and organization) structure of agro-ecological systems and the extent and complexity of the AE interface can have direct consequences for pathogen eco-evolutionary dynamics. Seasonal fluctuations in environmental conditions can affect pathogen spread and persistence (Altizer et al. 2006). In particular, periodic host absence in agricultural landscapes forces plant pathogens to survive on volunteer plants (wild relatives, alternative hosts, seedlings) or to have specific life-history strategies (saprophyte, free-living stages) that may lead to drastic reductions in pathogen population size [see Suffert et al. (2011) for an example on *Septoria tritici* blotch]. In addition, such pulses in host density have important consequences for pathogen evolution (van den Berg et al. 2011; Hamelin et al. 2011; Zhan et al. 2014). On the other hand, the diversity and spatial arrangement of host genotypes have been shown to influence disease spread and persistence at the field scale in variety mixtures (Mundt and Leonard 1986; Skelsey et al. 2005; Mundt et al. 2011) and to extend its influence at the landscape scale by modifying pathogen habitat connectivity (Zhu et al. 2000; Skelsey et al. 2010; Papaïx et al. 2014c). Pathogen evolution is also sensitive to the spatial heterogeneity of agricultural landscapes (Zhan et al. 2002; Stukenbrock and McDonald 2008; Sommerhalder et al. 2011) with large uniform areas facilitating the evolution of pathogen specialization and genotypes that are better adapted to crop hosts (Débarre and Gandon 2010; Papaïx et al. 2013).

Despite the broad range of work published on pathogen evolution in agricultural landscapes (Stukenbrock and McDonald 2008; Burdon et al. 2014), the potential consequences of spatio-temporal variation in the structure of the agro-ecological interface for the eco-evolutionary dynamics of plant diseases have been poorly understood (Burdon and Thrall 2008; Alexander et al. 2014). Indeed, most previous studies did not consider the role of remnant wild vegetation but focused on the cultivated crop. To our knowledge, only one modelling study considers both wild and cultivated hosts but do not take into account the explicit spatial configuration of the landscape (Fabre et al. 2012). The study involved a gene-for-gene system for virus epidemics in a landscape composed of a susceptible cultivar, a resistant cultivar and a wild reservoir. The reservoir was assumed to be selectively neutral and allowed the pathogen to survive during the off season. They found that epidemic intensity was the main factor explaining resistance breakdown. The landscape composition (cropping ratio between the susceptible and resistant cultivars) was also found to be crucial but even its influence was determined by epidemic intensity. However, the assumption that the wild reservoir was selectively neutral and the presence of a susceptible crop that increases pathogen population size prevented Fabre et al. (2012) highlighting any role resulting from viral dynamics in the reservoir.

Finding novel resistance genes and integrating them into new crop varieties is a long and costly process. Hence, the development of effective strategies for the deployment of resistant genotypes that increase resistance gene durability

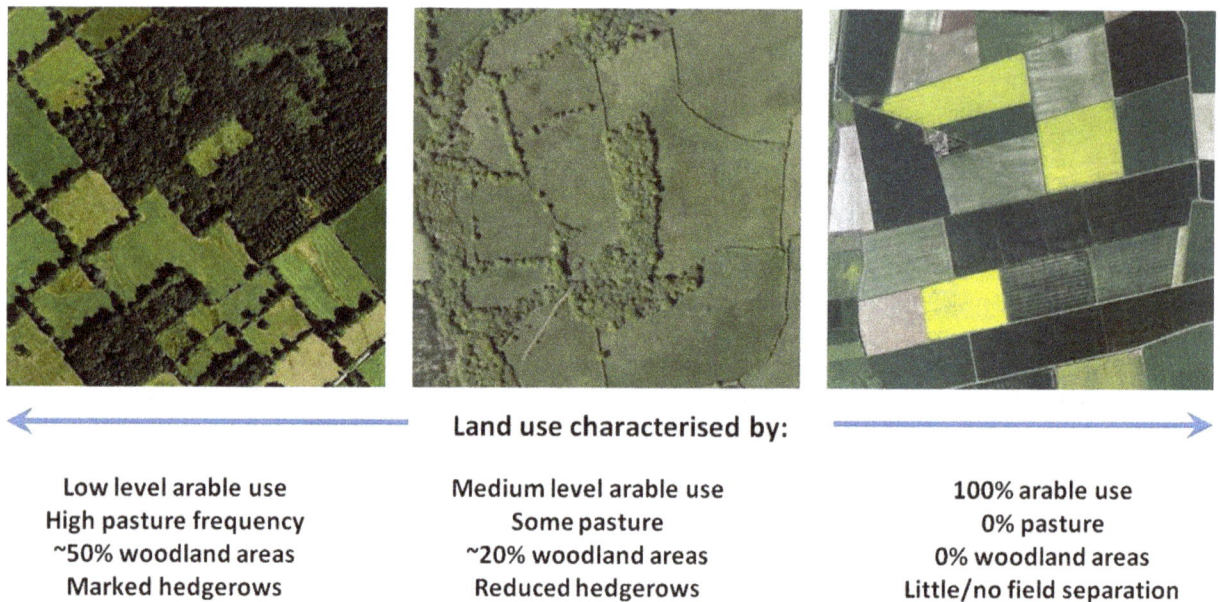

Land use characterised by:

Low level arable use	Medium level arable use	100% arable use
High pasture frequency	Some pasture	0% pasture
~50% woodland areas	~20% woodland areas	0% woodland areas
Marked hedgerows	Reduced hedgerows	Little/no field separation

Figure 1 Spatial structure of the agro-ecological interface across different farming systems (Imagery ©2014 Cnes/Spot Image, Digital Globe; Map data ©2014 Google).

should be a central goal of epidemiological studies of crop host–pathogen interactions (Burdon et al. 2014). Developing such strategies requires that we better understand how different agro-ecological landscapes (considering both wild and cultivated elements) might influence the spread and evolution of crop diseases in order to provide predictive tools to evaluate management practices (Shennan 2008; Thrall et al. 2011). Indeed, the shift of land-use patterns in agricultural landscapes through intensification or extensification of agricultural production systems modifies the interface between the agro-ecological elements, and we need to understand the potential consequences of such modifications for a better management of landscapes and control of crop pathogens.

Here, we develop a demographic–genetic simulation model to study how the spatial and temporal distribution of remnant wild vegetation patches embedded in an agricultural landscape can drive the emergence of a crop pathogen and its subsequent specialization on the crop host. In this study, we take a first step towards modelling more complex situations by considering a relatively simple system where the crop and the wild plant are each represented by only one genotype. We first present the model and the simulation experiment. Then, we study the evolutionary trajectory of the pathogen population in different agro-ecological landscapes. We characterized the landscape structures (spatial configurations and duration of the cropping season) that favoured crop pathogen emergence and studied how different life-history traits (dispersal ability and trade-off in aggressiveness) of the pathogen could affect

these outcomes. Finally, we discuss the implications for the management of remnant elements in agricultural landscapes.

Model and methods

Model overview

Our model describes the numerical dynamics and the evolutionary changes in the genetic composition of a pathogen population thriving on two different hosts: a wild host and a crop. The wild host inhabits patches of remnant vegetation embedded in an agricultural landscape composed of paddocks where the crop is sown. The approach we used allows us to control the spatial aggregation of wild patches and the area they covered. The wild host disperses among the patches of remnant vegetation and is present all year round. In contrast, the crop grows locally and is present only during the cropping season. As a consequence, the pathogen depends upon the wild host to bridge the off season, when the crop has been harvested. The pathogen disperses passively across the whole landscape (e.g. through wind dispersed propagules) regardless of the host type from which it is dispersed or the one on which it lands.

Pathogen genotypes are characterized by their aggressiveness on the host types. Aggressiveness is used here to describe the quantitative interaction between a pathogen genotype and a host type reflecting, for example, differences in spore infection efficacy, lesion development rate, time from infection to sporulation and the abundance of spores produced. Thus, aggressiveness is a composite trait directly linked to the fitness of a pathogen genotype on a given host type and to the potential amount of disease the host suffers in the presence of that pathogen genotype. Importantly, we consider a trade-off in aggressiveness on the two host types. Such trade-offs would reflect the constraint for the pathogen of simultaneously investing in different traits: the allocation of limited resources in one trait

has a negative impact on another trait (Pariaud et al. 2009; García-Arenal and Fraile 2013; Laine and Barrès 2013). At the beginning of a simulation, the pathogen population is only adapted to the wild host and cannot attack the crop. However, new pathogen genotypes can arise through mutation resulting first in the emergence of genotypes able to attack the crop and then in a gradual increase in aggressiveness on the crop.

Using this modelling framework, we quantify the impacts of the area and spatial configuration of the wild remnant patches as well as the duration of the cropping season on the emergence of crop pathogen and its subsequent specialization on the cultivated host over time (Fig. 2). We also investigate how the life history of the pathogen (i.e. strength of the trade-off in aggressiveness and dispersal ability) mediates the effect of spatial and temporal habitat variability on crop pathogen emergence (Fig. 2).

Population dynamics and pathogen evolution

Spatial structure

The spatial structure of the system within which hosts and pathogens interact is represented as a two-dimensional metapopulation composed of patches of two types: cultivated and wild (Fig. 3). The plant hosts are represented by two species: the crop is present in cultivated fields and the wild plant in wild remnant patches. Cultivated areas are

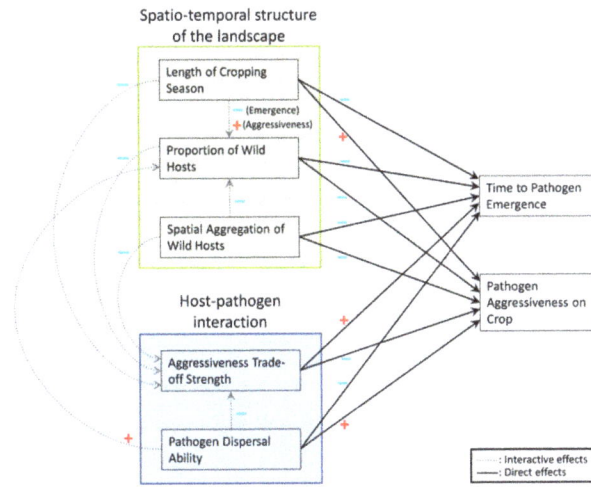

Figure 2 Summary of the direct (black solid arrows) and interactive (grey dashed arrows) effects of the input factors on the variables describing the pathogen evolution.

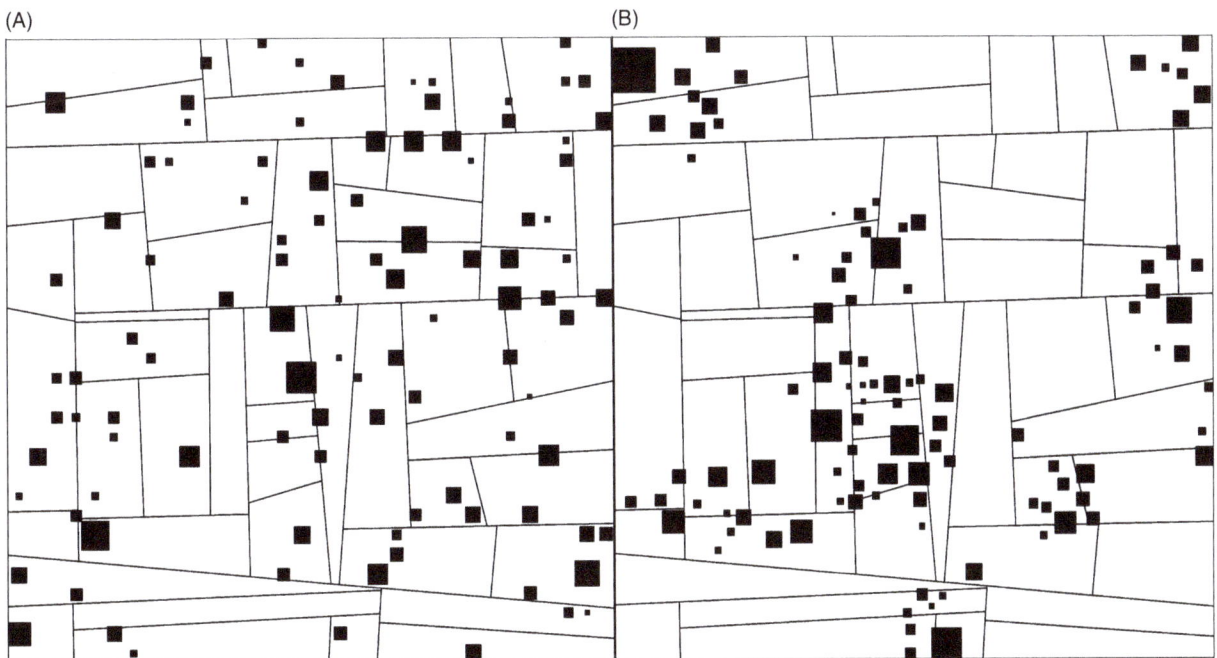

Figure 3 Examples of simulated landscapes composed of 49 paddocks (solid lines) and 100 wild patches (black squares). The wild host population covers 5% of the total landscape surface, and wild patches are randomly distributed (A) or clustered (B) in space.

generated by simulating a set of crop paddocks using a T-tessellation algorithm that makes it possible to control the size, number and shape of paddocks. This algorithm is based on the Metropolis–Hastings–Green principle that makes it possible to generate several landscape replicates sharing the same characteristics (Kiêu et al. 2013; Papaïx et al. 2014a). Wild patches are then positioned on that agricultural landscape with spatial aggregation (random and clustered – Fig. 3A,B) and the total surface covered (2.5%, 5% and 10%) determined as model inputs

Table 1. Definitions of the main terms and parameters used in modelling.

Symbols	Description	Value
(1) Spatial and temporal structure		
–	Number of paddocks	52, 49, 51, 48 and 52
–	Number of wild patches	100
q	Proportion of the agricultural landscape covered by the wild metapopulation	2.5%, 5% and 10%
N	Number of years	50
Y	Number of days in a year	360
T	Number of days in a cropping season	Between 60 and 300
(2) Crop and wild host dynamics		
δ^c	Crop growth rate	0.1
K^c	Carrying capacity of cultivated patches	Proportional to the paddock surface
$S_{0,c}$	Number of susceptible plants initiating the cropping season	10% of K^c
r^w	Wild host reproduction rate	1 by susceptible plant by day
K^w	Carrying capacity of wild patches	Proportional to the wild patch surface
d^w	Wild host death rate	0.1
$distM^w$	Wild host mean dispersal distance	10% of the landscape scale
(3) Pathogen population and evolutionary dynamics		
β	Shape of the trade-off function	0.6, 1 and 1.4
e	Pathogen infection efficiency	Varying according to the trade-off function. The maximal value for e is 0.4.
r^p	Pathogen reproduction rate	2 by infected plant by day
d^I	Infected plant death rate	0.1
m	Pathogen mutation rate	0.002 towards the 2 adjacent genotypes and 0.004 for the fully specialized genotypes
$distM^p$	Pathogen mean dispersal distance	2.5%, 10% and 25% of the landscape scale

(Table 1). Each wild patch centre was first located, and then surfaces were drawn from a log-normal distribution to obtain the desired percentage of landscape coverage by wild vegetation. For each combination of wild patch aggregation and surface coverage values, five different landscapes composed of 100 wild patches and, respectively, 52, 49, 51, 48 and 52 paddocks were constructed and used as landscape replicates.

Temporal structure

The epidemiological model we used is characterized by a temporal cycle (i.e. a year) composed of two time periods: the cropping season and the off season (Fig. 4). During the cropping season both the crop and wild hosts are present in the landscape. Conversely, during the off season, only the wild host is present. The seasons are separated by discrete events such as crop harvest or planting. Epidemics were simulated over $N = 50$ years composed of $Y = 360$ time steps (i.e. days). The cropping season forms the first T days of the year, whereas the off season is represented by the remaining $Y - T$ days of the year. The cropping season duration, T, was varied from 60 to 300 days at 60-day intervals. We assumed here that, within a given simulation run, the spatial structure remains fixed across years.

Eco-evolutionary dynamics

We present here a semi-discrete (Mailleret and Lemesle 2009) and deterministic version of the susceptible-infectious (SI) model used in the simulation experiment, which describes the dynamics of the densities of susceptible (S) and infectious (I) host plants in each local patch within the metapopulation. Let the subscripts i and n indicate the patch and year respectively, the subscripts c and w the crop and wild host species, respectively, and the subscript p the pathogen genotype. Let $\dot{S} = dS/dt$, and t^- and t^+ denote the time intervals immediately before and after time t, respectively. We also define C and W as the set of cultivated and wild patches, respectively. Table 1 summarizes the main terms and parameters used.

During the cropping season, both hosts are present and the dynamics for the cultivated patches are as follows:

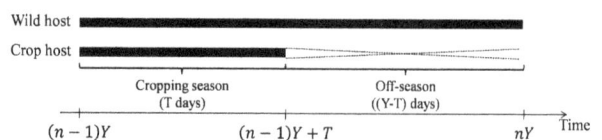

Figure 4 Schematic representation of the model temporal structure for one year. While the wild host is present all year round, the crop is cultivated only during the cropping season. n, index of the current year; Y, number of days during a year; T, cropping season duration.

$$\begin{cases} \dot{S}_i = \delta^c S_i \left(1 - \frac{S_i}{K_i^c}\right) - \pi_i^P(x) r^P \sum_p e_{p,c} \sum_{i' \in C \cap W} \mu_{i',i}^P \sum_{p'} m_{p',p} I_{i',p'}, \\ \dot{I}_{i,p} = \pi_i^P(x) r^P e_{p,c} \sum_{i' \in C \cap W} \mu_{i',i}^P \sum_{p'} m_{p',p} I_{i',p'} - d^I I_{i,p}, \end{cases} \quad (1)$$

for all t between $(n-1)Y$, and $(n-1)Y + T$, $n = 1, 2, \ldots$, for all $i \in C$. δ^c is the growth rate of the crop, K_i^c is the carrying capacity of patch i and is proportional to the area of that patch, r^P is the number of pathogen propagules produced by one infected plant per day, $e_{p,c}$ is the infection efficiency of pathogen genotype p for the crop, $\mu_{i',i}^P$ is the pathogen dispersal rate from patch i' to patch i, $m_{p',p}$ is the pathogen mutation rate from genotype p' to genotype p, and d^I is the death rate of an infected plant. $\pi_i^P(x)$ is the proportion of pathogen propagules that come into contact with a susceptible plant. This is an increasing function of $x = \frac{S_i}{S_i + I_i}$, the proportion of susceptible plants in the patch. We consider the following sigmoid function for $\pi_i^P(\cdot)$:

$$\pi_i^P(x) = 1 - \frac{\exp(-5.33x^3) - \exp(-5.33)}{1 - \exp(-5.33)} \quad (2)$$

giving an inflection point for $x \approx 0.5$. The function $\pi_i^P(x)$ ensures that, in patch i, $\pi_i^P(x) = 1$ if all the plants are susceptible and $\pi_i^P(x) = 0$ if there are no susceptible plants. In contrast to the crop, the wild host can disperse among remnant patches, which leads to the following dynamics for the host–pathogen interaction in wild patches:

$$\begin{cases} \dot{S}_i = \pi_i^w r^w \sum_{i' \in W} \mu_{i',i}^w S_{i'} - d^w S_i - \pi_i^P(x) r^P \sum_p e_{p,w} \sum_{i' \in C \cap W} \\ \qquad \mu_{i',i}^P \sum_{p'} m_{p',p} I_{i',p'}, \\ \dot{I}_{i,p} = \pi_i^P(x) r^P e_{p,w} \sum_{i' \in C \cap W} \mu_{i',i}^P \sum_{p'} m_{p',p} I_{i',p'} - d^I I_{i,p}. \end{cases}$$

$$(3)$$

for all $i \in W$. The rate of plant propagule establishment is represented by $\pi_i^w = 1 - (S_i + I_i)/K_i^w, i \in W$, $i \in W$, where K_i^w is the carrying capacity of patch i. It is equal to 0 if there is no available space ($S_i + I_i = K_i^w$), and to 1 if the patch is unoccupied. r^w is the number of plant propagules produced by one susceptible plant per day, $\mu_{i',i}^w$ is the wild host dispersal rate from patch i' to patch i, d^w is the death rate of susceptible plants, and $e_{p,w}$ is the infection efficiency of pathogen genotype p for the wild host.

The transition between the cropping and the off season was carried out by removing all of the crop in cultivated areas, but keeping the state of the wild patches unchanged. During the off season, only the wild host is present and its dynamics are the same as during the cropping season (eqn 3 – for all t between $(n-1)Y + T^+$ and nY). Finally, the transition between the off- and the new cropping season is described by:

$$\begin{cases} S_i(nY^+) = S_{0,c}, \\ I_{i,p}(nY^+) = 0, \end{cases} \quad (4)$$

for all $i \in C$. $S_{0,c}$ is the number of susceptible plants of the crop initiating the cropping season. The state of the wild patches is kept unchanged.

Dispersal
While the crop can only grow locally (i.e. where it has been sown) the wild plant and the pathogen can both disperse. Dispersal rates among patches are computed from an individual dispersal function. The propagule density emitted from a given source point z and arriving at a given reception point z' is given by:

$$g(\|z - z'\|) = \frac{(a-2)(a-1)}{2\pi b^2} \left(1 + \frac{\|z - z'\|}{b}\right)^{-a},$$
$$(5)$$

where $\|z - z'\|$ is the Euclidean distance between z and z', $b > 0$ is a scale parameter and $a > 2$ determine the length of the dispersal tail: the lower value of a, the longer the dispersal tail, and the more probable long distance dispersal events are. The mean dispersal distance travelled by a propagule is defined only when $a > 3$ as $distM = 2b/(a-3)$. The expression of $distM$ made it possible to vary the mean dispersal distance while keeping the probability of long distance dispersal events (parameter a) fixed. We thus considered three values for the mean dispersal distance of the pathogen ($distM^P = 2.5\%$, $distM^P = 10\%$ and $distM^P = 25\%$ of the landscape scale) by varying the scale parameter b and fixing a at 3.4. The mean dispersal distance of the wild host was fixed at $distM^w = 10\%$ of the landscape scale.

From eqn 5, the probability of a propagule dispersing from patch i to patch j is computed by performing the integration of the dispersal function $g(\cdot)$ between pairs of points that belong to the areas A_i and A_j of patches i and j, respectively (Bouvier et al. 2009):

$$\mu_{ij} = \frac{\int_{A_i} \int_{A_j} g(\|z - z'\|) dz dz'}{A_i} \quad (6)$$

In eqn 6, the integral of $g(\cdot)$ is divided by the area of the originating patch to ensure $0 \le \mu_{ij} \le 1$.

Pathogen population genetic structure
Pathogen genotypes are characterized by their aggressiveness on each host species. The aggressiveness of pathogen genotype p on host species h is defined by its nonspatial basic reproductive number in a monomorphic host population of plant species h, $R_{0,loc}^{ph}$. In epidemiology, the basic reproductive number is a classical measure of pathogen fitness. It represents the number of secondary infections arising from a single infected individual in a fully susceptible

host population and thus reflects the potential disease severity the pathogen can cause to its host. Note that the pathogen population goes extinct if its basic reproductive number is below unity. In our system, the aggressiveness of pathogen genotype p on host species h is thus defined by

$$R_{0,loc}^{ph} = e_{p,h} r^P / d^I.$$

We assume a trade-off in pathogen aggressiveness on the two host species: a gain in pathogen aggressiveness on the crop has a cost in terms of reduced aggressiveness on the wild host (and vice-versa). Thus for a generalist pathogen strain, performance on any particular host species is less than that achieved by the pathogen genotype specifically specialized to that host but is greater than the performance of the pathogen genotype specialized on the other host species. Gain and cost are linked through the relationship:

$$\text{cost} = 1 - \left(1 - \text{gain}^{\frac{1}{\beta}}\right)^{\beta}. \tag{7}$$

The parameter β determines the global concavity of the trade-off curve: the curve is concave when β is below unity, linear when $\beta = 1$ and convex otherwise. We will refer hereafter to concave curves as weak trade-offs, because they correspond to cases where the cost of being a generalist is low. Similarly, convex curves will be called strong trade-offs (Ravigné et al. 2009). In the simulation experiment, we fixed $r^P = 2$, $d^I = 0.1$ and the maximal infection efficiency at 0.4. The infection efficiencies of the other pathogen genotypes were computed from eqn 7, by varying the gain in aggressiveness between 0% and 100% and by considering three values for the trade-off shape, $\beta = 0.6$, $\beta = 1$ and $\beta = 1.4$.

The pathogen population is initially composed of one genotype but other pathogen genotypes can arise through mutation. We assume that the pathogen population evolves gradually: a new genotype arises from closely related genotypes by mutation with small gains or losses in $R_{0,loc}$. The probability that a pathogen propagule was of the same genotype as its parental individual was set to $m_{pp} = 0.996$, and then, we set $m_{p(p-1)} = m_{p(p+1)} = 0.002$. Exceptions were the pathogen genotypes with the highest aggressiveness on either the crop or the wild host – these mutated towards less specialized genotypes with a probability of 0.004 to keep their overall mutation rate equal to that of other genotypes.

Simulation experiment and statistical analysis
Experimental design
Simulations of the model described in Section Eco-evolutionary dynamics were carried out using discrete time intervals (one time step equalled to one day) and adding stochastic steps to account for possible drift when genotypes are at low frequencies. For each simulation, the path-

ogen population was initially composed of the wild host specialist which cannot infect the crop ($e_w = 0.4$ and $e_c = 0$). Epidemics were initiated by assuming that wild hosts were randomly infected with a probability of 0.01.

There were five input factors of interest (Table 1): duration of the cropping season (5 values – 60, 120, 180, 240 and 300 days), the proportion (3 values – 2.5%, 5% and 10%), the aggregation level (2 values – random and aggregated) of the wild patches, the trade-off shape (3 values – 0.6, 1 and 1.4) and the mean dispersal distance of the pathogen (3 values – 2.5%, 10% and 25%). We set up a complete factorial design by considering each combination of the values of the five input factors. For each of these conditions, 20 replicates were simulated as follows: 5 landscape replicates by 4 model replicates. This led to a total of 5400 simulations. Note that more than 4 model replicates would be necessary for a refined study of the effects due to model stochasticity in a specific landscape structure. That was not the aim of this study, and we preferred to pool the model and landscape replicates to have sufficient degrees of freedom for the statistical analyses.

Outputs
Local dynamics were aggregated to give global evolutionary trajectories in the pathogen population (Fig. A1). From these trajectories, we first characterized whether the emergence of a crop pathogen was successful through the variable E that equals 1 if a pathogen genotype that can infect the crop emerged and persisted in the simulation and 0 otherwise. Then, in the simulations where emergence occurred, we estimated the number of years that emergence required, T^E, as the first year for which the proportion of healthy plants in the cultivated area dropped by 5%. Finally, we characterized mean aggressiveness of the crop pathogen population ($\bar{R}_{0,loc}^c$) by averaging the aggressiveness $R_{0,loc}^{pc}$ over the pathogen genotypes p in the pathogen population that developed on the cultivated host at equilibrium.

Statistical models
The effects of the input factors (cropping season duration, proportion and aggregation level of wild patches, trade-off shape and pathogen mean dispersal distance) on the descriptors of the pathogen evolutionary trajectory (E, T^E, and $\bar{R}_{0,loc}^c$) were assessed by fitting generalized linear models (GLM) with various link functions using the R software (R Core Team 2014) (Table 2). As the number of successful emergences ($E = 1$) was very high within each combination of factors, it was not possible to estimate interactions for this variable. The GLMs used adequately fitted the data set as indicated by deviance residuals (Fig. A2). In addition, they explained 77.2%, 97.6% and 99.4% of the deviance for E, T^E, and $\bar{R}_{0,loc}^c$, respectively.

Table 2. Summary of the generalized linear models used for the analysis of the three descriptors of the global pathogen evolutionary trajectories: E, emergence of a crop pathogen ($E_s = 1$ if emergence was successful in simulation s and 0 otherwise); T^E, number of years required for the crop pathogen to emerge; $\bar{R}^c_{0,loc}$, mean aggressiveness of the pathogen population on the crop at equilibrium.

Descriptor	Distribution	Link function	Linear predictor
Emergence (E)	Binomial	Logit	-1 + LOC + PROP + BETA + CROP + DISP
Time before emergence (T^E)	Gamma	Log	-1 + LOC + LOC:(PROP + BETA + CROP + DISP + PROP:BETA + PROP:CROP + PROP: DISP + BETA:CROP + BETA:DISP + CROP:DISP)
Pathogen aggressiveness on the crop ($\bar{R}^c_{0,loc}$)	Normal	Natural	-1 + LOC + LOC:(PROP + BETA + CROP + DISP + PROP:BETA + PROP:CROP + PROP: DISP + BETA:CROP + BETA:DISP + CROP:DISP)

LOC, two-level factor of wild patches spatial aggregation (random or clustered); PROP, scaled wild host proportion; BETA, scaled parameter of the trade-off function (β); CROP, scaled cropping season duration; DISP, scaled pathogen mean dispersal distance; :, interactions.

Results

Observed outcomes

Over the entire simulation data set, four outcomes were observed: extinction of the pathogen population; persistence of the wild host specialist only; coexistence of one crop specialist and one wild host specialist; selection for one pathogen generalist (Fig. 5). Extinction of the pathogen population only occurred in 1.8% of the simulations and corresponded to cases where pathogen dispersal ability was the highest and the proportion of wild hosts the lowest. These extinctions occurred at the very beginning of the simulations and were due to a small pathogen population size combined with demographic stochasticity. Selection for generalist pathogens was generally observed when the aggressiveness trade-off was weak and pathogen dispersal ability was at its highest values but was also possible with a strong trade-off when the cropping season was very long. These two contrasting situations led, respectively, to a generalist with low (in the first case) and high (in the latter case) aggressiveness on the crop. In all other contexts, pathogen evolution resulted in coexistence of a crop specialist and a wild host specialist when the agricultural season was long enough and the trade-off not too strong and to unsuccessful emergence of a crop pathogen otherwise.

Effect of spatial configuration

The spatial configuration of the landscape was described by two of the five input factors: the aggregation level among wild patches and the proportion of area covered by the wild metapopulation. An increase in the aggregation of wild patches resulted in larger wild pathogen population sizes during the off season. Consequently, more new mutations were produced in more aggregated landscapes which facilitated the emergence of new pathogen genotypes able to thrive on the crop. Thus, more aggregated landscape patterns increased the number of successful emergences

(Table A1, intercept values) and decreased the time required for the crop pathogen to emerge (Table A2, intercept values). However, selection during the off season was stronger and, over longer evolutionary time scales, increases in the aggregation level of wild host patches led to a small but significant lower mean aggressiveness of the pathogen population on the crop at equilibrium (Table A2, intercept values).

Increases in the proportion of the landscape covered by patches of the wild host also increased pathogen population size because of increased wild host abundance. Thus, as for aggregation, this facilitated crop pathogen emergence (i.e. increased the number of successful emergences and decreased the time required for emergence to occur – Fig. 6A and Tables A1 and A2). At the same time, as its proportion increased, the wild host became an important habitat for the pathogen even during the agricultural season. As a consequence, increases in the proportion of the landscape covered by patches of the wild host decreased the mean aggressiveness of the pathogen population on the crop at equilibrium (Fig. 6B and Table A2). In addition, the effects of the proportion of the agricultural landscape covered by wild hosts on T^E and $\bar{R}^c_{0,loc}$ were more important when wild patches were randomly distributed in space (Table A2). Indeed, as its spatial aggregation increased, the wild host metapopulation sustained a larger pathogen population which made the pathogen less sensitive to variations in wild host abundance.

Effect of temporal heterogeneity

The longer the cropping season, the greater the time period during which the cultivated host was available which decreased pathogen dependency on the wild host for survival. As a consequence, increased duration of the cropping season meant that more crop pathogen emergences were successful (Table A1), the time required to observe crop

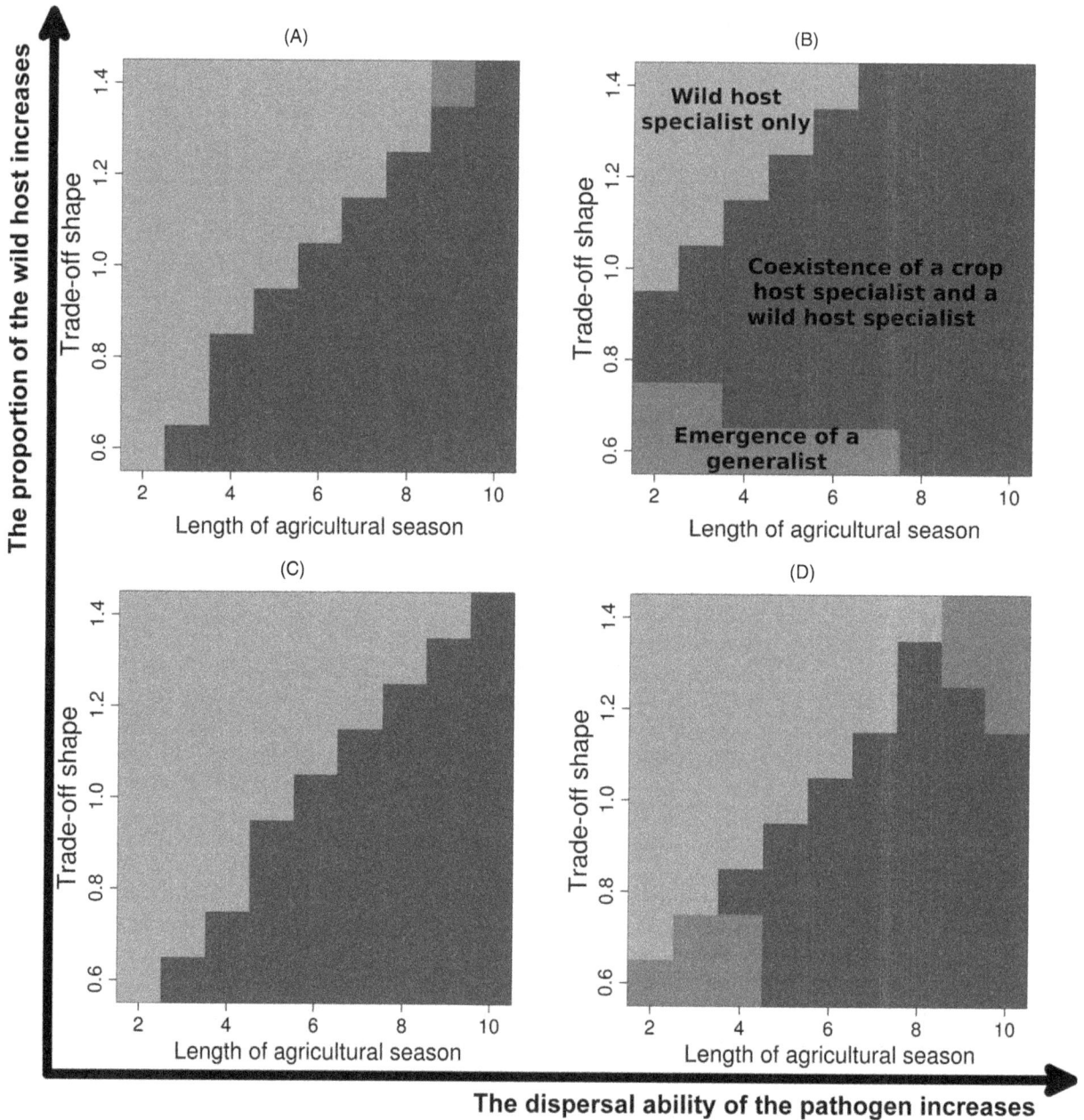

Figure 5 Patterns of pathogen diversity. The different plots represent different conditions in the proportion covered by the wild host and the pathogen dispersal ability: (A) high proportion of wild host (q = 10%) and low pathogen dispersal ($distM^P$ = 2.5%); (B) high proportion of wild host (q = 10%) and high pathogen dispersal ($distM^P$ = 25%); (C) low proportion of wild host (q = 2.5%) and low pathogen dispersal ($distM^P$ = 2.5%); and (D) low proportion of wild host (q = 2.5%) and high pathogen dispersal ($distM^P$ = 25%). These graphs represent the prediction of multinomial logistic regression models fitted to the simulated data set.

pathogen emergence was shorter (Fig. 7A and Table A2) and the mean aggressiveness of the pathogen population on the crop at equilibrium was greater (Fig. 7B and Table A2). These effects on the pathogen evolutionary trajectory were similar regardless of whether the spatial distribution of wild host patches was random or aggregated (Table A2). However, temporal heterogeneity mediated effects of the landscape spatial configuration (Table A2). In fact, increasing the length of the cropping season resulted in both a decreased effect of the proportion of wild hosts in the landscape on T^E (time to emergence) and an increased influence of the proportion of wild hosts in the landscape on the mean aggressiveness of the crop pathogen population, $\bar{R}^c_{0,loc}$.

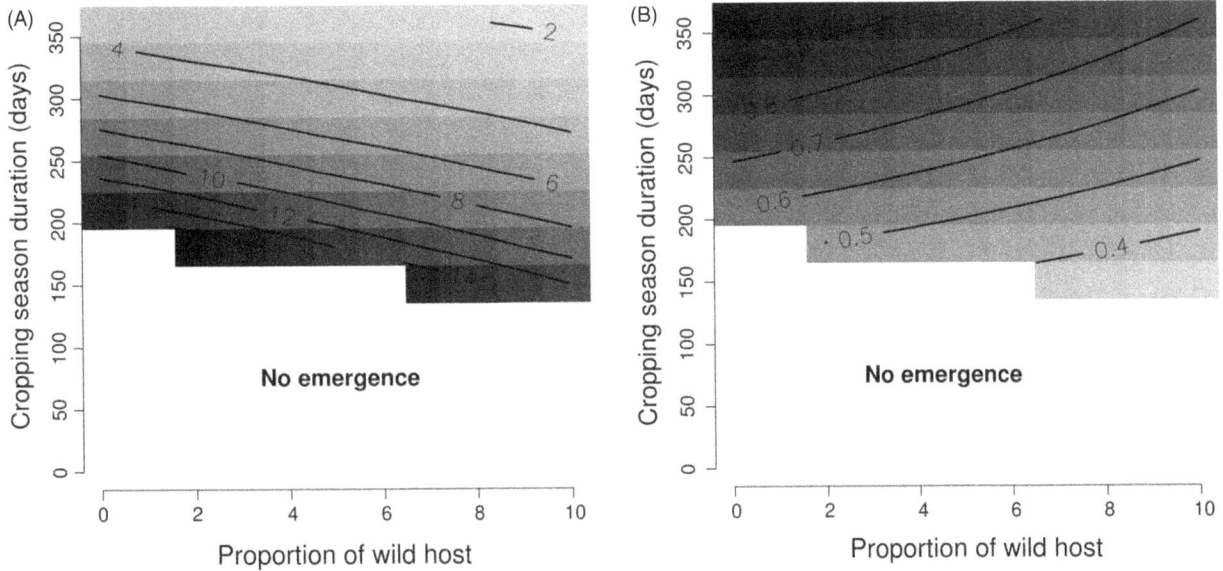

Figure 6 The effect of cropping season duration and the proportion of wild host in the agricultural landscape: (A) predicted values of the time (number of years) required to observe the emergence of a crop pathogen; and (B) relative mean aggressiveness of the pathogen population on the crop at equilibrium (0, the aggressiveness is minimal; 1 the aggressiveness is maximal). White: no successful emergence of a crop pathogen predicted by the model. Values for the grey scale are indicated by the contour lines.

Effects of life-history traits

Trade-off in aggressiveness

Pathogen aggressiveness is a measure of pathogen fitness and reflects the potential disease severity a pathogen can cause to its hosts. The strength of the trade-off in aggressiveness on the two hosts (β) reflected the difficulty experienced by pathogen genotypes able to thrive on the crop with regard to survival during the off season on the wild host. Thus, increases in the strength of the trade-off (i.e. higher values of β) acted in the same way on the three variables describing the global pathogen evolutionary trajectory: a stronger trade-off reduced the chance (fewer emergences were successful) and increased the time of emergence (Fig. 7A and Tables A1 and A2) and decreased the mean aggressiveness of the pathogen population on the crop at equilibrium (Fig. 7B and Table A2). The effects of the trade-off in aggressiveness on T^E and $\bar{R}_{0,loc}^c$ depended on the spatial configuration of the landscape and on the duration of the cropping season. Increases in the aggregation of wild patches, in the proportion of the agricultural landscape covered by wild hosts and in cropping season duration, resulted in a decreased impact of the trade-off shape on T^E and $\bar{R}_{0,loc}^c$ (Table A2).

Dispersal

The larger the pathogen mean dispersal distance, the more pathogen propagules originating from a wild patch landed in crop fields. As a consequence, selection pressure posed by the crop increased with the pathogen's ability to disperse

which facilitated the emergence of a crop pathogen (Table A1), decreased the time required for crop pathogen emergence to occur (Fig. 8A and Table A2) and increased the mean aggressiveness of the pathogen population on the crop at equilibrium (Fig. 8B and Table A2). The spatial distribution of wild host patches did not change these effects on the pathogen evolutionary trajectory (Table A2). Finally, increases in the scale of pathogen dispersal made the results more sensitive to changes in the proportion of the wild host in the landscape (Table A2).

Discussion

Role of landscape spatial configuration

The interface between agricultural and natural elements of agro-ecological landscapes can vary tremendously in different systems (Fig. 1), with the potential to significantly modify the way pathogens and their hosts interact. Modelling the spatial structure of the AE interface explicitly (described by the spatial aggregation of wild patches and the proportion of landscape they cover) enabled us to assess its effects on the emergence of a new crop pathogen and the potential for evolution towards increased aggressiveness (Fig. 2). Interestingly, we found a trade-off between the managements of pathogen emergence and aggressiveness. Landscape structures that promoted larger pathogen populations on the wild host (high proportion and aggregation of remnant wild patches) facilitated the emergence of a crop pathogen by both increasing emergence event and decreasing emergence time. However, such

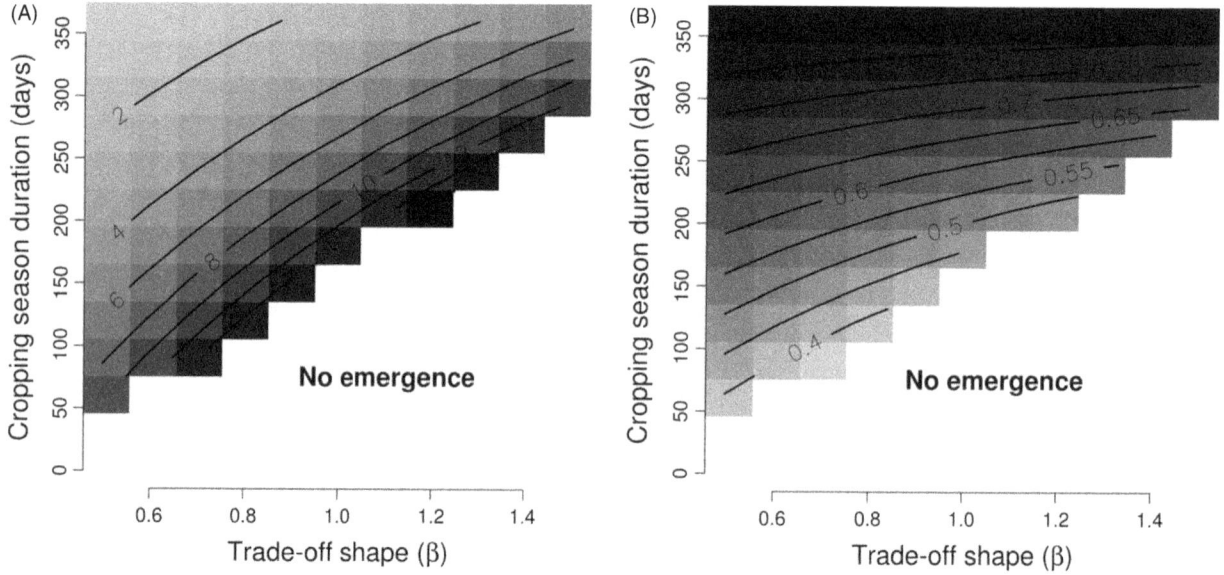

Figure 7 The effect of cropping season duration and trade-off shape: (A) predicted values of the time (number of years) required to observe the emergence of a crop pathogen, and (B) relative mean aggressiveness of the pathogen population on the crop at equilibrium (0, the aggressiveness is minimal; 1 the aggressiveness is maximal). Values for the grey scale are indicated by the contour lines.

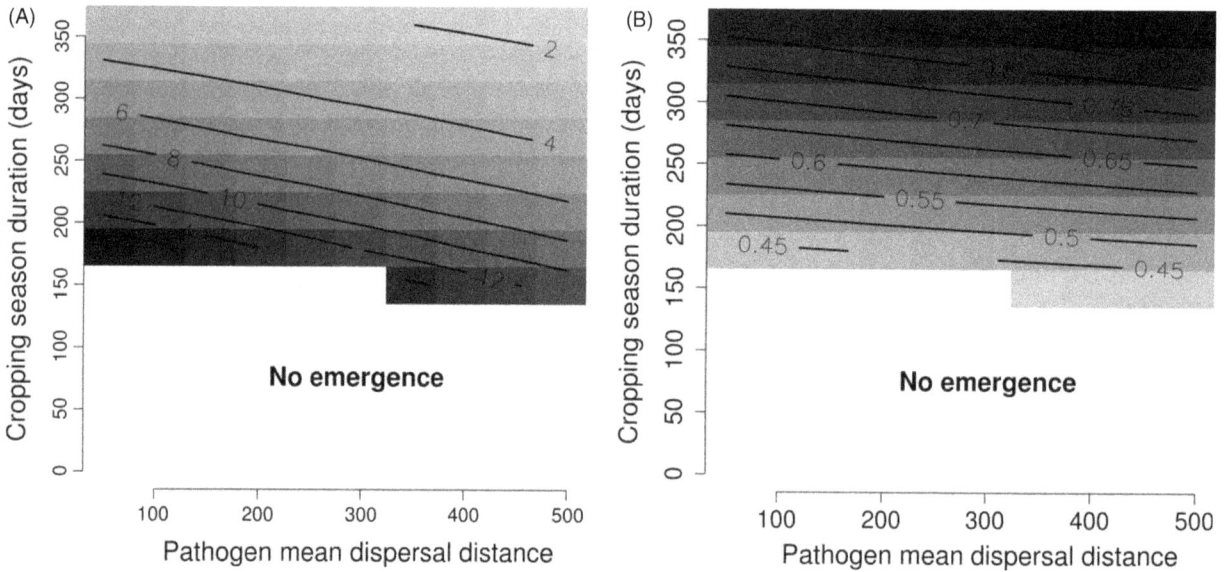

Figure 8 The effect of cropping season duration and pathogen mean dispersal distance: (A) predicted values of the time (number of years) required to observe the emergence of a crop pathogen, and (B) relative mean aggressiveness of the pathogen population on the crop at equilibrium (0, the aggressiveness is minimal; 1 the aggressiveness is maximal). White: no successful emergence of a crop pathogen predicted by the model. Values for the grey scale are indicated by the contour lines.

landscape structures also decreased the selection pressure posed by the crop which resulted, over longer time scales, in reduced aggressiveness of the pathogen population on the crop at equilibrium. Note that these results hold for a relatively low proportion (<10%) of wild hosts in the land-scape. For higher proportion and lower spatial aggregation, remnant wild patches may hamper disease spread at the landscape scale by decreasing the level of connectivity among crop fields (Papaïx et al. 2014c). More generally, wild patches can act as stepping stones or refugia for parasitoids and predators of vectors and pests to provide better biological control at the landscape scale (Woltz et al. 2012; MacFadyen and Müller 2013).

Trade-offs between the ecosystem services provided by remnant wild patches in agro-ecosystems are generally the rule as ecosystem services are not independent from each other (Millennium Ecosystem Assessment 2005). For example, Alexander et al. (2014) noted in the context of plant viruses that, while reduction in the abundance of both weeds and wild host plants can be beneficial for controlling viral diseases in crops, large-scale removal of noncrop plants also reduces heterogeneity in agro-ecological landscapes. This can influence the evolution and spread of viruses and thus potentially has negative as well as positive management consequences. As pointed out by Power (2010), these trade-offs should be considered in terms of spatial and temporal scales for management purposes. The spatial scale of emergence is generally local but depends on the spatial configuration of wild patches. Indeed, an aggregated distribution of local patches within the wild metapopulation could lead to more localized emergences than when patches are randomly distributed as the AE interface is more concentrated. In addition, emergence is a short-term evolutionary process. Once the pathogen has successfully shifted onto the crop, further mutations towards more aggressive pathotypes are more difficult to manage as they occur at larger spatial and longer temporal scales. Thus, with respect to the example we provide here, increasing the aggregation of wild vegetation within agricultural landscapes may be a useful approach to hinder pathogen evolution towards aggressive pathotypes at larger spatial and longer time scales while additional management strategies may be needed at local spatial scales and over shorter timeframes to control emergence (e.g. monitoring of pathogen populations).

Role of temporal heterogeneity

As expected, we found that longer agricultural seasons facilitated the emergence of a crop pathogen and increased its level of adaptation on the crop. Indeed, the perennial wild host enables the pathogen to survive during the off season for the crop. This situation generates a tension between the advantage (to the pathogen) of increased ability to attack crop hosts, but the potential disadvantage (when there are trade-offs) of a reduced ability to persist on wild hosts (which can represent an important refuge in the off season). Obviously, as the time period during which the crop was available increased, the pathogen's dependency on the wild host for survival decreased, which resulted in a better adaptation of the pathogen to the crop. Life history is, however, critical here: if a pathogen is less constrained (e.g. no trade-off or the pathogen can survive off-host as a saprophyte) then presumably generality (in

host range) will be more likely to emerge. Consistent with this, van den Berg et al. (2010, 2011) showed that a longer period of host absence selected for higher transmission rates in the presence of a trade-off between transmission and virulence but lower transmission rates in the presence of a trade-off between transmission and off-season survival. The role of the duration of the cropping season has direct consequences for the management of agricultural landscapes. For example, an increase in crop presence using crop varieties sown during the fall in temperate agricultural systems could increase the risk of disease emergence and the global adaptation of pathogen populations to crop hosts. More generally, crop rotations over time, that is inter- or intraseasonal changes in the crops, are known to impact disease dynamics in the long term and are recommended to provide disease breaks (Bennett et al. 2012). However, the efficiency of such rotations could be limited in regions where the crop is cultivated over large areas and for an aerial initial inoculum.

A key question concerning the off season for the crop that was not addressed here is how homogeneous the vegetation gap is across the landscape (Alexander et al. 2014). Indeed, different crops or crop cultivars can be planted and harvested at different times leading to a mosaic of host presence. In addition, self-sown volunteers and remnant plants can help obligate biotroph pathogens to bridge the gap when most crops are harvested. For example, volunteer wheat plants heavily infected with leaf rust (*Puccinia triticina*) are commonly observed (Burleigh et al. 1969; Mehta and Zadoks 1970; Moschini and Pérez 1999; Singh et al. 2004) indicating that the pathogen can survive locally and bridge the off season, resulting in the early appearance of rust infection in newly planted fields (Eversmeyer and Kramer 2000; Goyeau et al. 2006). Eversmeyer and Kramer (2000) suggested that the destruction of volunteer wheat would significantly reduce primary inoculum sources and disease severity as severe damage in wheat fields is observed when scattered plants are left when the new crop is planted. However, no information is available on the role of such volunteer plants in determining the year-to-year genetic structure of pathogen populations, and, in particular, in accelerating or hampering the fixation of new mutations or the rate of resistance breakdown.

In the same way, the demographic and genetic processes acting during the off season in the remnant wild vegetation patches of agro-ecological systems have received almost no consideration and we thus have little understanding of the role of the off season in shaping disease dynamics and pathogen genetic structure. Abiotic conditions during the off season are known to be important in determining pathogen survival and in-season epidemics (Marçais et al. 1996; Penczykowski et al. 2015). Biotic conditions can also play a critical role as exemplified by the case of barberry and stem rust discussed in the Introduction (Roelfs 1982). In addition, different pathogen genotypes may vary in their response to the off-season environment due to differential selection between pathogen life-history stages (Sommerhalder et al. 2011). This led Tack and Laine (2014) to extend the classic disease triangle to the off season. They developed their view with a meticulous investigation of the off-season dynamics of powdery mildew *Podophaera plantaginis* of *Plantago lanceolata* in which they found that pathogen survival during that time was affected by both environmental and spatial factors. Their results also suggested the presence of local adaptation by the pathogen to its local off-season environment. This opens the way for a better understanding and integration of off-season mechanisms in pathogen evolution and disease dynamics.

Role of life histories

Plant pathogens exhibit a diverse array of dispersal mechanisms which suggests that different host–pathogen interactions may occur over a broad range of spatial scales (Thrall and Burdon 1997). This aspect of pathogen life history thus has the potential to strongly mediate the effects of spatial structure on both disease epidemiology and pathogen evolutionary trajectories [e.g. Thrall and Burdon (1999, 2002)]. We found here that increases in the mean dispersal distance of the pathogen facilitated its shift from the wild host to the crop as well as its evolution on the crop towards more aggressive pathogen genotypes. In addition, increases in the scale of pathogen dispersal made the results more sensitive to changes in the proportion of the wild host in the landscape which is consistent with the literature (Papaïx et al. 2013). These results can be explained by the fact that greater pathogen dispersal distances increase the rate of propagule exchange between cropping and wild elements of agricultural systems. Indeed, the first pathogen genotypes arriving on the crop from the wild host can cause mild symptoms on the crop but cannot develop their own epidemic because their growth rate on the crop is negative ($\bar{R}_{0,loc}^{c} < 1$). Spore dispersal generates source–sink dynamics between wild and crop hosts. As pathogen dispersal distance increases, a larger pathogen population can be maintained on the crop due to increased spillover from the wild host (Holt 1993) which accelerates the evolution and the emergence of pathogen genotypes specialized on the crop.

Equally important are trade-offs between different life-history traits or in the ability to infect different hosts. The role of such a trade-offs has been studied using an adaptive dynamics theory approach in a nonspatial context (van den Berg et al. 2010, 2011; Hamelin et al. 2011). van den Berg et al. (2010, 2011) did not found situations in which coexistence of different pathogen genotypes was possible. Con-

versely, Hamelin et al. (2011) found that evolutionary branching in the pathogen population was possible due to the appearance of negative density dependence in the season-to-season dynamics but required that the trade-off between transmission and off-season survival has a concave shape (weak trade-off). Consistent with this, we found that when the trade-off between the adaptation to the wild host and the crop was weak, emergence and subsequent coexistence of two pathogen genotypes was the rule over a range of values for the length of the cropping season. Nevertheless, under a strong trade-off (convex shape), coexistence between a crop specialist and a wild host specialist was also observed but required a longer cropping season. This difference was probably due to the explicit consideration of the population spatial structure that favours the maintenance of diversity by means of several mechanisms (Sasaki et al. 2002; Salathé et al. 2005; Abrams 2006; Brown and Tellier 2011; Débarre and Lenormand 2011; Tellier and Brown 2011; Zhan and McDonald 2013). In addition, we have also shown that the spatio-temporal pattern of the landscape influences the speed of pathogen evolution and the level of adaptation of the pathogen population at equilibrium.

The analysis we reported here focused on the role of pathogen dispersal ability and on the existence of possible trade-offs in the ability to infect different hosts. Obviously, predictions for how crop composition and landscape structure affect pathogen interactions with (and reliance on) wild and cultivated hosts are also dependent on life-history traits other than dispersal and trade-offs, including for example mating system, transmission mode and the presence of saprophytic stages – all have the potential to modify pathogen persistence, population size and/or rates of evolution. For example, the ability of the wild host to disperse was fixed but the spatial scale of dispersal of both the host and the pathogen directly influence disease dynamics and plant pathogen co-evolution in the wild metapopulation. Low host and pathogen dispersal abilities imply a high level of asynchrony in disease dynamics among local populations with frequent local extinction and recolonization events and can result in a greater host and pathogen diversity (Thrall and Burdon 2002; Papaïx et al. 2014b). Conversely, as the host (respectively, the pathogen) dispersal ability increases severe boom and bust dynamics dominate due to a high level of synchrony among local populations, resulting in maladaptation of the pathogen (respectively, the host) population (Gandon 2002; Thrall and Burdon 2002; Papaïx et al. 2014b). Such different patterns of disease and host–pathogen co-evolutionary dynamics in wild patches are likely to have consequences for further pathogen evolution on the crop and need specific consideration to better predict the emergence of new crop pathogens.

Limitations: consideration of genetic diversity in host communities

As a first approximation, in this study, we considered a genetically homogeneous crop. The use of crop diversity through spatial diversification schemes and rotations is an active field of research focused on reducing disease severity and the evolutionary potential of pathogen populations (Zhan et al. 2014). From a management perspective, the spatio-temporal structure of crop diversity could be used in two different ways. First, crop diversity directly influences connectivity among cropping components and can delay the spread of the pathogen population (Skelsey et al. 2010; Papaïx et al. 2014c). This can result in a smaller pathogen population size on the crop with a lower survival probability during the off season (Suffert et al. 2011) and a lower evolutionary potential (Zhan et al. 2014). Second, crop diversity could directly influence the level of adaptation of the pathogen population to different crop cultivars and its genetic diversity (Marshall et al. 2009; Papaïx et al. 2011), potentially impacting its interaction with the wild host.

Heterogeneity within remnant wild vegetation should also be considered as wild plant communities are frequently very complex, being composed of native species, crop relatives and weedy exotics that can impose different selection pressures on pathogen populations (Mitchell and Power 2006; Thrall et al. 2007; Moore and Borer 2012; Seabloom et al. 2013). In addition, wild plant community dynamics differ from crops significantly in terms of population size, density and spatial distribution, genetic variability, and population continuity or predictability through time (Burdon 1993). Environmental conditions are also less stable potentially resulting in drastic fluctuations in population sizes and suboptimal growth conditions at least some of the time. All together these generally result in a marked stochasticity of pathogen dynamics at a local scale with repeated extinction/recolonization in individual demes even if disease dynamics appear stable at broader spatial scales. From an evolutionary perspective, the wild part of the AE interface is far from a homogeneous landscape (Burdon and Thrall 2014) and is better characterized as a mosaic of selection forces and intensity with different demes representing coevolutionary hot and cold spots (Thompson 1999; Smith et al. 2011). The spatial scale of local adaptation of both pathogen and plant populations can also be highly variable (Laine 2005; Jousimo et al. 2014) depending on other life-history attributes. Hence, it would be interesting to extend the present approach to consider more diverse situations for the wild elements (e.g. two wild hosts with some heterogeneity among host patches).

Literature cited

Abrams, P. A. 2006. Adaptive change in the resource-exploitation traits of a generalist consumer: the evolution and coexistence of generalists and specialists. Evolution 60:427–439.

Alexander, H. M., K. E. Mauck, A. E. Whitfield, K. A. Garrett, and C. M. Malmstrom 2014. Plant-virus interactions and the agro-ecological interface. European Journal of Plant Pathology 138:529–547.

Altizer, S., A. Dobson, P. Hosseini, P. Hudson, M. Pascual, and P. Rohani 2006. Seasonality and the dynamics of infectious diseases. Ecology Letters 9:467–484.

Bennett, A. J., G. D. Bending, D. Chandler, S. Hilton, and P. Mills 2012. Meeting the demand for crop production: the challenge of yield decline in crops grown in short rotations. Biological Reviews 87:52–71.

van den Berg, F., C. A. Gilligan, D. J. Bailey, and F. van den Bosch 2010. Periodicity in host availability does not account for evolutionary branching as observed in many plant pathogens: an application to *Gaeumannomyces graminis* var. *tritici*. Phytopathology 100:1169–1175.

van den Berg, F., N. Bacaer, J. A. J. Metz, C. Lannou, and F. van den Bosch 2011. Periodic host absence can select for higher or lower parasite transmission rates. Evolutionary Ecology 25:121–137.

Bianchi, F. J. J. A., C. J. H. Booij, and T. Tscharntke 2006. Sustainable pest regulation in agricultural landscapes: a review on landscape composition, biodiversity and natural pest control. Proceedings of the Royal Society of London. Serie B, Biological Sciences 273:1715–1727.

Bommarco, R., D. Kleijn, and S. G. Potts 2013. Ecological intensification: harnessing ecosystem services for food security. Trends in Ecology & Evolution 28:230–238.

Bouvier, A., K. Kiêu, K. Adamczyk, and H. Monod 2009. Computation of the integrated flow of particles between polygons. Environmental Modelling & Software 24:843–849.

Brown, J. K. M., and A. Tellier 2011. Plant-parasite coevolution: bridging the gap between genetics and ecology. Annual Review of Phytopathology 49:345–367.

Burdon, J. J. 1993. The structure of pathogen populations in natural plant communities. Annual Review of Phytopathology 31:305–323.

Burdon, J. J., and P. H. Thrall 2008. Pathogen evolution across the agro-ecological interface: implications for disease management. Evolutionary Applications 1:57–65.

Burdon, J. J., and P. H. Thrall 2014. What have we learned from studies of wild plant-pathogen associations? – the dynamic interplay of time, space and life-history. European Journal of Plant Pathology 138:417–429.

Burdon, J. J., L. G. Barrett, G. Rebetzke, and P. H. Thrall 2014. Guiding deployment of resistance in cereals using evolutionary principles. Evolutionary Applications 7:609–624.

Burleigh, J. R., A. A. Schulze, and M. G. Eversmeyer 1969. Some aspects of the summer and winter ecology of wheat rust fungi. Plant Disease Reporter 53:648–651.

Chaplin-Kramer, R., M. E. O'Rourke, E. J. Blitzer, and C. Kremen 2011. A meta-analysis of crop pest and natural enemy response to landscape complexity. Ecology Letters 14:922–932.

Débarre, F., and S. Gandon 2010. Evolution of specialization in a spatially continuous environment. Journal of Evolutionary Biology 23:1090–1099.

Débarre, F., and T. Lenormand 2011. Distance-limited dispersal promotes coexistence at habitat boundaries: reconsidering the competitive exclusion principle. Ecology Letters 14:260–266.

Eversmeyer, M. G., and C. L. Kramer 2000. Epidemiology of wheat leaf and stem rust in the central Great Plains of the USA. Annual Review of Phytopathology 38:491–513.

Fabre, F., E. Rousseau, L. Mailleret, and B. Moury 2012. Durable strategies to deploy plant resistance in agricultural landscapes. New Phytologist 193:1064–1075.

Gandon, S. 2002. Local adaptation and the geometry of host-parasite coevolution. Ecology Letters 5:246–256.

García-Arenal, F., and A. Fraile 2013. Trade-offs in host range evolution of plant viruses. Plant Pathology 62:2–9.

Garry, G., G. A. Forbes, A. Salas, M. Santa Cruz, W. G. Perez, and R. J. Nelson 2005. Genetic diversity and host differentiation among isolates of *Phytophthora infestans* from cultivated potato and wild solanaceous hosts in Peru. Plant Pathology 54:740–748.

Goyeau, H., R. Park, B. Schaeffer, and C. Lannou 2006. Distribution of pathotypes with regard to host cultivars in French wheat leaf rust populations. Phytopathology 96:264–273.

Hamelin, F. M., M. Castel, S. Poggi, D. Andrivon, and L. Mailleret 2011. Seasonality and the evolutionary divergence of plant parasites. Ecology 92:2159–2166.

Holt, R. D. 1993. Ecology at the mesoscale: the influence of regional processes on local communities. In R. Ricklefs, and D. Schluter, eds. Species Diversity in Ecological Communities, pp. 77–88. University of Chicago Press, Chicago, USA.

Jones, R. A. C. 2009. Plant virus emergence and evolution: origins, new encounter scenarios, factors driving emergence, effects of changing world conditions, and prospects for control. Virus Research 141:113–130.

Jousimo, J., A. J. M. Tack, O. Ovaskainen, T. Mononen, H. Susi, C. Tollenaere, and A.-L. Laine 2014. Ecological and evolutionary effects of fragmentation on infectious disease dynamics. Science 344:1289–1293.

Kiêu, K., K. Adamczyk-Chauvat, H. Monod, and R. S. Stoica 2013. A completely random T-tessellation model and Gibbsian extensions. Spatial Statistics 6:118–138.

Kolmer, J. A., Y. Jin, and D. L. Long 2007. Wheat leaf and stem rust in the United States. Australian Journal of Agricultural Research 58:631–638.

Laine, A.-L. 2005. Spatial scale of local adaptation in a plant-pathogen metapopulation. Journal of Evolutionary Biology 18:930–938.

Laine, A. L., and B. Barrès 2013. Epidemiological and evolutionary consequences of life-history trade-offs in pathogens. Plant Pathology 62 (Suppl. 1):96–105.

Lannou, C. 2012. Variation and selection of quantitative traits in plant pathogens. Annual Review of Phytopathology 50:319–338.

Lê Van, A., C. E. Durel, B. Le Cam, and V. Caffier. 2011. The threat of wild habitat to scab resistant apple cultivars. Plant Pathology 60:621–630.

Lebeda, A., I. Petrželová, and Z. Maryška 2008. Structure and variation in the wild-plant pathosystem: *Lactuca serriola-Bremia lactucae*. European Journal of Plant Pathology 122:127–146.

Leroy, T., B. Le Cam, and C. Lemaire 2014. When virulence originates from non-agricultural hosts: new insights into plant breeding. Infection, Genetics and Evolution 27:521–529.

MacFadyen, S., and W. Müller 2013. Edges in agricultural landscapes: species interactions and movement of natural enemies. PLoS ONE 8: e59658.

Mailleret, L., and V. Lemesle 2009. A note on semi-discrete modelling in the life sciences. Philosophical Transactions of the Royal Society of London. Series B, Biological Sciences 367:4779–4799.

Marçais, B., F. Dupuis, and M. L. Desprez-Loustau 1996. Modelling the influence of winter frosts on the development of the stem canker of

red oak, caused by *Phytophthora cinnamomi*. Annals of Forest Science 53:369–382.

Marshall, B., A. C. Newton, and J. Zhan 2009. Evolution of pathogen aggressiveness under cultivar mixtures. Plant Pathology 58:378–388.

McDonald, B. A., and C. Linde 2002. Pathogen population genetics, evolutionary potential, and durable resistance. Annual Review of Phytopathology 40:349–379.

Mehta, Y. R., and J. C. Zadoks 1970. Uredospore production and sporulation period of *Puccinia recondita* f. sp. *triticina* on primary leaves of wheat. Netherlands Journal of Plant Pathology 76:267–276.

van der Merwe, N. A., E. T. Steenkamp, C. Rodas, B. D. Wingfield, and M. J. Wingfield 2013. Host switching between native and non-native trees in a population of the canker pathogen *Chrysoporthe cubensis* from Colombia. Plant Pathology 62:642–648.

Millennium Ecosystem Assessment 2005. Ecosystems and Human Wellbeing: Synthesis. Island Press, Washington, DC.

Mitchell, C. E., and A. G. Power 2006. Disease dynamics in plant communities. In S. K. Collinge, and C. Ray, eds. Disease Ecology: Community Structure and Pathogen Dynamics, pp. 58–72. Oxford University Press, Oxford, UK.

Moore, S. M., and E. T. Borer 2012. The influence of host diversity and composition on epidemiological patterns at multiple spatial scales. Ecology 93:1095–1105.

Moschini, R. C., and B. A. Pérez 1999. Predicting wheat leaf rust severity using planting date, genetic resistance, and weather variables. Plant Disease 83:381–384.

Mundt, C. C., and K. J. Leonard 1986. Effect of host unit area on development of focal epidemics of bean rust and common maize rust in mixtures of resistant and susceptible plants. Phytopathology 76:895–900.

Mundt, C. C., K. E. Sackett, and L. D. Wallace 2011. Landscape heterogeneity and disease spread: experimental approaches with a plant pathogen. Ecological Applications 21:321–328.

Papaïx, J., H. Monod, H. Goyeau, P. du Cheyron, and C. Lannou 2011. Influence of cultivated landscape composition on variety resistance: an assessment based on wheat leaf rust epidemics. New Phytologist 191:1095–1107.

Papaïx, J., O. David, C. Lannou, and H. Monod 2013. Dynamics of adaptation in spatially heterogeneous metapopulations. PLoS ONE 8: e54697.

Papaïx, J., K. Adamczyk-Chauvat, A. Bouvier, K. Kiêu, S. Touzeau, C. Lannou, and H. Monod 2014a. Pathogen population dynamics in agricultural landscapes: the Ddal modelling framework. Infection, Genetics and Evolution 27:509–520.

Papaïx, J., J. J. Burdon, C. Lannou, and P. H. Thrall 2014b. Evolution of pathogen specialisation in a host metapopulation: joint effects of host and pathogen dispersal. Plos Computational Biology 10:e1003633.

Papaïx, J., S. Touzeau, H. Monod, and C. Lannou 2014c. Can epidemic control be achieved by altering landscape connectivity in agricultural systems? Ecological Modelling 284:35–47.

Pariaud, B., V. Ravigné, F. Halkett, H. Goyeau, J. Carlier, and C. Lannou 2009. Aggressiveness and its role in the adaptation of plant pathogens. Plant Pathology 58:409–424.

Penczykowski, R. M., E. Walker, S. Soubeyrand, and A.-L. Laine 2015. Linking winter conditions to regional disease dynamics in a wild plant-pathogen metapopulation. New Phytologist 205:1142–1152.

Power, A. G. 2010. Ecosystem services and agriculture: tradeoffs and synergies. Philosophical Transactions of the Royal Society of London. Series B, Biological Sciences 365:2959–2971.

R Core Team 2014. R: A Language and Environment for Statistical Computing. R Foundation for Statistical Computing, Vienna, Austria. URL http://www.R-project.org.

Ravigné, V., U. Dieckmann, and I. Olivieri 2009. Live where you thrive: joint evolution of habitat choice and local adaptation facilitates specialization and promotes diversity. The American Naturalist **174**: E141–E169.

Roelfs, A. P. 1982. Effects of barberry eradication on stem rust in the United States. Plant Disease **66**:177–181.

Salathé, M., A. Scherer, and S. Bonhoeffer 2005. Neutral drift and polymorphism in gene-for-gene systems. Ecology Letters **8**:925–932.

Sasaki, A., W. D. Hamilton, and F. Ubeda 2002. Clone mixtures and a pacemaker: new facets of red-queen theory and ecology. Proceedings of the Royal Society of London. Serie B, Biological Sciences **269**:761–772.

Seabloom, E. W., E. T. Borer, C. Lacroix, C. E. Mitchell, and A. G. Power 2013. Richness and composition of niche-assembled viral pathogen communities. PLoS ONE **8**:e55675.

Shennan, C. 2008. Biotic interactions, ecological knowledge and agriculture. Philosophical Transactions of the Royal Society of London. Series B, Biological Sciences **363**:717–739.

Singh, R. P., J. Huerta-Espino, W. Pfeiffer, and P. Figueroa-Lopez 2004. Occurrence and impact of a new leaf rust race on durum wheat in northwestern Mexico from 2001 to 2003. Plant Disease **88**:703–708.

Skelsey, P., W. A. H. Rossing, G. J. T. Kessel, J. Powell, and W. van der Werf 2005. Influence of host diversity on development of epidemics: an evaluation and elaboration of mixture theory. Phytopathology **95**:328–338.

Skelsey, P., W. A. H. Rossing, G. J. T. Kessel, and W. van der Werf 2010. Invasion of *Phytophthora infestans* at the landscape level: how do spatial scale and weather modulate the consequences of spatial heterogeneity in host resistance? Phytopathology **100**:1146–1161.

Smith, D. L., L. Ericson, and J. J. Burdon 2011. Co-evolutionary hot and cold spots of selective pressure move in space and time. Journal of Ecology **99**:634–641.

Sommerhalder, R. J., B. A. McDonald, F. Mascher, and J. Zhan. 2011. Effect of hosts on competition among clones and evidence of differential selection between pathogenic and saprophytic phases in experimental populations of the wheat pathogen *Phaeosphaeria nodorum*. BMC Evolutionary Biology **11**:188.

Stukenbrock, E. H., and B. A. McDonald 2008. The origins of plant pathogens in agro-ecosystems. Annual Review of Phytopathology **46**:75–100.

Suffert, F., I. Sache, and C. Lannou 2011. Early stages of *Septoria tritici* blotch epidemics of winter wheat: build-up, overseasoning, and release of primary inoculum. Plant Pathology **60**:166–177.

Tack, J. M., and A.-L. Laine 2014. Ecological and evolutionary implications of spatial heterogeneity during the off-season for a wild plant pathogen. New Phytologist **202**:297–308.

Tellier, A., and J. K. M. Brown 2011. Spatial heterogeneity, frequency-dependent selection and polymorphism in host-parasite interactions. BMC Evolutionary Biology **11**:319–334.

Thompson, J. N. 1999. Specific hypotheses on the geographic mosaic of coevolution. The American Naturalist **153**:S1–S14.

Thrall, P. H., and J. J. Burdon 1997. Host-pathogen dynamics in a metapopulation context: the ecological and evolutionary consequences of being spatial. Journal of Ecology **85**:743–753.

Thrall, P. H., and J. J. Burdon 1999. The spatial scale of pathogen dispersal: consequences for disease dynamics and persistence. Evolutionary Ecology Research 1:681–701.

Thrall, P. H., and J. J. Burdon 2002. Evolution of gene-for-gene systems in metapopulations: the effect of spatial scale of host and pathogen dispersal. Plant Pathology **51**:169–184.

Thrall, P. H., M. E. Hochberg, J. J. Burdon, and J. D. Bever 2007. Coevolution of symbiotic mutualists and parasites in a community context. Trends in Ecology & Evolution **22**:120–126.

Thrall, P. H., J. G. Oakeshott, G. Fitt, S. Southerton, J. J. Burdon, A. Sheppard, R. J. Russell et al. 2011. Evolution in agriculture: the application of evolutionary approaches to the management of biotic interactions in agro-ecosystems. Evolutionary Applications **4**:200–215.

Tscharntke, T., Y. Clough, S. A. Bhagwat, D. Buchori, H. Faust, D. Hertel, D. Holscher et al. 2011. Multifunctional shade-tree management in tropical agroforestry landscapes – a review. Journal of Applied Ecology **48**:619–629.

Tscharntke, T., Y. Clough, T. C. Wanger, L. Jackson, I. Motzke, I. Perfecto, J. Vandermeer et al. 2012. Global food security, biodiversity conservation and the future of agricultural intensification. Biological Conservation **151**:53–59.

Wang, B., C. L. Brubaker, and J. J. Burdon 2004. *Fusarium* species and Fusarium wilt pathogens associated with native *Gossypium* populations in Australia. Mycological Research **108**:35–44.

Wang, B., C. L. Brubaker, B. A. Summerell, P. H. Thrall, and J. J. Burdon. 2010. Local origin of two vegetative compatibility groups of *Fusarium oxysporum* f. sp *vasinfectum* in Australia. Evolutionary Applications **3**:505–524.

Webster, C. G., B. A. Coutts, R. A. C. Jones, M. G. K. Jones, and S. J. Wylie 2007. Virus impact at the interface of an ancient ecosystem and a recent agroecosystem: studies on three legume-infecting potyviruses in the southwest Australian floristic region. Plant Pathology **56**:729–742.

Wisler, G. C., and R. E. Norris 2005. Interactions between weeds and cultivated plants as related to management of plant pathogens. Weed Science **53**:914–917.

Woltz, J. M., R. Isaacs, and D. A. Landis 2012. Landscape structure and habitat management differentially influence insect natural enemies in an agricultural landscape. Agriculture Ecosystems and Environment **152**:40–49.

Zhan, J., and B. A. McDonald 2013. Experimental measures of pathogen competition and relative fitness. Annual Review of Phytopathology **51**:131–153.

Zhan, J., C. C. Mundt, M. E. Hoffer, and B. A. McDonald 2002. Local adaptation and effect of host genotype on the rate of pathogen evolution: an experimental test in a plant pathosystem. Journal of Evolutionary Biology **15**:634–647.

Zhan, J., P. H. Thrall, and J. J. Burdon 2014. Achieving sustainable plant disease management through evolutionary principles. Trends in Plant Science **19**:570–575.

Zhu, Y., H. Chen, J. Fan, Y. Wang, Y. Li, J. Chen, J. X. Fan et al. 2000. Genetic diversity and disease control in rice. Nature **406**:718–722.

Appendix A Statistical results

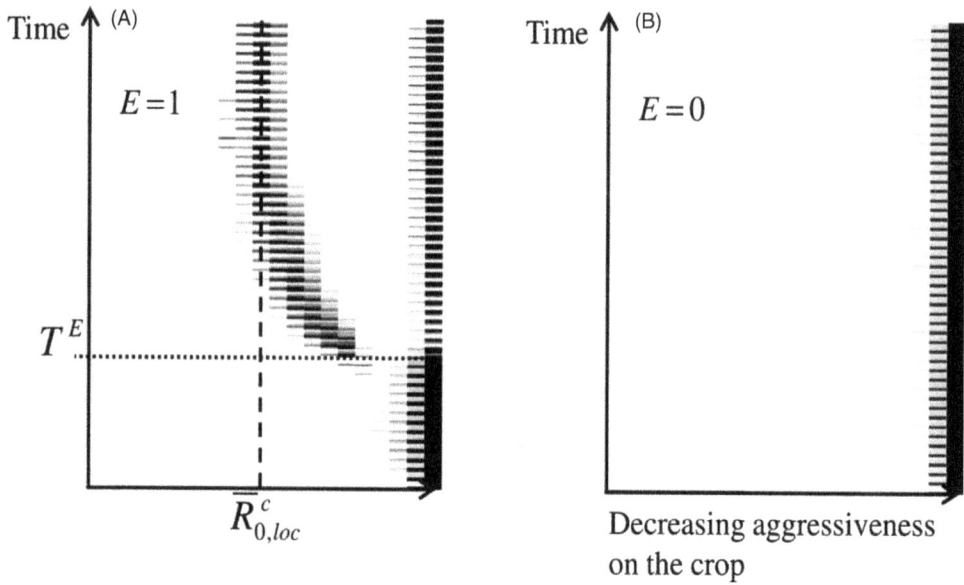

Figure A1 Examples of global evolutionary trajectories in the pathogen population when the emergence of a crop pathogen is successful (A, $E = 1$) or not (B, $E = 0$). In A, the time that the emergence required (T^E) and the mean aggressiveness of the stable pathogen population on the crop ($\bar{R}^c_{0,loc}$) are also displayed. The grey intensity indicates the frequency of pathogen genotypes (x-axis) across time (y-axis), white: the frequency is equal to 0, black: the frequency is equal to 1.

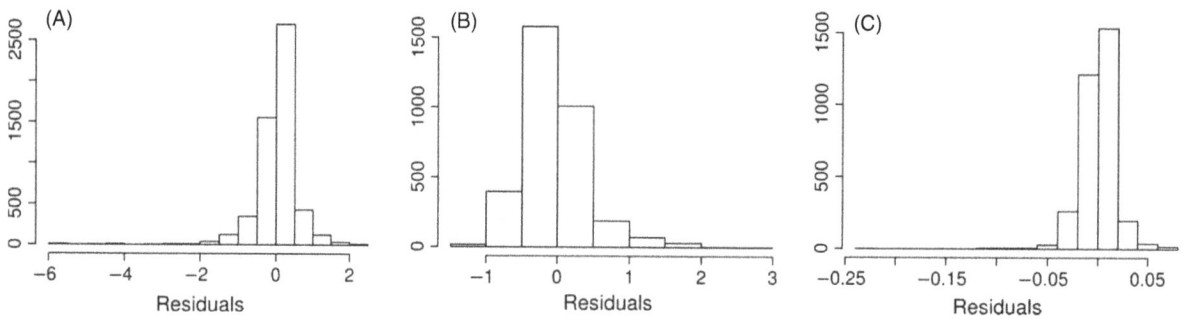

Figure A2 Histograms of deviance residuals for E (A), T^E (B) and $\bar{R}^c_{0,loc}$ (C). The models are specified in the Table 2 of the main text.

Table A1. Estimated effects of the proportion and the aggregation level of the wild patches, the trade-off shape, the cropping season duration and the pathogen mean dispersal distance on the emergence of a crop pathogen (E – Table 2 of the main text).

Effect	Mean (Confidence interval at 95%)
Intercept	
Random	1.76 (1.56,1.97)
Clustered	2.61 (2.37,2.86)
Wild host proportion (q)	0.96 (0.83,1.10)
Trade-off shape (β)	−4.91 (−5.27, −4.56)
Cropping season duration (Y)	5.25 (4.89,5.63)
Pathogen mean dispersal distance ($distM^P$)	1.09 (0.95,1.24)

Table A2. Estimated effects of the proportion and the aggregation level of the wild patches, the trade-off shape, the cropping season duration, the pathogen mean dispersal distance and interactions on the time required for the emergence of a crop pathogen (T^E) and the mean aggressiveness of the stable pathogen population on the crop ($\bar{R}^c_{0,loc}$ – Table 2 of the main text).

Effect	Mean (Confidence interval at 95%)			
	Time before emergence (log scale)		Mean aggressiveness ($*10^{-1}$)	
	Random	Clustered	Random	Clustered
Intercept	1.78 (1.75,1.82)	1.67 (1.64,1.70)	2.20 (2.19,2.21)	2.13 (2.12,2.14)
Wild host proportion (q)	−0.24 (−0.26, −0.21)	−0.17 (−0.19, −0.14)	−0.25 (−0.26, −0.24)	−0.21 (−0.22, −0.20)
Trade-off shape (β)	0.62 (0.58,0.65)	0.55 (0.52,0.59)	−0.21 (−0.22, −0.20)	−0.17 (−0.18, −0.16)
Cropping season duration (Y)	−0.66 (−0.69, −0.62)	−0.68 (−0.71, −0.65)	0.59 (0.58,0.60)	0.58 (0.57,0.59)
Pathogen mean dispersal distance ($distM^P$)	−0.25 (−0.28, −0.23)	−0.24 (−0.27, −0.22)	0.06 (0.05,0.07)	0.05 (0.04,0.06)
$q \times \beta$	−0.06 (−0.09, −0.04)	−0.06 (−0.09, −0.04)	0.04 (0.03,0.05)	0.02 (0.01,0.03)
$q \times Y$	0.03 (0.01,0.06)	0.04 (0.01,0.07)	−0.08 (−0.09, −0.07)	−0.07 (−0.08, −0.06)
$q \times distM^P$	−0.04 (−0.07, −0.01)	−0.05 (−0.08, −0.03)	−0.03 (−0.04, −0.03)	−0.01 (−0.02,0.00)
$\beta \times Y$	−0.10 (−0.14, −0.07)	−0.11 (−0.15, −0.08)	0.11 (0.10,0.12)	0.10 (0.09,0.11)
$\beta \times distM^P$	−0.13 (−0.16, −0.10)	−0.12 (−0.15, −0.09)	0.06 (0.05,0.07)	0.06 (0.05,0.07)
$Y \times distM^P$	0.03 (0.00,0.05)	0.06 (0.03,0.08)	0.03 (0.03,0.04)	0.01 (0.00,0.02)

×, denotes interactions.

11

Gene flow and natural selection shape spatial patterns of genes in tree populations: implications for evolutionary processes and applications

Victoria L. Sork[1,2]

1 Department of Ecology and Evolutionary Biology, University of California, Los Angeles, CA, USA
2 Institute of Environment and Sustainability, University of California, Los Angeles, CA, USA

Keywords
gene flow, landscape genomics, local adaptation, phylogeography, pollen dispersal, population differentiation, *Quercus*, seed dispersal.

Correspondence
Victoria L. Sork, Department of Ecology and Evolutionary Biology, University of California, Box 957239, Los Angeles, CA 90095-7239, USA.

e-mail: vlsork@ucla.edu

Abstract

A central question in evolutionary biology is how gene flow and natural selection shape geographic patterns of genotypic and phenotypic variation. My overall research program has pursued this question in tree populations through complementary lines of inquiry. First, through studies of contemporary pollen and seed movement, I have studied how limited gene movement creates fine-scale genetic structure, while long-distance gene flow promotes connectivity. My collaborators and I have provided new tools to study these processes at a landscape scale as well as statistical tests to determine whether changes in landscape conditions or dispersal vectors affect gene movement. Second, my research on spatial patterns of genetic variation has investigated the interacting impacts of geography and climate on gene flow and selection. Third, using next-generation genomic tools, I am now studying genetic variation on the landscape to find initial evidence of climate-associated local adaptation and epigenetic variation to explore its role in plant response to the climate. By integrating these separate lines of inquiry, this research provides specific insight into real-world mechanisms shaping evolution in tree populations and potential impacts of landscape transformation and climate change on these populations, with the prospective goal of contributing to their management and conservation.

Introduction

"In considering the distribution of organic beings over the face of the globe, the first great fact which strikes us is, that neither the similarity nor the dissimilarity of the inhabitants of various regions can be accounted for by their climatal and other physical conditions... The dissimilarity of inhabitants of different regions may be attributed to modification through natural selection... The degree of dissimilarity will depend on the migration of the more dominant forms of life from one region into another... Thus, the high importance of barriers comes into play by checking migration; as does time for the slow process of modification through natural selection". (Darwin 1859)

Understanding the relative contributions of gene flow and natural selection to geographical patterns of natural variation has been a recurring theme of evolutionary biology. Darwin goes back and forth between the two processes in *The Origin of Species*, and those who have followed (Fisher, Wright, Mayr, Stebbins—to name a few) have continued to grapple with the roles of these countervailing forces. Gene flow and natural selection are two central, and usually opposing, evolutionary forces: gene flow distributes, homogenizes, and maintains genetic variation that can act as the 'stuff of evolution', while natural selection reduces genetic variation to the variants that favor survival and reproduction. For tree species, which are often highly outcrossed with extensive gene flow, genetic drift plays a lesser role in their population genetics, except in small or substantially isolated populations (Ledig 1988, 1992; Ellstrand and Elam 1993). Effective dispersal mechanisms maintain connectivity of populations, except where they encounter a barrier. Nonetheless, within the species range, gene flow has a decay rate that will create genetic isolation by distance, setting up a pattern of genetic variation (genetic structure) within and among populations (Wright 1943; Slatkin 1985). Once a gene moves and a plant becomes established, it must survive where it lands, and long-lived plants will experience selection throughout life, with local ecological forces creating spatial genetic structure across the landscape (Bradshaw 1972). Endler (1973) proposed that selection is often so strong that the 'swamping out' effects of gene flow should be negligible. Of course, the relative impact of selection versus gene flow will vary across loci, depending on whether or not they experience selection.

One way to understand how the dynamics of gene movement create genetic structure is to analyze contemporary gene movement through pollen and seeds (Levin 1981; Hamrick 1987; Sork et al. 1999) and assess their impacts on local neighborhood size, which is the effective number of interbreeding individuals (Wright 1946; Crawford 1984). Numerous studies have used paternity analysis to assess the distance of pollen-mediated gene dispersal and the shape of the dispersal kernel (e.g., Adams et al. 1992; Smouse et al. 1999; Streiff et al. 1999; Burczyk et al. 2002; Slavov et al. 2005; Oddou-Muratorio et al. 2006; Meagher 2007; Pluess et al. 2009). Alternatively, one can use the genetic composition of the pollen pool, such as TwoGener and related methods (e.g., Smouse et al. 2001; Austerlitz et al. 2004; Smouse and Sork 2004; Smouse and Robledo-Arnuncio 2005; Robledo-Arnuncio et al. 2006), to document contemporary pollen-mediated gene dispersal, even when paternity analysis is not possible. Studies using both approaches often illustrate how well pollen flow promotes gene flow. In contrast, studies documenting seed-mediated gene dispersal using maternity analysis and alternative methods generally demonstrate more restricted gene movement (e.g., Hamrick et al. 1993; Godoy and Jordano 2001; Garcia et al. 2005; Grivet et al. 2005), with notable exceptions (e.g., Nason and Hamrick 1997; Aldrich and Hamrick 1998; Dick et al. 2003; Karubian et al. 2010). Various types of pollen and seed dispersal vectors have differential impacts on fine-scale genetic structure that can be assessed through spatial autocorrelation analysis (Sokal and Oden 1979; Loiselle et al. 1995; Epperson 2003; Vekemans and Hardy 2004). Resulting patterns are consistent with observations of contemporary gene movement, such that high movement reduces fine-scale structure and vice versa (Hamrick and Loveless 1986; Hardy et al. 2006; Dick et al. 2008). A few studies have been able to examine both processes simultaneously and show that the contribution of seed dispersal in particular plays a critical role in the genetic structure of the seedlings populations (e.g., Garcia et al. 2007; Grivet et al. 2009; Ottewell et al. 2012; Sork et al. 2015), thus shaping the opportunity for the natural selection to create local adaptation.

Until recently, studying adaptive genetic variation in natural populations has not been as feasible as studying gene flow. However, next-generation sequencing now makes it possible to identify the single-nucleotide polymorphisms (SNPs) that flag candidate genes underlying traits experiencing selection created by local environments, even for nonmodel species without reference genomes (Davey et al. 2011; Sork et al. 2013). For locally adapted genetic variation, allele frequencies at these SNPs might exhibit significantly higher population differentiation than other sampled SNPs, after controlling for demographic history (Davey et al. 2011). All the same, the proper designs and statistical tests to avoid false positives are still a work in progress (Lotterhos and Whitlock 2014, 2015). Another way to obtain evidence for local adaptation would be to examine the association between allele frequencies with environmental variables (Luikart et al. 2003; Nielsen 2005; Coop et al. 2010; Strasburg et al. 2012). These findings can be combined with other genomic and quantitative genetic approaches to determine the gene functions or phenotypes that might be under selection (Stapley et al. 2010; Neale and Kremer 2011; Andrew et al. 2013; Sork et al. 2013). For evolutionary ecologists and conservation biologists, it is exciting that we now have genomic tools to study the adaptive genetic variation in natural populations.

Understanding the interaction of gene flow and selection in natural populations is fundamental to many applied evolutionary questions. The field of landscape genetics has emerged in response to a need to understand the impact of fragmentation and other landscape changes on gene flow and the preservation of genetic variation (Sork et al. 1999; Manel et al. 2003; Sork and Waits 2010). Much of population genetic theory assumes null environments and equilibrium conditions, but the rapidly changing landscapes created by human perturbations make those assumptions invalid. Moreover, conservation scientists focus on the impact of those changes (Ledig 1992; Ellstrand and Elam 1993; Young et al. 1996; Nason and Hamrick 1997; Aguilar et al. 2008), thus, requiring new methods of estimating their genetic consequences at a landscape scale. In addition, to establish the credibility of conservation science, we have needed improved empirical tests to determine whether human disturbance has significantly affected populations. The concern of environmental scientists about the impact of anthropogenic landscape change on genetic diversity and gene flow has now been extended to apprehension about the effect of climate change on the sustainability of populations (Parmesan 2006; Aitken et al. 2008; Sork et al. 2010; Anderson et al. 2012; Kremer et al. 2014). Through models of future climate (e.g., Cayan et al. 2008), it will be possible to identify the geographic regions with a high probability of climate change in order to specify the areas of conservation priority and then examine populations within these regions to assess their vulnerability.

Over the last twenty plus years, my research has focused on gene flow and geographical patterns of genetic variation at different temporal and spatial scales. Several tree species in different ecosystems have been studied with the goals of understanding the fundamental evolutionary processes of gene flow and selection and applying that knowledge to the development of conservation and resource management practices and strategies. To generate a general framework for trees, I selected the iconic California endemic oak, *Quercus lobata* Née, as our focal study system (Fig. 1). *Q. lobata* is a foundation species for biodiversity in the ecosystems it defines, and it has experienced significant reduction in its geographic coverage since the arrival of Europeans (Kelly et al. 2005; Whipple et al. 2011). Pavlik et al. (1995) described valley oak as the 'monarch of California oaks by virtue of its size, age, and beauty'. This research has generated empirical data and new methods to address key concepts through three lines of inquiry:

1 *Contemporary gene flow at a landscape scale.* Studying ongoing pollen- and seed-mediated gene movement provides valuable insight into the intrinsic properties of the dispersal vectors and the influences of human disturbance on the distribution of genetic variation within and among populations.
2 *Geographic patterns of genetic variation.* Spatial patterns of genetic variation across the species range, especially when combined with environmental data, can clarify the contributions of gene flow, historical demographic events, and the environment on the distribution of genetic variation.

3 *Landscape genomics of adaptive genetic variation.* Landscape genomics provides the basis for studying adaptive genetic variation by geolocating sequence data, flagging outliers associated with climate, and identifying candidate genes. This approach provides initial evidence of local adaptation in climate-associated genes that can be complemented with studies of phenotypes in common gardens and experimental studies of gene expression in response to environmental treatments.

Background and initial drivers

Early in my career as a plant biologist, I was inspired by the classic work of Clausen et al. (1947) who showed that plants grew best when put near their native environments and that they grew better in those sites than plants derived from sites at other elevations. That work provided elegant proof of local adaptation, but, of course, the origins of that geographical differentiation could be explained by both natural selection on local populations and restricted gene flow from populations at different elevations. Several papers based on populations at the boundaries of mines with toxic metals showed that short-lived plant populations could differentiate rapidly due to strong selection (Antonovics and Bradshaw 1970; Antonovics 1971), but it was less clear whether highly outcrossed tree populations could differentiate on such a local scale. After reading Slatkin's papers on gene flow (1973, 1985, 1987), I wondered whether selection could be strong enough to overcome gene flow from adjacent populations of highly outcrossing tree species. Forest geneticists have utilized provenance studies to illustrate regional differences across tree

Figure 1 (A) Photo of adult *Quercus lobata* Née. (Photo by Andy Lentz ©) (B) Landscape view of typical oak savanna habitat with scattered large trees of *Q. lobata* (leafless canopies) and *Q. agrifolia* (green canopies) taken at UC Santa Barbara Sedgwick Reserve, Santa Ynez Valley, Santa Barbara Co., California, USA. (Photo by Delphine Grivet ©).

populations (Mátyás 1996). However, I questioned whether local adaptation could occur in tree populations on such a fine geographic scale as the mine boundaries or whether gene flow would swamp adaptive divergence. To test this, I set up a field experiment in Missouri where I planted acorns from adult northern red oaks growing in adjacent microhabitats using a quantitative genetic design (Sork et al. 1993). Among many differences found among the first-year seedlings, it was particularly noteworthy that percent leaf damage, an indicator of resistance to herbivores, was always lowest for seedlings grown on the microhabitat of their maternal seed tree (Fig. 2). Moreover, this difference among subpopulations also showed up when seedlings were planted in a common garden setting (Stowe et al. 1994). Although we cannot rule out maternal effects or other factors, these strong phenotypic differences on a scale less than a few hundred meters indicate the local selection pressures in tree populations could be sufficiently intense to overcome gene flow, or gene movement is sufficiently restricted that adaptation can evolve on a local scale.

A second major driver of my research program twenty years ago was to understand the impact of landscape alteration on genetic diversity of forest tree species. This work was stimulated by a multi-investigator study of the impact of forest ecosystem management on biodiversity, called the Missouri Ozark Forest Ecosystem Project (MOFEP), and administered through the Missouri Department of Conservation (Brookshire and Hauser 1993). Given evidence that landscape fragmentation and reductions in population size could jeopardize gene flow and connectivity (for example, see Hamrick et al. 1991; Ledig 1992; Templeton and Georgiadis 1996; Nason and Hamrick 1997), my role was to assess whether forest management practices would have an impact on genetic diversity of tree populations in 200+ ha plots. In contrast to the other MOFEP studies that looked at immediate impacts of forest-cutting practices on insect, bird, and mammal biodiversity (Brookshire and Hauser 1993; Brookshire et al. 1997), because of the longevity of trees, our project would not be able to measure changes in diversity for at least 50 years. Instead, we chose to investigate the impacts on contemporary pollen flow because any changes in gene flow at the present time would shape future diversity. However, at that time, gene movement was studied largely through paternity and maternity analysis (e.g., Devlin et al. 1988; Adams et al. 1992; Streiff et al. 1999), which was not feasible for continuously distributed tree populations on a landscape scale (Sork et al. 1999; Smouse and Sork 2004). This applied conservation science investigation required a method of studying contemporary gene flow that could be conducted across a large area and that could test specific hypotheses about the impact of landscape change on future genetic diversity. Fortunately,

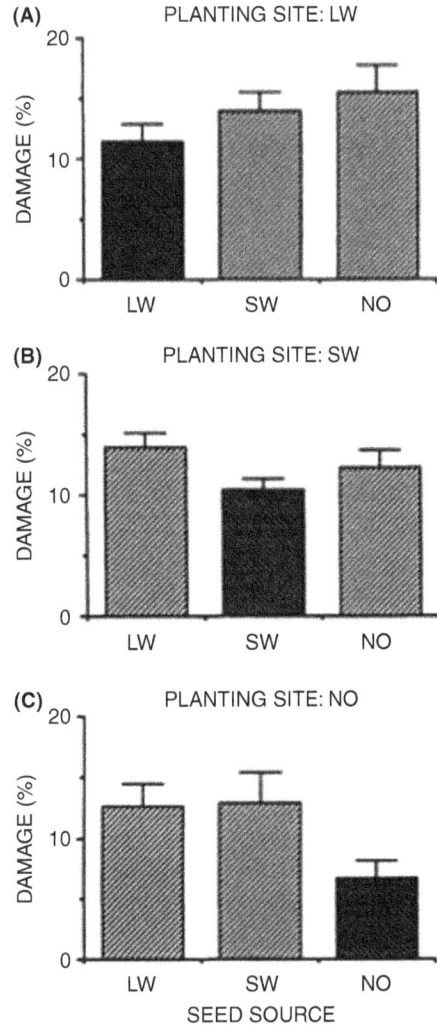

Figure 2 Percent leaf damage by herbivores on *Quercus rubra* seedlings grown from seed derived from three adjacent slopes in a Missouri oak–hickory forest and reciprocally planted in those sites. Microhabitats consisted of lower west-facing slope (LW), south-facing slope (SW), and north-facing slope (NO). (Taken from Sork et al. 1993. American Naturalist).

during my sabbatical at the National Center for Ecological Analysis and Synthesis, I was able to organize an NCEAS-funded working group of population genetics experts that would address this issue.

Contemporary gene flow in trees at a landscape scale

Evolutionary biology has long inferred historical gene flow through genetic structure statistics (Slatkin 1985, 1987; Neigel 1997), but the study of contemporary gene flow allows an examination of the process of gene movement. Short distance gene movement creates neighborhoods (*sensu* Wright 1946), which sets up the opportunity for both genetic drift and natural selection to reduce the genetic variation. Long-distance gene movement, on the other hand, maintains connectivity among tree populations. If gene flow is impeded by landscape changes, then a population becomes isolated and can decrease in effective size, which increases risk of inbreeding depression and loss of adaptive genetic variation through drift (Ledig 1992; Ellstrand and Elam 1993; Aguilar et al. 2008). The effects of landscape change may be mitigated in plant species with long dispersal kernels with fat tails, where gene flow will promote genetic diversity within local populations and connectivity among populations (Austerlitz et al. 2004; Smouse and Sork 2004; Sork and Smouse 2006). Indeed, the study of dispersal kernels has provided valuable detail about the processes of gene movement away from focal plants for many species (e.g., Austerlitz et al. 2004; Oddou-Muratorio et al. 2005; Robledo-Arnuncio and Gil 2005). However, my interest has been to test specific hypotheses about the genetic consequences of gene movement in different landscape contexts when locating all potential parents was not feasible. For this goal, new statistical approaches were developed through my collaboration with Peter Smouse (Rutgers University) and a series of talented younger scientists, including Frederic Austerlitz, Cyril Dutech, Rodney Dyer, Delphine Grivet, Jordan Karubian, Andrea Pluess, Juan Jose Robledo-Arnuncio, and Douglas Scofield. Building on earlier work in Missouri Ozark Forests studying oak, pine, and flowering dogwood (Sork et al. 1997, 2005; Gram and Sork 1999, 2001; Dyer and Sork 2001; Apsit et al. 2002), we eventually focused our studies on the pollen and seed dispersal vectors of California populations of valley oak, *Quercus lobata* (Fig. 3). The overarching goals of this research are both to understand the processes of gene movement themselves from an evolutionary perspective and to provide a way to assess the extent to which they are influenced by landscape conditions.

Pollen movement

For most tree populations, gene flow occurs primarily through pollen. When I first began this line of research, the primary method of studying pollen movement was to identify the paternity of progeny sampled from a maternal source (Meagher and Thompson 1987; Devlin et al. 1988; Adams et al. 1992; Streiff et al. 1999). While providing a very precise description of the pollen dispersal kernel, average dispersal distance, and the extent of long-distance dispersal, its application is often limited to a single location, to a single year, and to a population where all of the adult parents could be identified completely or at least within a certain radius around a focal seed tree. For species with

long-distance dispersal or in difficult-to-sample habitats, paternity analysis is often not feasible (Sork et al. 1999; Smouse and Sork 2004; Sork and Smouse 2006). Moreover, the costs of replicating studies of paternity analysis where large numbers of progeny are needed relative to the number of pollen sources (Sork et al. 1999) make replication across sites and years difficult, feasibility issues aside. To find an alternative method, we changed our line of thinking: *What if we used the genetic structure of the pollen pool itself to make inferences about the effective number of pollen donors, the mean effective dispersal distance, and the effective neighborhood size (sensu Wright), and didn't worry about finding the pollen sources?* To do this, we developed a two-generation procedure, dubbed TwoGener, which combines the simplicity of a population structure approach with the parent-/offspring-deductive aspects of the parentage approach (Smouse et al. 2001). Our approach emphasizes the genetic consequences of pollen movements rather than the description of each movement (Smouse and Sork 2004). A key parameter is the effective number of pollen donors (N_{ep}). This number is typically much smaller than the total number of pollen donors because a few local pollen sources dominate, with rare long-distance pollen flow from multiple different sources. From an evolutionary perspective, this result explains how local adaptation is compatible with long-distance gene flow. From a conservation standpoint, estimating N_{ep} focuses on the diversity of the pollen pool as a critical element for conservation strategies because it reflects risk of genetic drift.

In our first application of the methods, we found that the mean *effective* pollen dispersal distance of the white oak, *Quercus alba*, in the heavily vegetated Ozark forest was restricted to about 17 m and the effective number of pollen donors was equivalent to 8 individuals (Smouse et al. 2001). Applying the same analysis to *Quercus lobata* in a savannah setting, we estimated that the mean effective pollen dispersal distance was almost four times greater than the distance for *Q. alba* (~65 m), and only about two effective pollen donors (Sork et al. 2002). We recognize that these initial estimates underestimated the effective number of pollen sources and dispersal distance, but the breakthrough was the introduction of a parameter that would allow comparison of the relative impacts of forest versus savanna settings. Initially, these paradoxically small estimates surprised some molecular ecologists because paternity studies demonstrate long-distance gene flow and many pollen sources. However, none of these findings are inconsistent with long-distance dispersal or multiple matings; they simply reflect the high frequency of the short distance movements in leptokurtic kernels and the fact that our measure is an 'effective number' rather than a total number of all pollen sources, as we later illustrate (Pluess et al. 2009).

Figure 3 Stages of plant reproduction of wind-pollinated and animal-dispersed *Quercus lobata* involved in gene flow. (A) Emergence of male flowers (catkins) in February–April. (B) Subsequent emergence of separate female flowers with conspicuous stigma for pollen receipt. (C) The California ground squirrel, *Otospermophilus beecheyi*, is one of the several rodent seed predators that occasionally disperse acorns locally. (D) Postdispersal seedling establishment. (E) Acorn removal from tree by acorn woodpecker, *Melanerpes formicivorus*, followed by transport to granary for storage. (F) Long-distance dispersal of acorn by western scrub jay, *Aphelocoma californica*, resulting in caching beneath soil. (Photos A–D by Andy Lentz ©; photos E–F by Marie Read ©).

An advantage of the TwoGENER framework was that it introduced a tool for testing hypotheses about gene movement that was previously lacking for conservation genetic studies. For example, in an investigation of the impact of landscape change on pollen dispersal in flowering dogwood, *Cornus florida*, we presented tests that demonstrated that the pollen pool structure of forest sites differed among three forest-cutting regimes and that forest thinning promoted pollen movement (Sork et al. 2005). Following these initial studies, the TwoGENER framework has evolved over time. We began by improving how we model the dispersal kernel (Austerlitz et al. 2004), and then we developed a new way to estimate effective pollen dispersal distance and effective number of pollen donors (Robledo-Arnuncio et al. 2006, 2007). We have emphasized effective number of donors because this parameter provides a useful assessment of the opportunity for evolutionary change, the risks of genetic drift, and other measures of interest to evolutionary and conservation biologists. This parameter, which is related to the effective population size of population genetics and akin to the species diversity index of ecology, can be used to test for statistical differences across years, landscapes, or species.

Seed movement at a landscape scale

Because the study of seed dispersal is hindered by the difficulties in tracking seeds when the parents cannot be located, we wanted to extend our pollen pool structure approach to seed dispersal. In parallel with our treatment of male parentage in pollen dispersal studies (Smouse and Robledo-Arnuncio 2005), we have articulated the consequences of seed movement for the effective number of maternal parents (N_{EM}) within local patches and for the local site (Grivet et al. 2005). Taking advantage of the fact that maternal tissue in seeds allows us to unambiguously identify the maternal genotypes (Godoy and Jordano 2001), we illustrated our efforts using the study system of acorn woodpeckers at the UC Santa Barbara Sedgwick Reserve (see Fig. 3E). Although woodpeckers are more likely to acts as seed consumers than dispersers, they transport acorns to 'granaries', which are storage sites often in the barks of trees, and can be considered to be seed pools (Grivet et al. 2005). Woodpeckers can fly kilometers on any given day, but our analyses showed that the effective number of maternal trees found in any single granary (each containing the combined collection from a single woodpecker family's efforts) is 2–3 local trees (Scofield et al. 2010; Scofield et al. 2011), as would be predicted by optimal foraging theory (Thompson et al. 2014).

In the acorn woodpecker study, maternal oak genotypes had been geolocated, so we knew not only who they were, but also where they were. However, even where the maternal locations remain opaque, maternal identification is still useful. A prime example can be illustrated by our study of seed dispersal by long-wattled umbrellabirds in an Ecuadorian tropical forest (Karubian et al. 2010). Umbrellabirds are characterized by a lek mating system and feed preferentially on palm fruits during the lekking season. The male birds forage in groups and fly at increasingly long distances to find fruits before returning to their lek where the seeds pass through their system. Even without locating the seed source, by genotyping seeds collected in seed traps in leks, this project was able to demonstrate that the birds visited many trees in the region and went long distances to do so.

Eventually, we translated our effective number of parent parameters into $\alpha-$ and $\gamma-$ diversity statistics so that evolutionary biologists and ecologists could talk the same language. In the first phase of development, we described genetic diversity within a pool or patch (α) and genetic diversity across patches within the area (γ). We presented an elaborated treatment of diversity statistics in an *American Naturalist* article (Scofield et al. 2012) where we added measures of divergence (δ) in seed sources among patches as an alternative to the β-parameter used in ecological diversity studies and introduced more statistical tests that permit comparison of different dispersal agents. This study illustrated the new measures by comparing seed transport by acorn woodpeckers with umbrellabirds, two species with very different social organizations and foraging patterns. The analyses showed that the territorial woodpeckers visited significantly fewer maternal source trees than did lekking males and that the seed pools of different woodpecker families were significantly more divergent from each other than were those created by males in different adjacent leks. These findings suggest that umbrellabirds foraged in many palm trees, and that different males visited the same trees before bringing palm fruits back to their respective lekking sites.

Whether we refer to the number of effective seed sources per patch as N_{EM} or α-diversity, it is valuable to be able to complement this information with direct documentation of seed movement. Taking advantage of the fact that first-year valley oak seedlings remain attached to the maternally inherited acorn tissue and that we have a well-mapped study site at the UC Santa Barbara Sedgwick Reserve, with every adult genotype identified, we were able to track seed dispersal into valley oak seedling patches away from an adult canopy. We observed that most of the seedlings came from nearby trees, but many were dispersed from a seed tree 1–4 kilometers away, putatively by scrub jays (see Fig. 3F). Although the local diversity of those seedling patches was small, those rare long-distance events demonstrate that seed dispersal can also contribute to long-distance gene flow in oaks. In fact, in a separate study of the Mexican oak *Quercus castanea*, we found evidence of high seed-mediated connectivity in a fragmented landscape (Herrera-Arroyo et al. 2013). Such studies have illustrated both our novel methods and specific insight about gene flow in bird-dispersed tree species.

Joint impacts of pollen and seed movement

Because gene flow in plant populations is sequential, involving pollen dispersal followed by seed dispersal, each can have a separate contribution to the genetic diversity of the seedling population. However, seed dispersal has a disproportionate impact on the local effective size of recruiting populations because it moves both maternal and paternal alleles, whereas pollen only moves the male alleles (Crawford 1984; Hamilton and Miller 2002; Garcia and Grivet 2011). Estimating the separate contributions of pollen and seeds to gene flow has been elusive because many studies compare the two processes using different sets of individuals or markers. To overcome that limitation, we compared haploid pollen and diploid seed dispersal using maternally inherited tissue in the acorn seed coats attached to the seedlings and biparentally inherited leaf tissue from the same seedlings (Grivet et al. 2009), which can be performed even for incomplete genotypes resulting from DNA degradation in buried acorns (Smouse et al. 2012). To assess the respective contributions of pollen and seed flow, we introduced a novel indirect assessment of the separate male and female gametic contributions to total effective parental size (N_e), based on kinship coefficients (Grivet et al. 2009; Robledo-Arnuncio et al. 2012). Here, we demonstrated that the effective number of parents of dispersed seedlings ($N_e = 6.7$) was greater than nondispersed seedlings ($N_e = 3.6$), largely because the nondispersed seedlings had only one maternal source. Thus, we provided important new evidence that seed dispersal has a substantial impact on the local neighborhood size of newly established seedlings despite the homogenizing contribution of the pollen donors (Grivet et al. 2009).

The next development in our contemporary gene flow studies was to assess the genetic diversity of the male and female haploid gametic contributions to the seedling pools (Sork et al. 2015). This study utilized the diversity indices introduced in Scofield et al. (2012) to compare allelic diversity resulting from pollen and seed movement rather than estimates of the effective number of pollen or seed sources. Allelic diversity is estimated at the patch level (α–diversity) and site level (γ-diversity). β-diversity (=γ/α) represents the turnover in diversity across patches and is an estimate of the effective number of patches for the site. When the allelic composition of patches is similar across the site, β-diversity is low, while divergent allelic composition among patches leads to high β-diversity. As before, we found that α–diversity resulting from seed dispersal was much less than that from pollen dispersal within patches. A key difference between this studies with the Grivet et al. (2009) study of effective parental size is not the switch to diversity parameters, but the direct comparison of the haploid genetic contribution of pollen sources with the haploid contribution of the seed sources. Here, we show that female gametes still contribute less to the γ-diversity of seedling populations than do male gametes, without the confounding inclusion of the diploid genotype.

Both studies demonstrate similar patterns of contemporary gene flow in oaks, but the diversity indices allow more dimensional analysis of the consequences of gene movement. First, despite long-distance seed dispersal in valley oak, female gametes do not contribute as much to seedling genetic diversity as do wind-dispersed male gametes. The difference is likely due to the fact that jays move acorns from a limited number of trees, even when they move them long distances. Second, the dispersal of male gametes maintains both the local diversity and connectedness of the regional gene pool. Third, the examination of γ-diversity adds an important element to the assessment of the genetic consequences of dispersal, because it provides an effective way of summarizing accumulated genetic diversity across seedling pools within an area.

An additional contribution of these papers is the introduction of statistical tests to assess differences in diversity indices between sites or conditions. In general, this approach provides a framework to test the impact of changes in ecological circumstances created by a loss of a pollinator or dispersal agent or landscape alteration. There are three possible cases and the use of statistical tests will determine whether the α-/γ-diversity parameters differ significantly within each: (i) γ-diversity does not change, but the α− versus γ-diversity does; (ii) the α- and β-diversity values are not affected, but the overall γ-diversity is changed; (iii) α-, β-, and γ-diversity measures are all affected. If statistical tests show that a disturbance affects cases (ii) or (iii), one can reliably conclude that the disruption is jeopardizing genetic diversity.

Geographic patterns of genetic variation

The geographical patterns of genetic variation in a species are the outcome of demographic history of population expansion and contraction, migration history, and the impact of natural selection both spatially and temporally. By putting genes on a map, it is possible to extract inference on these processes from the geographical patterns of allele frequencies. The analysis of spatial patterns is a form of discovery-based science that provides insight into the environmental context of the organism and its evolution. With my early ecological training, understanding the environmental context of my study systems is inherent in how I think about evolution. My research has always utilized a spatially explicit framework, and the landscape context can play such an important role in the distribution of genetic variation. This work has been done in collaboration with a geographer and plant ecologist, Frank Davis, a forest geneticist, Robert Westfall, postdocs, Cyril Dutech, and especially Delphine Grivet and Paul Gugger. Our studies fall into three types of approaches: spatial autocorrelation, geographic and climate associations of genetic gradients, and phylogeography.

Spatial autocorrelation

The simplest form of spatial analysis of genetic variation is to examine the scale of spatial autocorrelation of alleles, particularly appropriate for examination of the extent of gene dispersal. Our early work in this arena involved a study of seed dispersal by tropical avian frugivores (Loiselle et al. 1995), in which John Nason developed a clever statistical parameter that transforms the autocorrelation value across distance classes into a kinship estimate. In a separate study of fine-scale genetic structure of adult valley oak (Dutech et al. 2005), we found significant spatial structure in adults that was on a scale greater than contemporary gene flow, which provided evidence that the local populations have sufficient spatial structure to allow the evolution of local adaptation. In a more recent study (Sork et al. 2015), we were able to dissect the male and female gametic contributions to the fine-scale genetic structure of oak seedlings to demonstrate that the male gametes reduce structure while the female gametes produce it. Because it is not always possible to directly observe gene dispersal, spatial autocorrelation analyses provide valuable insight about the genetic consequences of propagule movement. Most plant populations show some degree of spatial genetic structure (SGS), suggesting this short distance gene flow creates opportunity for the survival of locally adapted genetic variation or that local adaptation might even promote SGS.

Geographic and environmental associations with genetic gradients

A key limitation of spatial autocorrelation analysis is that it does not incorporate spatially explicit environmental factors associated with geographic location. When a species is widely distributed, its evolutionary history can result in a geographic mosaic of genotypes that reflects the movement of genes, the impact of genetic bottlenecks, and the effects of local selection pressures. We can use that mosaic to infer evolutionary history and the processes that have shaped the geographic patterns (Avise 2000) or for conservation purposes (Moritz 1995). For example, Richard Rayburn, a former resource manager for California State Parks, and Craig Moritz proposed a conservation strategy of prioritizing the protection of regions that are 'evolutionary hotspots' (Rayburn and Moritz 2006). To identify the regions of conservation importance for valley oak, we used a multivariate approach that we had previously used to analyze the environmental correlations with geographic patterns of genetic variation in Missouri Ozark forests (Gram and Sork 1999, 2001). Multivariate models, such as canonical correspondence analysis (CCA), are useful because they detect small genetic differences across loci that reveal cumulative patterns of differentiation that are not apparent with analyses of one locus at a time (Conkle and Westfall 1984; Kremer and Zanetto 1997) and provide signatures of genetic differences that can be correlated with environmental gradients or associated with geographic regions. In one study, we used the multilocus genotypes of valley oak as indicators of unique genetic composition and demonstrated how one could design a reserve network to maximize the preservation of unique genetic diversity for the species (Sork et al. 2009). In another study, we analyzed the geographic patterns of multilocus chloroplast and nuclear microsatellite genetic markers of individuals distributed widely across the valley oak range. This allowed us to identify the areas with

distinctive histories and genetic composition that should be given priority in reserve network design (Grivet et al. 2008). Sadly, two of the regions we identified lay in areas of high human development and little protection where habitat loss has already been severe. Thus, this conservation genetic analysis revealed that without a careful preservation plan, valuable evolutionary information will be lost for valley oak and that efforts to preserve populations in this area will be particularly important.

Given concerns about the impact of rapid climate change on long-lived species, the association of genetic variation with climate can provide a map of regional differences in genetic variation that could indicate how regions might respond differently to changes in climate. This topic is particularly relevant for trees that were established under climate conditions several hundred years ago. In a species-wide analysis of genetic variation in valley oak, we found the geographic structure of multivariate genotypes is highly correlated with climate variables (Sork et al. 2010), which was initially surprising given that putatively neutral markers were used. However, the patterns make sense if climate shapes evolutionary processes, such as migration or population expansion and contraction, which in turn affect genome-wide patterns of variation. This study employed climate niche modeling (Maximum Entropy, MAXENT) based on downscaled historical (1971–2000) and future (2070–2100) climate grids to illustrate how geographic genetic variation could interact with regional patterns of 21st century climate change. Our models predicted that future climate niches for valley oak will shift in location due to changes in climate. In some regions, the climate niches of local populations might show local displacement of only a few meters to hundreds of meters, while in others, they might shift 60–100 kilometers. Thus, for several populations, it is unlikely that the movement of genes through pollen dispersal or colonization events through seed dispersal will be great enough to allow the spread of valley oak to more suitable sites. This work predicts that the late 21st century climate conditions could lead to regional extinctions for many valley oak populations unless they have preexisting tolerance for warmer, drier, and more seasonal climate environments. In fact, for some tree species, interventions, such as assisted migration, may be necessary (Aitken and Whitlock 2013).

Phylogeography

Phylogeography is the study of geographic patterns of genetic lineages with the goal of understanding how the physical landscape and climate cycles shape the evolutionary history of species, the emergence of new species, and the extinction of others (Avise 2000). Through my team's phylogeographic studies of valley oak and other species in California (Grivet et al. 2006; Sork et al. 2010; Gugger et al. 2013; Ortego et al. 2014; Sork and Werth 2014; Werth and Sork 2014), we have shown that the lack of recent glaciation in much of California and the overwhelming majority of the valley oak species range has allowed the accumulation of high genetic diversity and long-term persistence of local populations. For example, we learned that valley oak has much greater chloroplast diversity than European oaks, whose range has experienced recent glaciation (Grivet et al. 2006). In Gugger et al. (2013), we found genetic divergence between eastern and western subpopulations of valley oak that took place approximately 150 000 years ago during the last interglacial warming. This estimate means that contemporary oak populations are at least several hundred thousand years old, which would explain why they have more chloroplast haplotype diversity than European oaks. We have also shown that these haplotypes are highly localized (Grivet et al. 2006; Gugger et al. 2013), indicating that colonization via seeds has remained local over extended periods. In fact, Gugger et al. (2013) found that the current distribution of genetic variation retains a signature of climate association with the time of the Last Glacial Maximum, less than 20 000 years ago. Whether that residual pattern represents adaptation remains an open question. Overall, these findings provide valuable background on the evolutionary history of valley oak. In addition, our work on valley oak and other California species contributes to the growing body of literature documenting California as a hotspot of biodiversity (Raven and Axelrod 1978; Calsbeek et al. 2003; Rissler et al. 2006; Lancaster and Kay 2013).

Landscape genomics of adaptive genetic variation

For populations of valley oak, a tree found in the rainy northwestern part of its range is likely to have a different genetic composition than one from the warm, dry south. This pattern could be true due to isolation by distance factors alone. However, I am especially interested in the question: To what extent are trees locally adapted and, in particular, at what spatial scale is valley oak adapted to its local climate environment? Given that the generation time of a valley oak is on the order of 100 years, it seems unlikely that local populations will evolved to keep pace with the current rapid rate of climate change. However, some populations may have preexisting tolerance to hotter, drier climates with more seasonality, while other populations may not. Thus, identifying geographic patterns of adaptive genetic variation to prioritize regions for conservation is a topic of concern to many forest geneticists (Neale 2007; Aitken et al. 2008; Alberto et al. 2013; Sork et al. 2013; Kremer et al. 2014).

With the emergence of next-generation sequencing for nonmodel systems, our research group is characterizing geographic patterns of adaptive genetic variation associated with response to climate. Recently, I have participated in working groups that have emphasized genomic research as a high priority for understanding the evolutionary processes in natural populations (Andrew et al. 2013) and for developing new genomic tools for managing forest tree populations (Sork et al. 2013), especially oaks (Petit et al. 2013). This new phase of my research over the last five years has benefitted from discussions with forest geneticists Sally Aitken, David Neal, and from collaborations with Andrew Eckert, Matteo Pellegrini, Alex Platt, Jessica Wright and in particular, my postdoc Paul Gugger. Our joint long-term goal is to understand the basic evolutionary biology of local adaption in tree populations, particularly in response to climate, and the extent to which the genetic composition of tree populations will shape their response to climate. I have multiple projects underway using several genomic approaches, but here, I will present two initial studies. In the first, we utilized DNA sequence data from specific genes with functions related to climate response and applied a landscape genomic approach to associate genetic gradients with climate gradients as evidence for selection (Sork et al. 2016). The second study, we compared genetic and epigenetic differences across populations to see whether methylated DNA sequences could be important in shaping phenotypic variation in response to climate (Platt et al. 2015).

Climate-associated candidate genes

In designing my first study of adaptive genetic variation, I was concerned that gene flow in oaks was so extensive that the signal of genes under selection would be weak. So, we identified 40 genes within four functional categories for response to climate that could be potential candidates under selection: bud burst and flower timing genes, growth genes, osmotic stress genes, and temperature stress genes (Sork et al. 2016). Then, we sequenced individuals from thirteen localities throughout the species range. Based on 195 single-nucleotide polymorphisms (SNPs) across the 40 genes, we identified outlier SNPs likely to be under selection through several approaches. First, we found that the top 5% of F_{ST} estimates ranged from 0.25 to 0.68, while the mean $F_{ST} = 0.03$. Second, in correlation analyses of the 195 SNP frequencies with climate gradients, we found three SNPs within bud burst and flowering genes and two SNPs within temperature stress genes that were significantly associated with mean annual precipitation (Fig. 4). Finally, given that adaptive variation is often polygenic and climate niche is multivariate, we used a canonical correlation multivariate model to examine this highly complex story for

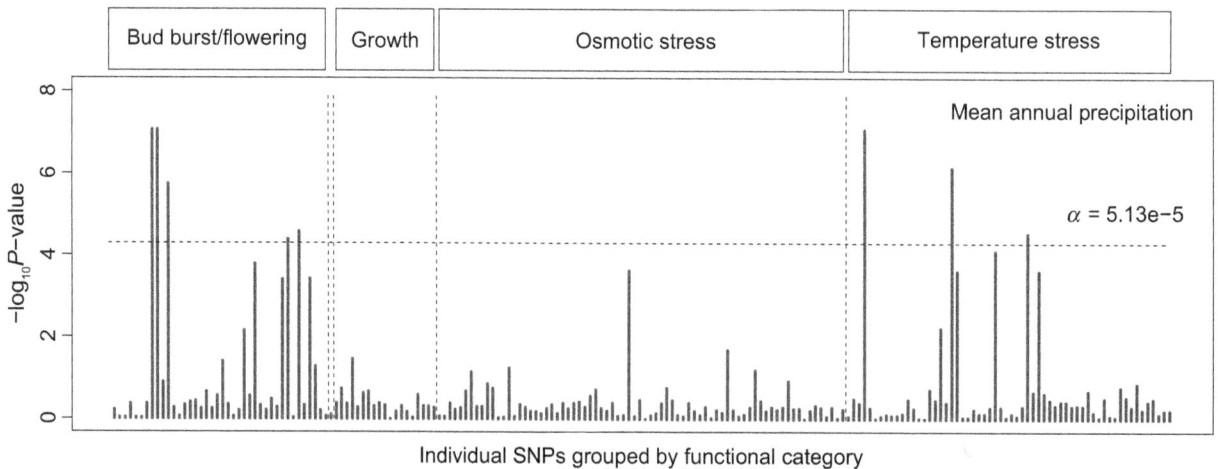

Figure 4 Manhattan plot showing environmental association analysis of 195 *Q. lobata* SNPs found within four functional gene categories versus mean annual precipitation. Outlier SNPs with significant correlations with climate variables are detected by plotting the negative log probability of SNP–climate association. We found significant correlations with mean annual precipitation only. The dashed line indicates the significance threshold after correcting for multiple tests. (Modified from Sork et al. 2016. American Journal of Botany).

each of the four functional sets of genes. Many of the SNPs identified through the single-locus models above were again outliers in these multivariate models, but we also found additional SNPs of smaller effects in growth genes that may be under selection. These analyses revealed that 10 of the 40 candidate genes showed evidence of spatially divergent selection (Sork et al. 2016), an exciting finding because it demonstrates the potential of landscape genomic studies to detect adaptive genetic variation in a nonmodel systems.

Future landscape genomic work should include a random and much larger sample of SNPs that will allow for a good estimate of background genetic structure that controls for the confounding effects of demographic history (Coop et al. 2010; Lotterhos and Whitlock 2014, 2015). Ideally, landscape genomic studies should include as many individuals and localities as possible to increase the power to detect climate associations. The use of functional genes increases the efficiency of finding outliers, but it also presents an ascertainment bias by not testing all potential genes. Alternatively, one can use a genome-wide method to randomly subsample the genome, such as RADseq (Baird et al. 2008) or GBS (Elshire et al. 2011) that could find both neutral and adaptive variants, but this methods utilizes only a small percent of the genome (often <5%) and could miss more genes under selection. Some studies include both candidate genes to see whether they show spatially divergent selection and randomly selected SNPs to describe demographic history (Keller et al. 2012; De Kort et al. 2014). As the costs of sequencing decline, whole-genome sequencing will be an option for genome-wide association studies (GWAS) designed to identify the important

environmentally associated candidate genes that might be responsible for tree performance and survival.

Epigenetic differentiation in valley oak

Epigenetic variation has the potential to shape phenotypic plasticity and adaptation to the environment, each of which could be particularly important for the persistence of long-lived organisms such as trees (Franks and Hoffmann 2012; Braeutigam et al. 2013). This potential is illustrated by studies of natural populations of *Arabidopsis thaliana* that reveal patterns of DNA methylation which vary across environments and may provide a mechanism to enhance the adaptation to the local environment (Schmitz 2014). Evidence from tree populations illustrates one epigenetic mechanism, DNA methylation of cytosine residues, which has all the ingredients to be involved in evolution by natural selection in plants. It is sometimes heritable across many generations (Becker et al. 2011; Schmitz et al. 2011; Zhang et al. 2013), among individuals, and among populations (Herrera and Bazaga 2010, 2011; Schmitz et al. 2013). Mounting evidence suggests DNA methylation may also underlie trait variation by affecting gene expression when found in regulatory or genic regions. For example, flowering time in *Arabidopsis* (Soppe et al. 2000), floral symmetry in *Linaria* (Cubas et al. 1999), and fruit ripening in *Solanum* (Manning et al. 2006) are partially under epigenetic control by cytosine methylation.

To test whether epigenetic variation differs across populations of valley oak, we sampled individuals from three southern populations with divergent elevations, levels of rainfall, and seasonality of rainfall (Platt et al. 2015). We then sequenced bisulfite-treated reduced-representation 'genomes', generating 100 base pair sequence fragments from randomly selected loci across the genome and converting unmethylated cytosines to thymines. Cytosine methylation occurs in three DNA sequence contexts, CG, CHG, and CHH (where H is non-G), each of which has different regulatory pathways (Law and Jacobsen 2010) and thus potentially differing roles in adaptation. We found that CG sites, which are commonly found in genic regions (Cokus et al. 2008; Takuno and Gaut 2013), were the most highly methylated and, on average, showed much more differentiation across populations than either CHG or CHH sequences. To assess whether the SNP (single-nucleotide polymorphism) and SMP (single methylated site polymorphism) variation differs across populations, we calculated population differentiation, F_{ST}, at each of the polymorphic loci as a measure of the extent to which each polymorphism is distributed nonrandomly across populations. Typically, tree population genetic studies using putatively neutral, nuclear microsatellite markers report F_{ST} values in the range of 0.08–0.20 (Loveless and Hamrick 1984). Similarly, in a study of range-wide populations of valley oak, we found $F_{ST} = 0.12$ for microsatellite loci (Grivet et al. 2008). In contrast, Platt et al. (2015) found that mean epigenetic differentiation among three populations at variably methylated CG sites (CG-SMPs) was higher than that based on SNPs ($F_{ST} = 0.28$ versus $F_{ST} = 0.19$, respectively) and that the distribution of CG-SMP F_{ST} values included more values at the upper end of the distribution (Fig. 5). The findings of this study provide evidence that divergent selection and local adaptation are operating on portions of the valley oak genome marked by CG-SMPs. While we do not yet know the mechanisms creating such high levels of epigenetic differentiation among populations, recent work on Norway spruce shows that temperature during zygote formation appears to create an epigenetic 'memory' in the progeny that regulates bud phenology and cold acclimation later on in life (Yakovlev et al. 2011). It will be necessary to

combine studies that associate epigenetic variation with environmental and phenotypic variation with studies that look at the epigenetic effects on those phenotypes, if we are to assess the extent to which epigenetic processes play a role in either phenotypic plasticity or local adaptation. The role of DNA methylation in shaping phenotypes that enhance tolerance to local environments is a long way from being understood, but this study is a good start to assessing the extent of epigenetic differentiation in natural tree populations.

Summary and conclusions

1 The interaction of pollen and seed dispersal vectors with landscape features shapes contemporary gene flow. My research has shown that pollen movement in oaks is responsible for maintaining connectivity among populations, yet local pollination is sufficiently frequent that it is possible to create fine-scale genetic structure that could facilitate local adaptation. Seed dispersal is more likely to be restricted, so it accounts for much more of the fine-scale genetic structure than does local pollen movement. Dispersal agents, such as western scrub jays, facilitate long-distance acorn movement that would allow colonization of new sites, but because they visit few trees, the genetic diversity of patchy seedling populations is reduced during this phase from that introduced through pollen dispersal. The generality that has emerged from this work and that of others for many tree species is that localized gene movement of pollen and seeds will create sufficient genetic structure to allow the evolution of local adaptation, while long-distance gene flow will maintain high levels of genetic diversity and spread favorable alleles over large spatial scales.

2 The analysis of contemporary gene flow in plants provides valuable information on the extent to which human disturbances are jeopardizing future genetic diversity by isolating populations. However, conservation scientists need to test hypotheses about the impact of human disturbance. My collaborators and I have also introduced statistical tests for landscape scale questions to determine whether landscape change through fragmentation isolates or enhances gene movement or whether dispersal vectors differ in their impact on the genetic structure of populations and thus their loss would jeopardize genetic diversity. Thus, our studies are part of a growing body of literature in landscape genetics exploring whether plant populations are threatened by human activities.

3 The geographic patterns of genetic variation reflect the evolutionary history of a species that is the foundation for current responses to environmental conditions. Valley oak populations are likely to be several hundred

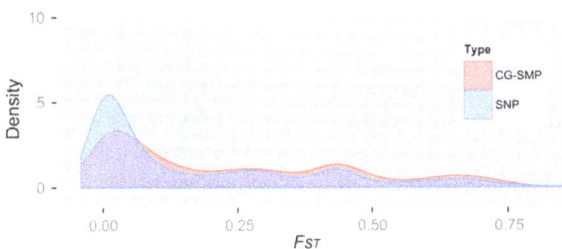

Figure 5 Densities of F_{ST} values for CG-single methylated polymorphisms (CG-SMP) and single-nucleotide polymorphism (SNP) across three populations of *Quercus lobata* are shown for the range of F_{ST} values. (Modified from Platt et al. 2015. Molecular Ecology).

Box 1: Up close and personal

A passion for science and four brothers explain a lot about who I am. By the age of ten, I had decided to be either a scientist or a math professor at UCLA—a scientist because I loved to do experiments and a math professor because I enjoyed doing problems in my head. The decision between science and math came during middle school when I was told girls were not good at math. If that was true, I thought, I had better focus on science. And as for the brothers—trying to keep up with three older and one younger and wanting to do whatever they could, motivated me to work harder. It helped that I had an older brother who thought I could do anything as well or better than anyone else. Everyone should have that kind of older brother. All my effort and his faith never got me onto the Little League baseball team I longed to join, but it did teach me perseverance and resiliency. At least having a skewed sex ratio in my family preadapted me for the world of science.

Throughout my undergrad years at UC Irvine, I had the pleasure of working in research labs on independent projects. After two academic quarters in field courses (Field Biology Quarter and Marine Biology Quarter), I realized my passion for understanding the natural world and switched my focus from microbiology and genetics to ecology and evolution. Under the mentorship of Professor Dick MacMillan, I had the world's most fun conducting my honor's project studying desert rodents and plants and camping a few times a month to collect data in the remote areas of southern California. The other great influence during my undergraduate years was my volunteer work with young students in the barrios of Santa Ana. The physical conditions of the school and the poor quality of their education made me realize a lack of a level playing field in society. My path to graduate school had a few bumps as several professors advised that academia was not a place for a woman and a potential faculty mentor unabashedly informed me that he did not accept women as graduate students. Not to be deterred, I continued onto the University of Michigan with great excitement to do science and little confidence that I belonged.

Graduate school was one of the best, but most challenging, periods of my life. I met some amazingly supportive professors and graduate students, but I also found the departmental atmosphere to be competitive and aggressive. As a teaching assistant, the female undergrads would seek me out as a role model—a role that I felt so unsuited for. With the support of professors such as Deborah Rabinowitz and John Vandermeer, and several fellow (or should I say sister) graduate students including Carol Augspurger, Ann Sakai, Elizabeth Lacey, Kay Dewey, and others, I developed a critical support network. Key to my success was participating as a graduate teaching assistant in the Women's Studies Program where the brilliant feminist graduate students and professors inspired me. There, I developed a critical analysis of why women felt out of place in science (and elsewhere outside the home) that gave me greater conviction that I was entitled to an academic career. Perhaps the reason I continued in academia is that I loved, and still do love, research and discovery of knowledge. My experiences during graduate school solidified my resolve that one of my tasks as a professor would be to ensure that the door was open and the playing field was level for all students (women, African Americans, Latinos, first-generation college students, and others who are currently in the minority in our classrooms).

As a faculty member at University of Missouri St. Louis, I had wonderful colleagues and the rewards of teaching an urban commuting student population that was largely first generation and very diverse in racial/ethnic composition and economic status. During that time, I learned to balance teaching, research, and service. Service included helping to establish the PhD program in Biology and being the founding director of the International Center for Tropical Ecology at UM-St. Louis in partnership with the Missouri Botanical Garden. Through research in the tropics and Missouri Ozark forests, I became aware of the fine distinction between basic and applied evolutionary biology. Through pivotal research leaves at the University of Chicago, University of California Berkeley, and the National Center for Ecological Analysis and Synthesis at UC Santa Barbara, I advanced my understanding of the evolutionary theory and statistical methods needed to test my ideas empirically. My thinking evolved from that of an ecologist who loves nature, to an evolutionary ecologist who appreciates population genetics, and then to an evolutionary ecologist who integrates these scientific disciplines with their applications to conservation and resource management.

After achieving tenure, I adopted two daughters at birth, three years apart, and began one of the most fulfilling parts of my life. I cannot say how many times I have been asked how I balance my career and being a mother. One answer is that it is not easy. It helps when you have—as my close friends tell me—almost unbounded energy and drive. The other answer is that when you love what you do nothing seems undoable. You just find a way to do it. Actually, I cannot imagine a career that is more family-friendly than being a university professor so it saddens me when female graduate students or postdocs tell me that they do not want to stay in academia because the work–life balance is too skewed. I suspect that the real problems are a lack of societal support for child care and the chilly climate for women and underrepresented minorities, where the majority group is often treated with more respect, opportunities, and advancement (there are data that show this bias but I won't go into them here). Through personal observation, I became aware that men with children and women with children were treated differently. For women, having children was interpreted as a lack of dedication to science while for men it was never noticed unless they were applauded when they missed an afternoon meeting to carpool their children. I hope that it is easier now for everyone to balance career and family (for example, there is now daycare at conferences), but I am sensitive that the climate for women and for faculty of color is not always welcoming. It is clear that those of us who have succeeded need to be actively part of the solution.

During the last fifteen years at UCLA, I have had dual roles—one as the enthusiastic researcher and professor with joy in pursuing science and teaching, and the other as an administrative leader with the responsibility to create an environment of excellence, thousand years old with high genetic diversity within populations and high gene flow among populations. While nuclear markers indicate extensive connectivity across the species range, chloroplast haplotypes indicate that colonization is local. This restricted dispersal would allow evolution of local adaptation. In addition, the species-wide spatial pattern of genetic variation reflects an influence of climate through either gene flow or selection against migrants into populations. Regardless, this pattern explains the prevalence of genetic and phenotypic differences among tree populations shown here and which is likely to be true for many forest tree species.

4 The emergence of next-generation sequencing tools now allows the direct analysis of candidate genes and epigenetic impacts that could shape phenotypic response to the environment. Our work demonstrates extensive genetic differentiation across populations for certain genes and a correlation between genetic and climate gradients that provides initial evidence of local adaptation. However, future work needs to also control for background genetic structure that may be confounded with environmental associations. Our evidence of significant epigenetic differentiation at methylated sites further provokes the question of the extent to which epigenetic processes might shape phenotypic response to local environments.

5 We have also demonstrated that the homogenizing influences of long-distance gene flow are not sufficient to 'swamp out' the impact of selection on tree populations. Instead, genes show a strong impact of selection likely due to both the strength of selection and the presence of fine-scale genetic structure.

6 Valley oak populations have already experienced a reduction in regional coverage due to habitat transformation and problems with regeneration. Restoration efforts will need to consider future climates as they consider the impact of climate change on the future environments of their restoration sites. It is my goal to develop and 'ground-truth' landscape genomic approaches that can be used to inform future conservation strategies when long-term common garden and reciprocal transplant experiments are not feasible. This information will not only help to preserve this and other California oaks, but also provide a case study for other tree species of conservation concern.

Future directions

My research efforts have taken place in the larger context of landscape genetics and genomics, plant evolutionary biology, forest genetics, and conservation science. My studies are built around the valley oak study system is poised to complement other exciting forest genetic research to address the emerging questions about the ability of long-lived species to face and survive anthropogenic challenges. Maintaining healthy tree populations is not only important to their survival, but also important to sustaining the ecological processes of the ecosystems they define (Hughes et al. 2008; Schoener 2011).

1 *Gene movement on the landscape.* Unlike studies of animal landscape genetics that have focused on the impact of landscape features on migration, landscape genetic studies of plant populations initially focused more on geographical and environmental associations of genetic variation (Manel et al. 2003, 2010; Storfer et al. 2007), than on gene movement (with some notable exceptions, such as McRae 2006; Dyer et al. 2012). With the availability of genomic tools for nonmodel systems, the next frontier is the multidisciplinary analysis of gene movement to detect the adaptive genetic variation using large numbers of SNPs that permit the simultaneous analysis of historical migration and demographic events (Sork and Waits 2010; Manel and Holderegger 2013; Sork et al. 2013). Using species-wide samples of valley oak, we

currently have a landscape genomic study underway based on GBS to assess the extent to which neutral versus selective processes are shaping the geographic structure of genetic variation.

2 *Evolution of local adaptation.* To understand how local adaptation evolves, we need first to identify candidate genes underlying phenotypic traits under selection and the fitness consequences of those traits (Mitchell-Olds et al. 2007; Stapley et al. 2010; Anderson et al. 2011; Strasburg et al. 2012). Landscape genomic analysis using F_{ST} outlier tests or environmental association analysis is one way to identify the candidate genes for local selection, when controlling for demographic history (for example, see Eckert et al. 2010; Holliday et al. 2010; Keller et al. 2012). For forest trees, the GWAS approach can be a useful way to detect candidate genes associated with phenotypes (Nichols et al. 2010). However, additional work in evolutionary and quantitative genetics may be needed to document the functions of those genes. Alternatively, experimental studies examining gene expression under different environmental conditions could be used to identify the number of genes and the gene networks responsible for phenotypes that might be under selection (Alberto et al. 2013; De Kort et al. 2014). Eventually, experiments will be necessary to test the adaptive consequences of individual alleles detected through these approaches (Barrett and Hoekstra 2011). In short, the breakthroughs in understanding local adaptation will come from an integrative approach. In our research program, we plan to integrate findings from a variety of approaches: environmental associational analysis of landscape genomic studies, studies of phenotypes measured a recently established long-term provenance study for valley oak (Delfino-Mix et al. 2015), studies of gene expression and physiological response in drought-treated oak seedlings (Peñaloza Ramírez, Gugger, Wright, Sork, unpublished data; Steele, Sweet, Davis, Sork, unpublished data) that utilize our reference transcriptome (Cokus et al. 2015), and the production of a high quality gene sequence of valley oak to facilitate our ability to identify the climate-associated genes and their genetic architecture (Sorel, Langley, Pellegrini, Sork, Salzberg, work in progress).

3 *Evolutionary and ecological response to climate change.* How plants are going to respond to rapid climate change is a critical issue (Parmesan 2006; Anderson et al. 2012). For long-lived trees that became established under very different climate conditions and have a generation time that hampers rapid evolution, this concern is particularly salient (Savolainen et al. 2007; Aitken et al. 2008; Kremer et al. 2012; Alberto et al. 2013). Are most tree spe-

cies resilient to a broad range of climate conditions (Hamrick 2004)? Is local adaptation sufficiently strong that we will need to assist migration across populations (Aitken and Whitlock 2013)? Will plant response require the evolution of new locally adapted phenotypes with existing or new genetic variation (Stapley et al. 2010)? Does epigenetic variation offer a mechanism for trees to respond ecologically to climate (Braeutigam et al. 2013)? My long-term goal is to use valley oak as a case study of how a long-lived tree species can respond to climate change. Besides the information generated for this particular species, we hope to assess the extent to which we can map specific genes and identify regional populations of concern to design management strategies for the maintenance of healthy forests.

In short, the integration of 21st century genomic approaches, quantitative and population genetics, and recently developed geographic methods of modeling environmental data provides exciting and unprecedented opportunities to understand the evolutionary and ecological processes of plants and important applications.

Acknowledgements

Major funding sources include National Science Foundation (NSF-DEB-0514956, NSF-DEB- 0089445), National Center for Ecological Analysis and Synthesis, Missouri Department of Conservation, USDA Forest Service, University of Missouri St. Louis, and UCLA. Field work was performed in part at the Missouri Ozark Forest Ecosystem Project study plots, the University of California Natural Reserve System Sedgwick Reserve, and other UC-NRS Reserves. I thank S. Rose, A. Sakai, and S. Weller for comments on this manuscript and their long-term support. S. Steele and K. Beckley helped with manuscript preparation. The alert, insightful, and patient referees and editor deserve my gratitude for their comments on an earlier draft. I express thanks to all the colleagues, postdocs, graduate students, field assistants, and friends who have enhanced my science and are too numerous to list here. I particularly recognize the scientific influence of the following people: Augspurger, Hamrick, Janzen, Raven, Schaal, Schemske, Slatkin, Templeton, and Vandermeer. In addition, I thank collaborators highlighted in this essay: Austerlitz, F. Davis, Dutech, Dyer, Eckert, Grivet, Gugger, Karubian, Nason, Pellegrini, Platt, Pluess, Robledo-Arnuncio, Scofield, Smouse, Werth, Westfall, and J. Wright. Finally, I acknowledge P. Smouse – delightful, stimulating, and responsive collaborator and friend. This essay is dedicated to my family and friends (see Box 1) and my supportive PhD advisor, Deborah Rabinowitz, who passed away many years ago all too young.

Literature cited

Adams, W. T., A. R. Griffin, and G. F. Moran 1992. Using paternity analysis to measure effective pollen dispersal in plant populations. American Naturalist 140:762–780.

Aguilar, R., M. Quesada, L. Ashworth, Y. Herrerias-Diego, and J. Lobo 2008. Genetic consequences of habitat fragmentation in plant populations: susceptible signals in plant traits and methodological approaches. Molecular Ecology 17:5177–5188.

Aitken, S. N., and M. C. Whitlock 2013. Assisted gene flow to facilitate local adaptation to climate change. Annual Review of Ecology, Evolution, and Systematics 44:367–388.

Aitken, S. N., S. Yeaman, J. A. Holliday, T. L. Wang, and S. Curtis-McLane 2008. Adaptation, migration or extirpation: climate change outcomes for tree populations. Evolutionary Applications 1:95–111.

Alberto, F. J., S. N. Aitken, R. Alia, S. C. Gonzalez-Martinez, H. Hanninen, A. Kremer, F. Lefevre et al. 2013. Potential for evolutionary responses to climate change: evidence from tree populations. Global Change Biology 19:1645–1661.

Aldrich, P. R., and J. L. Hamrick 1998. Reproductive dominance of pasture trees in fragmented tropical forest mosaic. Science 281:103–105.

Anderson, J. T., J. H. Willis, and T. Mitchell-Olds 2011. Evolutionary genetics of plant adaptation. Trends in Genetics 27:258–266.

Anderson, J. T., A. M. Panetta, and T. Mitchell-Olds 2012. Evolutionary and ecological responses to anthropogenic climate change. Plant Physiology 160:1728–1740.

Andrew, R. L., L. Bernatchez, A. Bonin, C. A. Buerkle, B. C. Carstens, B. C. Emerson, D. Garant et al. 2013. A road map for molecular ecology. Molecular Ecology 22:2605–2626.

Antonovics, J. 1971. Effects of a heterogeneous environment on genetics of natural populations. American Scientist 59:593–599.

Antonovics, J., and A. D. Bradshaw 1970. Evolution in closely adjacent plant populations. 8. Clinal patterns at a mine boundary. Heredity 25:349–362.

Apsit, V. J., V. L. Sork, and R. J. Dyer. 2002. Patterns of mating in an insect-pollinated tree species in the Missouri Ozark Forest Ecosystem Project (MOFEP). In S. R. Shifley, and J. M. Kabrick, eds. Proceedings of the Second Missouri Ozark Forest Ecosystem Symposium: Post-Treatment Results of the Landscape Experiment; 2000 October 17–18; St. Louis, MO. Gen. Tech. Rep. NC-227. U.S. Department of Agriculture, Forest Service, North Central Forest Experiment Station, St. Paul, MN.

Austerlitz, F., C. W. Dick, C. Dutech, E. K. Klein, S. Oddou-Muratorio, P. E. Smouse, and V. L. Sork 2004. Using genetic markers to estimate the pollen dispersal curve. Molecular Ecology 13:937–954.

Avise, J. C. 2000. Phylogeography: The History and Formation of Species. Harvard University Press, Cambridge, MA.

Baird, N. A., P. D. Etter, T. S. Atwood, M. C. Currey, A. L. Shiver, Z. A. Lewis, E. U. Selker et al. 2008. Rapid SNP discovery and genetic mapping using sequenced RAD markers. PLoS ONE 3:e3376.

Barrett, R. D. H., and H. E. Hoekstra 2011. Molecular spandrels: tests of adaptation at the genetic level. Nature Reviews Genetics 12:767–780.

Becker, C., J. Hagmann, J. Muller, D. Koenig, O. Stegle, K. Borgwardt, and D. Weigel 2011. Spontaneous epigenetic variation in the Arabidopsis thaliana methylome. Nature 480:245–249.

Bradshaw, A. D. 1972. Some of the evolutionary consequences of being a plant. Evolutionary Biology 5:25–47.

Braeutigam, K., K. J. Vining, C. Lafon-Placette, C. G. Fossdal, M. Mirouze, J. Gutierrez Marcos, S. Fluch et al. 2013. Epigenetic regulation of adaptive responses of forest tree species to the environment. Ecology and Evolution 3:399–415.

Brookshire, B., and C. Hauser. 1993. The Missouri Forest Ecosystem Project. In A. R. Gillespie, G. R. Parker, P. E. Pope, and G. Rink, eds. Proceedings of the 9th Central Hardwood Forest Conference. USDA, Forest Service, North Central Forest Experiment Station, St. Paul, MN. General Technical Report NC-161.

Brookshire, B. L., R. Jensen, and D. C. Dey. 1997. The Missouri Ozark Forest Ecosystem Project: past, present and future. In B. L. Brookshire, and S. R. Shifley, eds. Proceedings of the Missouri Ozark Forest Ecosystem Project symposium: an experimental approach to landscape research; 1997 June 3–5; St. Louis, MO General Technical Report NC-193. USDA Forest Service, North Central Forest Experimental Station, St. Paul, MN.

Burczyk, J., W. T. Adams, G. F. Moran, and A. R. Griffin 2002. Complex patterns of mating revealed in a Eucalyptus regnans seed orchard using allozyme markers and the neighbourhood model. Molecular Ecology 11:2379–2391.

Calsbeek, R., J. N. Thompson, and J. E. Richardson 2003. Patterns of molecular evolution and diversification in a biodiversity hotspot: the California Floristic Province. Molecular Ecology 12:1021–1029.

Cayan, D. R., A. L. Luers, G. Franco, M. Hanemann, B. Croes, and E. Vine 2008. Overview of the California climate change scenarios project. Climatic Change 87:S1–S6.

Clausen, J., D. D. Keck, and W. M. Hiesey 1947. Heredity of geographically and ecologically isolated races. American Naturalist 81:114–133.

Cokus, S. J., S. Feng, X. Zhang, Z. Chen, B. Merriman, C. D. Haudenschild, S. Pradhan et al. 2008. Shotgun bisulphite sequencing of the Arabidopsis genome reveals DNA methylation patterning. Nature 452:215–219.

Cokus, S. J., P. F. Gugger, and V. L. Sork. 2015. Evolutionary insights from de novo transcriptome assembly and SNP 2 discovery in California white oaks. BMC Genomics 6:552.

Conkle, M. T., and R. D. Westfall. 1984. Evaluating breeding zones for ponderosa pine in Califorrnia. In Progeny Testing: Proceedings of Servicewide Genetics Workshop, Charleston, SC, December 5-9, 1983. USDA Forest Service, Washington, DC.

Coop, G., D. Witonsky, A. Di Rienzo, and J. K. Pritchard 2010. Using environmental correlations to identify loci underlying local adaptation. Genetics 185:1411–1423.

Crawford, T. J. 1984. The estimation of neighborhood parameters for plant populations. Heredity 52:273–283.

Cubas, P., C. Vincent, and E. Coen 1999. An epigenetic mutation responsible for natural variation in floral symmetry. Nature 401:157–161.

Darwin, C. 1859. On the Origin of Species by Means of Natural Selection. J. Murray, London.

Davey, J. W., P. A. Hohenlohe, P. D. Etter, J. Q. Boone, J. M. Catchen, and M. L. Blaxter 2011. Genome-wide genetic marker discovery and genotyping using next-generation sequencing. Nature Reviews Genetics 12:499–510.

De Kort, H., K. Vandepitte, H. H. Bruun, D. Closset-Kopp, O. Honnay, and J. Mergeay 2014. Landscape genomics and a common garden trial reveal adaptive differentiation to temperature across Europe in the tree species Alnus glutinosa. Molecular Ecology 23:4709–4721.

Delfino-Mix, A., J. W. Wright, P. F. Gugger, C. Liang, and V. L. Sork. 2015. Establishing a range-wide provenance test in valley oak (Quercus lobata Née) at two California sites. In R. B. Standiford, and K. Purcell eds. Proceedings of the Seventh California Oak Symposium: Managing Oak Woodlands in a Dynamic World. November 3–6, 2014, Visalia, CA, USDA Forest Service General Technical Report PSW–GTR–XX. (in press).

Devlin, B., K. Roeder, and N. C. Ellstrand 1988. Fractional paternity assignment: theoretical development and comparison to other methods. Theoretical and Applied Genetics 76:369–380.

Dick, C. W., G. Etchelecu, and F. Austerlitz 2003. Pollen dispersal of tropical trees (Dinizia excelsa: Fabaceae) by native insects and African honeybees in pristine and fragmented Amazonian rainforest. Molecular Ecology 12:753–764.

Dick, C. W., O. J. Hardy, F. A. Jones, and R. J. Petit 2008. Spatial scales of pollen and seed-mediated gene flow in tropical rain forest trees. Tropical Plant Biology 1:20–33.

Dutech, C., V. L. Sork, A. J. Irwin, P. E. Smouse, and F. W. Davis 2005. Gene flow and fine-scale genetic structure in a wind-pollinated tree species Quercus lobata (Fagaceaee). American Journal of Botany 92:252–261.

Dyer, R. J., and V. L. Sork 2001. Pollen pool heterogeneity in shortleaf pine, Pinus echinata Mill. Molecular Ecology 10:859–866.

Dyer, R. J., D. M. Chan, V. A. Gardiakos, and C. A. Meadows 2012. Pollination graphs: quantifying pollen pool covariance networks and the influence of intervening landscape on genetic connectivity in the North American understory tree, Cornus florida L. Landscape Ecology 27:239–251.

Eckert, A. J., A. D. Bower, S. C. Gonzalez-Martinez, J. L. Wegrzyn, G. Coop, and D. B. Neale. 2010. Back to nature: ecological genomics of loblolly pine (Pinus taeda, Pinaceae). Molecular Ecology 19:3789–3805.

Ellstrand, N. C., and D. R. Elam 1993. Population genetics of small population size: implications for plant conservation. Annual Review of Ecology and Systematics 23:217–242.

Elshire, R. J., J. C. Glaubitz, Q. Sun, J. A. Poland, K. Kawamoto, E. S. Buckler, and S. E. Mitchell 2011. A robust, simple genotyping-by-sequencing (GBS) approach for high diversity species. PLoS ONE 6: e19379.

Endler, J. A. 1973. Gene flow and population differentiation. Science 179:243–250.

Epperson, B. K. 2003. Geographical Genetics (MPB-38). Princeton University Press, Princeton, NJ.

Frankham, R., J. D. Ballou, and D. A. Briscoe 2002. Introduction to Conservation Genetics. Cambridge University Press, Cambridge, UK.

Franks, S. J., and A. A. Hoffmann 2012. Genetics of climate change adaptation. Annual Review of Genetics 46:185–208.

Garcia, C., and D. Grivet 2011. Molecular insights into seed dispersal mutualisms driving plant population recruitment. Acta Oecologica-International Journal of Ecology 37:632–640.

Garcia, C., J. M. Arroyo, J. A. Godoy, and P. Jordano 2005. Mating patterns, pollen dispersal, and the ecological maternal neighbourhood in a Prunus mahaleb L. population. Molecular Ecology 14:1821–1830.

Garcia, C., P. Jordano, and J. A. Godoy 2007. Contemporary pollen and seed dispersal in a Prunus mahaleb population: patterns in distance and direction. Molecular Ecology 16:1947–1955.

Godoy, J. A., and P. Jordano 2001. Seed dispersal by animals: exact identification of source trees with endocarp DNA microsatellites. Molecular Ecology 10:2275–2283.

Gram, W. K., and V. L. Sork 1999. Population density as a predictor of genetic variation for woody plant species. Conservation Biology 13:1079–1087.

Gram, W. K., and V. L. Sork 2001. Association between environmental and genetic heterogeneity in forest tree populations. Ecology 82:2012–2021.

Grivet, D., P. E. Smouse, and V. L. Sork 2005. A novel approach to an old problem: tracking dispersed seeds. Molecular Ecology 14:3585–3595.

Grivet, D., M.-F. Deguilloux, R. J. Petit, and V. L. Sork 2006. Contrasting patterns of historical colonization in white oaks (Quercus spp.) in California and Europe. Molecular Ecology 15:4085–4093.

Grivet, D., V. L. Sork, R. D. Westfall, and F. W. Davis 2008. Conserving the evolutionary potential of California valley oak (Quercus lobata Née): a multivariate genetic approach to conservation planning. Molecular Ecology 17:139–156.

Grivet, D., J. J. Robledo-Arnuncio, P. E. Smouse, and V. L. Sork 2009. Relative contribution of contemporary pollen and seed dispersal to the effective parental size of seedling population of California valley oak (Quercus lobata, Née). Molecular Ecology 18:3967–3979.

Gugger, P. F., M. Ikegami, and V. L. Sork 2013. Influence of late Quaternary climate change on present patterns of genetic variation in valley oak, Quercus lobata Née. Molecular Ecology 22:3598–3612.

Hamilton, M. B., and J. R. Miller 2002. Comparing relative rates of pollen and seed gene flow in the island model using nuclear and organelle measures of population structure. Genetics 162:1897–1909.

Hamrick, J. L., and M. D. Loveless 1986. The influence of seed dispersal mechanisms on the genetic structure of plant populations. In A. Estrada, and T. H. Fleming, eds. Frugivores and Seed Dispersal, pp. 211–223. Dr. W. Junk, Dordrecht.

Hamrick, J. L. 1987. Gene flow and distribution of genetic variation in plant populations. In K. M. Urbanska, ed. Differentiation Patterns in Higher Plants. Harcourt Brace Jovanovich, London.

Hamrick, J. L. 2004. Response of forest trees to global environmental changes. Forest Ecology and Management 197:323–335.

Hamrick, J. L., M. J. W. Godt, D. A. Murawski, and M. D. Loveless 1991. Correlations between species traits and allozyme diversity – implications for conservation biology. In D. A. Falk, and K. E. Holsinger, eds. Genetics and Conservation of Rare Plants. Center for Plant Conservation, St. Louis, MO.

Hamrick, J. L., D. A. Murawski, and J. D. Nason 1993. The influence of seed dispersal mechanism on the genetic structure of tropical tree populations. Vegetatio 107:281–297.

Hardy, O. J., L. Maggia, E. Bandou, P. Breyne, H. Caron, M. H. Chevallier, A. Doligez et al. 2006. Fine-scale genetic structure and gene dispersal inferences in 10 Neotropical tree species. Molecular Ecology 15:559–571.

Herrera, C. M., and P. Bazaga 2010. Epigenetic differentiation and relationship to adaptive genetic divergence in discrete populations of the violet Viola cazorlensis. New Phytologist 187:867–876.

Herrera, C. M., and P. Bazaga 2011. Untangling individual variation in natural populations: ecological, genetic and epigenetic correlates of long-term inequality in herbivory. Molecular Ecology 20:1675–1688.

Herrera-Arroyo, M. L., V. L. Sork, A. Gonzalez-Rodriguez, V. Rocha-Ramirez, E. Vega, and K. Oyama 2013. Seed-mediated connectivity among fragmented populations of Quercus castanea (Fagaceae) in a Mexican landscape. American Journal of Botany 100:1663–1671.

Holliday, J. A., K. Ritland, and S. N. Aitken 2010. Widespread, ecologically relevant genetic markers developed from association mapping of climate-related traits in Sitka spruce (Picea sitchensis). New Phytologist 188:501–514.

Hughes, A. R., B. D. Inouye, M. T. J. Johnson, N. Underwood, and M. Vellend 2008. Ecological consequences of genetic diversity. Ecology Letters 11:609–623.

Karubian, J., V. L. Sork, T. Roorda, R. Duraes, and T. B. Smith 2010.

Destination-based seed dispersal homogenizes genetic structure of a tropical palm. Molecular Ecology 19:1745–1753.

Keller, S. R., N. Levsen, M. S. Olson, and P. Tiffin 2012. Local adaptation in the flowering-time gene network of balsam poplar, *Populus balsamifera* L. Molecular Biology and Evolution 29:3143–3152.

Kelly, P. A., S. E. Phillips, and D. F. Williams. 2005. Documenting ecological changes in time and space: the San Joaquin Valley of California. In E. A. Lacey, and P. Myers, eds. Mammalian Diversification: from Chromosomes to Phylogeography (a Celebration of the Career of James L. Patton), vol. **133**: pp. 57–78. UC Publication in Zoology, Berkeley, CA.

Kremer, A., and A. Zanetto 1997. Geographical structure of gene diversity in *Quercus petraea* (Matt.) Liebl. II: multilocus patterns of variation. Heredity **78**:476–489.

Kremer, A., O. Ronce, J. J. Robledo-Arnuncio, F. Guillaume, G. Bohrer, R. Nathan, J. R. Bridle et al. 2012. Long-distance gene flow and adaptation of forest trees to rapid climate change. Ecology Letters **15**:378–392.

Kremer, A., B. M. Potts, and S. Delzon 2014. Genetic divergence in forest trees: understanding the consequences of climate change. Functional Ecology **28**:22–36.

Lancaster, L. T., and K. M. Kay 2013. Origin and diversification of the California Flora: Re-examining classic hypotheses with molecular phylogenies. Evolution **67**:1041–1054.

Law, J. A., and S. E. Jacobsen 2010. Establishing, maintaining and modifying DNA methylation patterns in plants and animals. Nature Reviews Genetics **11**:204–220.

Ledig, F. T. 1988. The conservation of diversity in forest tress. BioScience **38**:471–479.

Ledig, F. T. 1992. Human impacts on genetic diversity in forest ecosystems. Oikos **63**:87–108.

Levin, D. A. 1981. Dispersal versus gene flow in plants. Annals of the Missouri Botanical Garden **68**:233–253.

Loiselle, B. A., V. L. Sork, J. Nason, and C. Graham 1995. Spatial genetic structure of a tropical understory shrub, *Psychotria officinalis* (Rubiaceae). American Journal of Botany **82**:1420–1425.

Lotterhos, K. E., and M. C. Whitlock 2014. Evaluation of demographic history and neutral parameterization on the performance of F-ST outlier tests. Molecular Ecology **23**:2178–2192.

Lotterhos, K. E., and M. C. Whitlock 2015. The relative power of genome scans to detect local adaptation depends on sampling design and statistical method. Molecular Ecology **24**:1031–1046.

Loveless, M. D., and J. L. Hamrick 1984. Ecological determinants of genetic structure in plant populations. Annual Review of Ecology and Systematics **15**:65–95.

Luikart, G., P. R. England, D. Tallmon, S. Jordan, and P. Taberlet 2003. The power and promise of population genomics: from genotyping to genome typing. Nature Reviews Genetics **4**:981–994.

Manel, S., and R. Holderegger 2013. Ten years of landscape genetics. Trends in Ecology & Evolution **28**:614–621.

Manel, S., M. K. Schwartz, G. Luikart, and P. Taberlet 2003. Landscape genetics: combining landscape ecology and population genetics. Trends in Ecology & Evolution **18**:189–197.

Manel, S., S. Joost, B. K. Epperson, R. Holderegger, A. Storfer, M. S. Rosenberg, K. T. Scribner et al. 2010. Perspectives on the use of landscape genetics to detect genetic adaptive variation in the field. Molecular Ecology **19**:3760–3772.

Manning, K., M. Tor, M. Poole, Y. Hong, A. J. Thompson, G. J. King, J. J. Giovannoni et al. 2006. A naturally occurring epigenetic mutation in a gene encoding an SBP-box transcription factor inhibits tomato fruit ripening. Nature Genetics **38**:948–952.

Mátyás, C. 1996. Climatic adaptation of trees: rediscovering provenance tests. Euphytica **92**:45–54.

McRae, B. H. 2006. Isolation by resistance. Evolution **60**:1551–1561.

Meagher, T. R. 2007. Paternity analysis in a fragmented landscape. Heredity **99**:563–564.

Meagher, T. R., and E. Thompson 1987. Analysis of parentage for naturally established seedlings of *Chamaelirium luteum* Liliaceae. Ecology **68**:803–812.

Mitchell-Olds, T., J. H. Willis, and D. B. Goldstein 2007. Which evolutionary processes influence natural genetic variation for phenotypic traits? Nature Reviews Genetics **8**:845–856.

Moritz, C. 1995. Uses of molecular phylogenies for conservation. Philosophical Transactions of the Royal Society of London Series B, Biological Sciences **349**:113–118.

Nason, J. D., and J. L. Hamrick 1997. Reproductive and genetic consequences of forest fragmentation: two case studies of neotropical canopy trees. Journal of Heredity **88**:264–276.

Neale, D. B. 2007. Genomics to tree breeding and forest health. Current Opinion in Genetics & Development **17**:539–544.

Neale, D. B., and A. Kremer 2011. Forest tree genomics: growing resources and applications. Nature Reviews Genetics **12**:111–122.

Neigel, J. E. 1997. A comparison of alternative strategies for estimating gene flow from genetic markers. Annual Review of Ecology and Systematics **28**:105–128.

Nichols, K. M., D. B. Neale, J. Gratten, A. J. Wilson, A. F. McRae, D. Beraldi, P. M. Visscher et al. 2010. Association genetics, population genomics, and conservation: revealing the genes underlying adaptation in natural populations of plants and animals. In J. A. DeWoody, J. W. Bickham, C. H. Michler, K. M. Nichols, O. E. Rhodes, and K. E. Woeste, eds. Molecular Approaches in Natural Resource Conservation and Management, pp. 123–168. Cambridge University Press, Cambridge, UK.

Nielsen, R. 2005. Molecular signatures of natural selection. Annual Reviews of Genetics **39**:197–218.

Oddou-Muratorio, S., E. K. Klein, and F. Austerlitz 2005. Pollen flow in the wildservice tree, *Sorbus torminalis* (L.) Crantz. II. Pollen dispersal and heterogeneity in mating success inferred from parent-offspring analysis. Molecular Ecology **14**:4441–4452.

Oddou-Muratorio, S., E. K. Klein, B. Demesure-Musch, and F. Austerlitz 2006. Real-time patterns of pollen flow in the wild-service tree, *Sorbus torminalis* (Rosaceae). III. Mating patterns and the ecological maternal neighborhood. American Journal of Botany **93**:1650–1659.

Ortego, J., P. F. Gugger, and V. L. Sork. 2014. Climatically stable landscapes predict patterns of genetic structure and admixture in the Californian canyon live oak. Journal of Biogeography **42**:328–338.

Ottewell, K., E. Grey, F. Castillo, and J. Karubian 2012. The pollen dispersal kernel and mating system of an insect-pollinated tropical palm, *Oenocarpus bataua*. Heredity **109**:332–339.

Parmesan, C. 2006. Ecological and evolutionary responses to recent climate change. Annual Review of Ecology Evolution and Systematics **37**:637–669.

Pavlik, B. M., P. C. Muick, S. G. Johnson, and M. Popp 1995. Oaks of California. Cachuma Press, Oakland.

Petit, R. J., J. Carlson, A. L. Curtu, M. L. Loustau, C. Plomion, A. Gonzalez-Rodriguez, V. Sork et al. 2013. Fagaceae trees as models to integrate ecology, evolution and genomics. New Phytologist **197**:369–371.

Platt, A., P. F. Gugger, M. Pellegrini, and V. L. Sork 2015. Genome-wide signature of local adaptation linked to variable CpG methylation in oak populations. Molecular Ecology 24:3823–3830.

Pluess, A. R., V. L. Sork, B. Dolan, F. W. Davis, D. Grivet, K. Merg, J. Papp et al. 2009. Short distance pollen movement in a wind-pollinated tree, *Quercus lobata* (Fagaceae). Forest Ecology and Management 258:735–744.

Raven, P. H., and D. I. Axelrod. 1978. Origin and Relationships of the California Flora. Vol. 72, *University of California Publications in Botany*. University of California Press, Berkeley and Los Angeles.

Rayburn, R., and C. Moritz. 2006. Using Evolutionary Hotspots to Identify Important Areas for Conservation in California. California State Department of Parks and Recreation, Sacramento, CA.

Rissler, L. J., R. J. Hijmans, C. H. Graham, C. Moritz, and D. B. Wake 2006. Phylogeographic lineages and species comparisons in conservation analyses: a case study of California herpetofauna. American Naturalist 167:655–666.

Robledo-Arnuncio, J. J., and L. Gil 2005. Patterns of pollen dispersal in a small population of *Pinus sylvestris* L. revealed by total-exclusion paternity analysis. Heredity 94:13–22.

Robledo-Arnuncio, J. J., F. Austerlitz, and P. E. Smouse 2006. A new method of estimating the pollen dispersal curve independently of effective density. Genetics 173:1033–1045.

Robledo-Arnuncio, J. J., F. Austerlitz, and P. E. Smouse 2007. POLDISP: a software package for indirect estimation of contemporary pollen dispersal. Molecular Ecology Notes 7:763–766.

Robledo-Arnuncio, J. J., D. Grivet, P. E. Smouse, and V. L. Sork 2012. PSA: software for parental structure analysis of seed or seedling patches. Molecular Ecology Resources 12:1180–1189.

Savolainen, O., T. Pyhajarvi, and T. Knurr 2007. Gene flow and local adaptation in trees. Annual Review of Ecology Evolution and Systematics 38:595–619.

Schmitz, R. J. 2014. The secret garden-epigenetic alleles underlie complex traits. Science 343:1082–1083.

Schmitz, R. J., M. D. Schultz, M. G. Lewsey, R. C. O'Malley, M. A. Urich, O. Libiger, N. J. Schork et al. 2011. Transgenerational epigenetic instability is a source of novel methylation variants. Science 334:369–373.

Schmitz, R. J., M. D. Schultz, M. A. Urich, J. R. Nery, M. Pelizzola, O. Libiger, A. Alix et al. 2013. Patterns of population epigenomic diversity. Nature 495:193–198.

Schoener, T. W. 2011. The newest synthesis: understanding the interplay of evolutionary and ecological dynamics. Science 331:426–429.

Scofield, D. G., V. L. Sork, and P. E. Smouse 2010. Influence of acorn woodpecker social behaviour on transport of coast live oak (*Quercus agrifolia*) acorns in a southern California oak savanna. Journal of Ecology 98:561–572.

Scofield, D. G., V. R. Alfaro, V. L. Sork, D. Grivet, E. Martinez, J. Papp, A. R. Pluess et al. 2011. Foraging patterns of acorn woodpeckers (*Melanerpes formicivorus*) on valley oak (*Quercus lobata* Née) in two California oak savanna-woodlands. Oecologia 166:187–196.

Scofield, D. G., P. E. Smouse, J. Karubian, and V. L. Sork 2012. Use of Alpha, Beta, and Gamma diversity measures to characterize seed dispersal by animals. American Naturalist 180:719–732.

Slatkin, M. 1973. Gene flow and selection in a cline. Genetics 75:733–756.

Slatkin, M. 1985. Gene flow in natural populations. Annual Review of Ecology and Systematics 16:393–430.

Slatkin, M. 1987. Gene flow and the geographic structure of natural populations. Science 236:787–792.

Slavov, G. T., G. T. Howe, A. V. Gyaourova, D. S. Birkes, and W. T. Adams 2005. Estimating pollen flow using SSR markers and paternity exclusion: accounting for mistyping. Molecular Ecology 14:3109–3121.

Smouse, P. E., and J. J. Robledo-Arnuncio 2005. Measuring the genetic structure of the pollen pool as the probability of paternal identity. Heredity 94:640–649.

Smouse, P. E., and V. L. Sork 2004. Measuring pollen flow in forest trees: an exposition of alternative approaches. Forest Ecology and Management 197:21–38.

Smouse, P. E., T. R. Meagher, and C. J. Kobak 1999. Parentage analyis in *Chamaelirium luteum* (L.) Gray (Liliaceae): why do some males have higher reproductive contributions? Journal of Evolutionary Biology 12:1069–1077.

Smouse, P. E., R. J. Dyer, R. D. Westfall, and V. L. Sork 2001. Two-generation analysis of pollen flow across a landscape. I. Male gamete heterogeneity among females. Evolution 55:260–271.

Smouse, P. E., V. L. Sork, D. G. Scofield, and D. Grivet 2012. Using seedling and pericarp tissues to determine maternal parentage of dispersed valley oak recruits. Journal of Heredity 103:250–259.

Sokal, R. R., and N. L. Oden 1979. Spatial autocorrelation in biology: I. Methodology. Biological Journal of Linnean Society 10:199–228.

Soppe, W. J. J., S. E. Jacobsen, C. Alonso-Blanco, J. P. Jackson, T. Kakutani, M. Koornneef, and A. J. M. Peeters 2000. The late flowering phenotype of *fwa* mutants is caused by gain-of-function epigenetic alleles of a homeodomain gene. Molecular Cell 6:791–802.

Sork, V. L., and P. E. Smouse 2006. Genetic analysis of landscape connectivity in tree populations. Landscape Ecology 21:821–836.

Sork, V. L., and L. Waits 2010. Contributions of landscape genetics — approaches, insights, and future potential. Molecular Ecology 19:3489–3495.

Sork, V. L., and S. Werth 2014. Phylogeography of *Ramalina menziesii*, a widely distributed lichen-forming fungus in western North America. Molecular Ecology 23:2326–2339.

Sork, V. L., K. A. Stowe, and C. Hochwender 1993. Evidence for local adaptation in closely adjacent subpopulations of northern red oak (*Quercus rubra* L.) expressed as resistance to leaf herbivores. American Naturalist 142:928–936.

Sork, V. L., A. Koop, de la Fuente M. A., P. Foster, and J. Raveill. 1997. Patterns of genetic variation in woody plant species in the Missouri Ozark Forest Ecosystem Project (MOFEP). In B. L. Brookshire, and S. R. Shifley, eds. Proceedings of the Missouri Ozark Forest Ecosystem Project Symposium: an Experimental Approach to Landscape Research, 1997 June 3–5, St. Louis, MO. U. S. Department of Agriculture Forest Service North Central Forest Experiment Station, St. Paul, MN.

Sork, V. L., J. Nason, D. R. Campbell, and J. F. Fernandez 1999. Landscape approaches to historical and contemporary gene flow in plants. Trends in Ecology and Evolution 14:219–224.

Sork, V. L., F. W. Davis, P. E. Smouse, V. J. Apsit, R. J. Dyer, J. F. Fernandez-M, and B. Kuhn 2002. Pollen movement in declining populations of California Valley oak, *Quercus lobata*: where have all the fathers gone? Molecular Ecology 11:1657–1668.

Sork, V. L., P. E. Smouse, V. J. Apsit, R. J. Dyer, and R. D. Westfall 2005. A two-generation analysis of pollen pool genetic structure in flowering dogwood, *Cornus florida* (Cornaceae), in the Missouri Ozarks. American Journal of Botany 92:262–271.

Sork, V. L., F. W. Davis, and D. Grivet. 2009. Incorporating genetic information into conservation planning for California valley oak. Paper read at Proceedings of the Sixth Symposium on Oak Woodlands: California's Oaks: Today's Challenges, Tomorrow's Opportunities. 2006 October 9–12, at Santa Rosa, CA.

Sork, V. L., F. W. Davis, R. Westfall, A. Flint, M. Ikegami, H. F. Wang, and D. Grivet 2010. Gene movement and genetic association with regional climate gradients in California valley oak (*Quercus lobata* Née) in the face of climate change. Molecular Ecology 19:3806–3823.

Sork, V. L., S. N. Aitken, R. J. Dyer, A. J. Eckert, P. Legendre, and D. B. Neale 2013. Putting the landscape into the genomics of trees: approaches for understanding local adaptation and population responses to changing climate. Tree Genetics and Genomics 9:901–911.

Sork, V. L., P. E. Smouse, D. Grivet, and D. G. Scofield. 2015. Impact of asymmetric male and female gamete dispersal on allelic diversity and spatial genetic structure in valley oak (*Quercus lobata* Née). Evolutionary Ecology In press.

Sork, V. L., K. Squire, P. F. Gugger, S. E. Steele, E. D. Levy, and A. J. Eckert. 2016. Landscape genomic analysis of candidate genes for climate adaptation in a California endemic oak, *Quercus lobata* Née (Fagaceae). American Journal of Botany In press.

Stapley, J., J. Reger, P. G. D. Feulner, C. Smadja, J. Galindo, R. Ekblom, C. Bennison et al. 2010. Adaptation genomics: the next generation. Trends in Ecology & Evolution 25:705–712.

Storfer, A., M. A. Murphy, J. S. Evans, C. S. Goldberg, S. Robinson, S. F. Spear, R. Dezzani et al. 2007. Putting the 'landscape' in landscape genetics. Heredity 98:128–142.

Stowe, K. A., V. L. Sork, and A. W. Farrell. 1994. Effect of water availability on the phenotypic expression of herbivore resistance in northern red oak seedlings (*Quercus rubra* L.). Oecologia 100:309–315.

Strasburg, J. L., N. A. Sherman, K. M. Wright, L. C. Moyle, J. H. Willis, and L. H. Rieseberg 2012. What can patterns of differentiation across plant genomes tell us about adaptation and speciation? Philosophical Transactions of the Royal Society of London Series B, Biological Sciences 367:364–373.

Streiff, R., A. Ducousso, C. Lexer, H. Steinkellner, J. Gloessl, and A. Kramer 1999. Pollen dispersal inferred from paternity analysis in a mixed oak stand of *Quercus robur* L. and *Q. petraea* (Matt.) Liebl. Molecular Ecology 8:831–841.

Takuno, S., and B. S. Gaut 2013. Gene body methylation is conserved between plant orthologs and is of evolutionary consequence. Proceedings of the National Academy of Sciences 110:1797–1802.

Templeton, A. R., and N. J. Georgiadis 1996. A landscape approach to conservation genetics: conserving evolutionary processes in the African Bovidae. In J. C. Avise, and J. L. Hamrick, eds. Conservation Genetics. Chapman and Hall, New York.

Thompson, P. T., P. E. Smouse, D. G. Scofield, and V. L. Sork. 2014. What seeds tell us about bird movement: a multi-year analysis of Acorn Woodpecker foraging patterns on two oak species. Movement Ecology 2:12. doi:10.1186/2051-3933-2-12.

Vekemans, X., and O. J. Hardy 2004. New insights from fine-scale spatial genetic structure analyses in plant populations. Molecular Ecology 13:921–935.

Werth, S., and V. L. Sork 2014. Ecological specialization in *Trebouxia* (Trebouxiophyceae) photobionts of *Ramalina menziesii* (Ramalinaceae) across six range-covering ecoregions of western North America. American Journal of Botany 101:1127–1140.

Whipple, A. A., R. M. Grossinger, and F. W. Davis 2011. Shifting baselines in a California oak savanna: nineteenth century data to inform restoration scenarios. Restoration Ecology 19:88–101.

Wright, S. 1943. Isolation by distance. Genetics 28:114–138.

Wright, S. 1946. Isolation by distance under diverse systems of mating. Genetics 31:39.

Yakovlev, I. A., D. K. A. Asante, C. G. Fossdal, O. Junttila, and O. Johnsen 2011. Differential gene expression related to an epigenetic memory affecting climatic adaptation in Norway spruce. Plant Science 180:132–139.

Young, A., T. Boyle, and T. Brown 1996. The population genetic consequences of habitat fragmentation for plants. Trends in Ecology & Evolution 11:413–418.

Zhang, Y. Y., M. Fischer, V. Colot, and O. Bossdorf 2013. Epigenetic variation creates potential for evolution of plant phenotypic plasticity. New Phytologist 197:314–322.

12

Drug-resistant HIV-1 protease regains functional dynamics through cleavage site coevolution

Nevra Özer,[1],* Ayşegül Özen,[2] Celia A. Schiffer[2] and Türkan Haliloğlu[1]

1 Polymer Research Center and Chemical Engineering Department, Bogazici University, Bebek, Istanbul, Turkey
2 Department of Biochemistry and Molecular Pharmacology, University of Massachusetts Medical School, Worcester, MA, USA
* Present address: Department of Bioengineering, Marmara University, Goztepe, Kadikoy, Istanbul, Turkey

Keywords
coevolution, elastic network model, fluctuations, HIV-1 protease.

Correspondence
Türkan Haliloğlu, Polymer Research Center and Chemical Engineering Department, Bogazici University, Bebek 34342 Istanbul, Turkey.

e-mail: halilogt@boun.edu.tr
Celia A. Schiffer, Department of Biochemistry and Molecular Pharmacology, University of Massachusetts Medical School, Worcester, MA, USA.

e-mail: celia.schiffer@umassmed.edu

Abstract

Drug resistance is caused by mutations that change the balance of recognition favoring substrate cleavage over inhibitor binding. Here, a structural dynamics perspective of the regained wild-type functioning in mutant HIV-1 proteases with coevolution of the natural substrates is provided. The collective dynamics of mutant structures of the protease bound to p1-p6 and NC-p1 substrates are assessed using the Anisotropic Network Model (ANM). The drug-induced protease mutations perturb the mechanistically crucial hinge axes that involve key sites for substrate binding and dimerization and mainly coordinate the intrinsic dynamics. Yet with substrate coevolution, while the wild-type dynamic behavior is restored in both p1-p6 ($^{LP1'F}$p1-p6$_{D30N/N88D}$) and NC-p1 (^{AP2V}NC-p1$_{V82A}$) bound proteases, the dynamic behavior of the NC-p1 bound protease variants (NC-p1$_{V82A}$ and ^{AP2V}NC-p1$_{V82A}$) rather resemble those of the proteases bound to the other substrates, which is consistent with experimental studies. The orientational variations of residue fluctuations along the hinge axes in mutant structures justify the existence of coevolution in p1-p6 and NC-p1 substrates, that is, the dynamic behavior of hinge residues should contribute to the interdependent nature of substrate recognition. Overall, this study aids in the understanding of the structural dynamics basis of drug resistance and evolutionary optimization in the HIV-1 protease system.

Introduction

Protein interactions mediate the function of biological systems, where the evolution of interactions is important to understand the functional mechanism in act (Juan et al. 2008; Lovell and Robertson 2010). Evolutionary signals are generated either by whole-sequence evolution or by site-specific coevolution (Lovell and Robertson 2010). Coevolution can be defined as a reciprocal change in one site affecting the selection pressure at another site allowing for adaptation (Thompson 1994). This can occur as either an intramolecular or an intermolecular process, where coevolution arises from the evolutionary interaction between sites within a single molecule in the former, and the latter is due to co-adaptation as a result of the evolutionary interaction between different molecules (Juan et al. 2008; Lovell and Robertson 2010).

Understanding evolution within the complex relationship between sequence, structure and function for a particular phenotype is quite limited (Xia and Levitt 2004; Tomatis et al. 2008). Selective pressures for evolvability should act at both structural and dynamics levels, where the sequence divergence is constrained by the conservation of structural features and further by the conservation of functional motion. Thus, the functional importance of protein dynamics should be credited for the sequence evolution (Maguid et al. 2008; Juan et al. 2008). The structural flexibility and plasticity of proteins are imperative in performing their biological functions, especially in molecular recognition (Teague 2003; Gerstein and Echols 2004; Marianayagam and Jackson 2005; Friedland et al. 2009; Ramanathan and Agarwal 2011; Mittal et al. 2012). The mechanism of this recognition between proteins and ligands is probably predefined by the rules encoded in the protein structure and dynamics, the detailed knowledge of which would be useful in design and engineering of drugs. Evolvability and natural variation are also correlated with drug resistance, which is a central problem in drug design for many diseases (Earl and Deem 2004; Creavin 2004; Berkhout and Sanders 2005; Nalam and Schiffer 2008). Influenza, tuberculosis, malaria, cancer, and HIV/AIDS are some of the important examples of diseases that confront drug resistance.

HIV-1 protease is an effective therapeutic target of the most effective antiviral drugs for the treatment of HIV-1 infection (Prabu-Jeyabalan et al. 2002). The protease is a symmetric homodimer containing a single active site formed at the dimer interface by two conserved catalytic aspartic acid residues, one from each monomer, and covered by two flexible flaps (Wlodawer and Erickson 1993). The enzyme recognizes substrate sites on the Gag and Gag-Pro-Pol polyproteins which are asymmetric in both size and charge around the cleavage site, while the currently prescribed inhibitors are relatively symmetric. Yet with the drug-resistant mutations on the protease, the affinity for inhibitors is lowered while efficient processing is still maintained and the structure reassumes a drastic asymmetry (Prabu-Jeyabalan et al. 2004). Thus, substrate specificity should be based on a conserved shape rather than a particular amino acid sequence. This is explained by a consensus volume of substrates recognized by the protease, defined as the 'substrate envelope', which not only explains specificity but also has significant implications for drug resistance and substrate coevolution (Kolli et al. 2009). The envelope is achieved by packing of the substrate residues which makes the substrate recognition interdependent (Prabu-Jeyabalan et al. 2002; King et al. 2004). This interdependency implies that a drug-resistant protease mutation that causes unfavorable interactions with one substrate position is often compensated by a mutation at another position within the substrate sequence. Two examples of this evolutionary interplay between substrate and enzyme are the coevolution of p1-p6 substrate cleavage site with D30N/N88D protease mutations (Bally et al. 2000) and the coevolution of NC-p1 substrate cleavage site with V82A protease mutation (Doyon et al. 1996). Furthermore, cleavages at these sites of Gag, which have the most significant polymorphism among all HIV-1 substrate sites, are known to be rate limiting steps in polyprotein processing (Tözser et al. 1991; Feher et al. 2002).

Most drug-resistant mutations within the protease active site occur where the inhibitors protrude beyond the substrate envelope and contact the protease (King et al. 2004). Those protease mutations may be associated with other mutations in the substrate sites that extend beyond the substrate envelope and/or with other mutations in the viable protease (Kolli et al. 2006). Substrate dynamics was later incorporated into the substrate envelope, where p1-p6 and NC-p1 are shown to be two of the three most dynamic substrates (Ozen et al. 2011). Dynamic substrates exhibit a worse fit within this dynamic substrate envelope as they sample a wider conformational space resulting in a greater deviation from the substrate envelope. Accordingly, the dynamic p1-p6 and NC-p1 substrates protrude beyond the dynamic envelope more than expected based on their molecular volume compared to the other substrates (Ozen et al. 2011). Thus, the compensatory mutations in these cleavage sites optimize the portion of the substrate volume that stays within the dynamic substrate envelope. In the presence of D30N/N88D mutations in the protease, the coevolutionary mutation LP1'F at the p1-p6 cleavage site provides a better fit within the dynamic substrate envelope. Similarly, in NC-p1, the compensatory AP2V substitution occurs in the presence of V82A mutation in the protease. The interdependency between the changes in the substrate sequence in response to the drug-induced protease mutations can be explained by the energetic fitness of the sequences to the structural space of HIV-1 protease. This fitness possibly implies a conservation for the dynamic behavior of the residues in the HIV-1 protease-substrate/inhibitor complex system. With this approach, the structural basis of drug resistance, that is, the effect of structural and dynamic constraints on the sequence evolution (Liu and Bahar 2012; Gerek et al. 2013), could further be clarified. In this study, we investigate the dynamics to understand the mechanism that results in the dynamic substrate envelope which was validated as the substrate recognition motif for HIV-1 protease.

Significant structural and functional features of biomolecular complexes can be elucidated by the detailed analysis of fluctuations around their native states (Bahar et al. 2010). When dynamics is decomposed into a collection of modes of motion, the cooperative low frequency/large amplitude modes have been shown to be significantly correlated with the biological function (Nicolay and Sanejouand 2006). The principal component analysis (PCA) is a computational approach to extract the collective behavior from the fluctuations observed in molecular dynamics (MD) trajectories (Tournier and Smith 2003). The cooperative motions can alternatively be studied by normal mode analysis (Ma 2005; Cui and Bahar 2005). Elastic network models have been well-accepted for studying the large-scale motion of protein structures in recent years (Chennubhotla et al. 2005; Nicolay and Sanejouand 2006; Bahar et al. 2010; Gniewek et al. 2012). Despite the simplicity of this approach, the application of the elastic network models such as the Gaussian Network Model (GNM) (Bahar et al. 1997; Haliloglu et al. 1997) and the Anisotropic Network Model (ANM) (Atilgan et al. 2001) to the HIV-1 protease system have also produced results that are highly in accord with those of both experimental studies and MD simulations (Bahar et al. 1998; Zoete et al. 2002; Kurt et al. 2003; Micheletti et al. 2004; Hamacher and McCammon 2006; Yang et al. 2008; Hamacher 2010). The computational studies on the structural dynamics of HIV-1 protease suggest that understanding the dynamic behavior of the enzyme is crucial for its intrinsic flexibility and function (Perryman et al. 2004; Hornak and Simmerling 2007; Ozer et al. 2010).

In protein-ligand interactions, the ligand prefers the conformations that best match its structural and dynamic behavior among those intrinsically accessible to the unbound protein (Bakan and Bahar 2009). The conformational changes experimentally observed in the enzymes by binding a broad range of ligands can be predicted by the most cooperative lowest frequency modes of motion by ANM, where the hinges are the key mechanistic regions of the structure that control the conformational ensemble. Hinge motion has been shown to be an important mechanism that underlies the functional conformational changes, and catalytic residues tend to be positioned near the hinge regions that are unique for particular architectures (Yang and Bahar 2005). The elastic distortions of these dynamically important hinge residues, which serve as central hubs, effectively trigger the correlated fluctuations of a large number of residues (Zheng and Brooks 2005). Thus, the dynamic behavior of the hinge axes should mainly determine the flexibility and intrinsic dynamics of the structure. In our recent work, the ANM analysis of the fluctuations of the bound HIV-1 protease structures demonstrated that the hinge residues of the most cooperative modes display variation in their fluctuations depending on the bound substrate (Ozer et al. 2010). Further, flexible substrates adapt to the conformational changes of the protease better than the conformationally and dynamically restricted inhibitors, implying the rationale for more diverse inhibitors. Here, to study the dynamic behavior that accompany coevolution in the HIV-1 protease system, fluctuations of the wild-type, mutant, and coevolved structures of two protease-substrate complexes, namely the p1-p6 and NC-p1 complexes, are analyzed by ANM. The results put forward the motive for the evolutionary optimization of the sequences of HIV-1 protease-substrate complex from a mechanistic functional dynamics perspective.

Materials and methods

Nomenclature

Throughout this article, the wild-type HIV-1 protease (WT), the HIV-1 protease mutants (D30N, D30N/N88D, or V82A), and the cleavage site (LP1$'$F or AP2V) variants in a protease-substrate complex are designated by a sub-

script and a superscript to the name of the cleavage site. For example, $^{LP1'F}$p1-p6$_{D30N}$ denotes a complex of D30N protease variant with the LP1$'$F mutant of the p1-p6 cleavage site, where LP1$'$F refers to a Leu-to-Phe mutation at P1$'$ position of the cleavage site. The substrate residues on the amino-terminal side of the scissile bond and the protease monomer on that side are termed as unprimed; whereas, those on the carboxy-terminal side of the scissile bond are termed as primed.

Structures

The wild-type structures are the crystal structures of HIV-1 protease in complex with its natural substrates downloaded from the Protein Data Bank (PDB) (Bernstein et al. 1977; Prabu-Jeyabalan et al. 2002, 2004). The structures of p1-p6$_{D30N}$, p1-p6$_{D30N/N88D}$, $^{LP1'F}$p1-p6$_{D30N/N88D}$, NC-p1$_{V82A}$, and ^{AP2V}NC-p1$_{V82A}$ variants are modeled *in silico* based on their wild-type structures and simulated by MD simulations. The details of the modeling and MD simulation protocols were described elsewhere (Ozen et al. 2012). The representative conformations are selected by clustering the MD sampled conformations.

Cluster analysis on MD simulation trajectories

A representative set of conformations of all wild-type protease-substrate complexes, p1-p6$_{D30N}$, p1-p6$_{D30N/N88D}$, $^{LP1'F}$p1-p6$_{D30N/N88D}$, NC-p1$_{V82A}$, and ^{AP2V}NC-p1$_{V82A}$, generated from MD simulation trajectories of eleven ns production phase are clustered separately to group the 'redundant' conformations and examine the unique conformers. For the modeled mutant structures that were run for fifteen *ns*, the clusters of the endmost snapshots of eleven ns are used to allow selection of MD-relaxed structures. The similarity measure to group the MD sampled conformations is root-mean-square deviation (RMSD) in this study. MMTSB Toolset's (Feig et al. 2004) *kclust* utility that uses the k-means clustering is used to perform conformational clustering. The convergence of the simulations is judged by the relative populations of clusters as even fairly long MD trajectories may not be converged for flexible systems (Lyman and Zuckerman 2006). The number of clusters depends on the cutoff value of RMSD (cluster radius); as RMSD cutoff increases, less number of clusters are found by the algorithm. With a total of 550 structures for each HIV-1 protease-substrate complex within eleven ns, the cluster radius is set as 1.3 Å after various trials. The number of clusters obtained hereby and the percentage of all the clusters can be seen on Table 1. The percentage of largest clusters varies between 35% and 86% while that of the largest two clusters in total varies between 67% and 100%. Then, the representative structures of the

Table 1. Percentage of the clusters of molecular dynamics (MD) sampled structures.

	p1-p6$_{WT}$	p1-p6$_{D30N/N88D}$	$^{LP1'F}$p1-p6$_{D30N/N88D}$	NC-p1$_{WT}$	NC-p1$_{V82A}$	^{AP2V}NC-p1$_{V82A}$	MA-CA$_{WT}$	CA-p2$_{WT}$	p2-NC$_{WT}$	RT-RH$_{WT}$	RH-IN$_{WT}$
Cluster 1	50	57	41	37	51	35	40	68	65	86	49
Cluster 2	25	36	32	34	39	32	29	23	23	14	42
Cluster 3	16	7	17	17	10	15	22	9	12		9
Cluster 4	5		10	12		12	9				
Cluster 5	4					6					

largest clusters are further analyzed by ANM. To test the convergence in the dynamic behavior observed, the representative members of the second largest clusters are also analyzed by ANM and the results are compared with those of the largest clusters.

Anisotropic Network Model

The simple elastic network models are originally introduced by Tirion (Tirion 1996), where the complex vibrational properties of macromolecular systems are reproduced by a model using a single uniform harmonic potential. GNM (Bahar et al. 1997; Haliloglu et al. 1997) and ANM (Atilgan et al. 2001) are the two models that have widely been used in recent years.

Anisotropic Network Model (Atilgan et al. 2001) predicts the directionalities and magnitudes of the motions of protein structures around their equilibrium states by a harmonic vibrational analysis. The conformations that describe the fluctuations of residues from the average in the principal directions of motion are generated using the elastic network formed by connecting all neighboring heavy atoms. The total potential energy for a system of N nodes is the summation over all harmonic interactions of close-neighboring (i, j) pairs calculated as:

$$V = (\gamma/2) \left[\sum_{i,j}^{N} h(r_c - R_{ij})(R_{ij} - R_{ij}^{0})^2 \right]$$

γ is the harmonic force constant, and R_{ij} is the instantaneous distance, and R_{ij}^0 is the equilibrium distance between sites i and j in the native structure. $h(r_c - R_{ij})$ is the Heaviside step function which is 1 if $(r_c - R_{ij}) \geq 0$ and zero otherwise. r_c, the cutoff distance, is taken as 9 Å, which has successfully been used to account for inter-residue interactions in the all-atom structure model of the HIV-1 protease system (Ozer et al. 2010).

The Hessian matrix $\boldsymbol{H}$ is a $3N \times 3N$ symmetric matrix, which holds the anisotropic information regarding the orientation of nodes i, j. $\boldsymbol{H}$ is composed of $N \times N$ super elements $\boldsymbol{H}_{ij}$ each of size 3×3 given by the second derivatives of the potential V. An orthogonal transformation of the real symmetric Hessian matrix gives the normal modes of the elastic network with $3N - 6$ nonzero eigenvalues λ_i and corresponding eigenvectors $\mathbf{u}_i$.

$$\mathbf{H}^{-1} = \sum_{i=1}^{3N-6} \frac{1}{\lambda_i} \mathbf{u}_i \mathbf{u}_i^T$$

The fluctuations of nodes are used to construct and explicitly view pairs of alternative conformations sampled in the individual modes, simply by adding the fluctuation vectors to the equilibrium position vectors in the respective modes. The mode shapes exhibit the distribution of mobility among residues driven by different frequency modes, where the minima correspond to hinge regions.

Orientational correlation analysis

The orientational correlation between the fluctuations in different structures is assessed by the calculation of the inner product of the fluctuation vectors. That is, the structures are superimposed, and cosine of the angle between their fluctuation vectors is evaluated by a dot product calculation. The normalized correlation values range between 1 (perfect correlation, where the angle between the fluctuation vectors of the superimposed structures are 0°) and −1 (maximum variation in fluctuation direction, where the angle between the fluctuation vectors of the superimposed structures are 180°). The orientational correlations are assessed on a residue basis in the two most cooperative modes of motion, to observe which residues' fluctuations differ between various protease-substrate complex structures. Absolute lower orientational correlation values indicate the residues that display larger variations in their fluctuations' directions.

Table 2. HIV-1 protease-substrate complex structures used in the analyses.

Wild-type structures (PDB code)	p1-p6$_{WT}$ (1kjf)
	NC-p1$_{WT}$ (1tsu)
	MA-CA$_{WT}$ (1kj4)
	CA-p2$_{WT}$ (1f7a)
	p2-NC$_{WT}$ (1kj7)
	RT-RH$_{WT}$ (1kjg)
	RH-IN$_{WT}$ (1kjh)
Mutant structures	p1-p6$_{D30N}$
	p1-p6$_{D30N/N88D}$
	NC-p1$_{V82A}$
Coevolved mutant structures	$^{LP1'F}$p1-p6$_{D30N/N88D}$
	^{AP2V}NC-p1$_{V82A}$

Results and discussion

The HIV-1 protease structures investigated in this study are listed on Table 2. Details of the nomenclature used throughout the article are described in the Materials and methods section.

Cooperative motion of protease and substrate

The slow modes describe the most cooperative global motion of the HIV-1 protease-peptide complex structures and are likely associated with the enzymatic function of the protease (Kurt et al. 2003; Yang and Bahar 2005; Yang et al. 2008). The residue fluctuation profiles in the slowest two modes share a similar trend in the wild-type and mutant complexes investigated here (Fig. 1), which is consistent with the earlier results on the wild-type, substrate- and inhibitor-bound protease structures (Ozer et al. 2010). The generality of this observation across substrate and inhibitor complexes of protease variants suggest that the hinge regions such as the dimerization region (res. 5–10), the active site (res. 25–27), the flap (res. 45–55), and the substrate cleft (res. 80–90) display the least fluctuations in their mean positions and coordinate the motion. The distribution of the mean-square fluctuations in p1-p6 and NC-p1 bound protease structures analyzed here displays that residues 56, 69, 78, 93 and residues 25–27, 49–51, 84, 97 are observed to be at the minima of the corresponding mode shapes, in the first (Fig. 1A,C) and second (Fig. 1B, D) slowest modes, respectively.

The motion of a ligand-bound HIV-1 protease is demonstrated for one representative structure in Fig. 2. In the slowest mode (Fig. 2A), the protease monomers rotate around two axes parallel to the Z direction, coupled with the peptide fluctuation along the Y direction. In second slowest mode (Fig. 2B), the monomers rotate around two different axes parallel to X and Z directions and the peptide motion is significant in the terminal residues. The hinge

axes in the corresponding modes are shown as dashed lines. As the functional conformational changes are mainly due to elastic distortions of the hinge residues that trigger the correlated fluctuations (Zheng and Brooks 2005), the dynamic behavior of the hinge axes, that is, the fluctuation of residues located at the hinge sites, should determine the flexibility and intrinsic dynamics of the bound protease. Here, the hinge regions suggested by the slowest mode mainly coordinate the intrachain cooperative motions along with the motion of the peptide, whereas those suggested by the second slowest mode are mostly responsible for the correlations across the dimerization interface (Fig. 2).

Deformation of hinge axes by protease mutations and coevolution: the orientational variations in the fluctuations of hinge residues

Among different complex structures of the wild-type protease, although the overall residue mobility profiles are similar (Fig. 1), some variations in the amplitude and the orientation of the residue fluctuation vectors might be possible. The substrate stabilizes one of the several conformations accessible to the protease functionally and energetically favorable (from both thermodynamic and kinetic aspects). For example, the protease's interaction with a specific substrate should corroborate the substrate kinetics, that is, the rate of cleavage. Structural variations across substrate complexes should be reflected in specific recognition of substrates, so should be the difference in their dynamic behavior. This presumable difference is therefore investigated further by the orientational correlations between the residue fluctuations in mutant structures.

In the ANM analysis of the dynamics of the wild-type HIV-1 protease structures bound to its natural substrates, the highest orientational differences in the residue fluctuations and the largest extent of the asymmetry of the residue fluctuations in the two protease monomers were observed primarily along the hinge axes (Ozer et al. 2010). This suggests that there is a substrate-specific behavior implicated therein in relevant modes (Fig. 2) by the fluctuations of hinge residues. Here, the premise is whether the specific dynamic interaction of each substrate with the protease should be conserved or not. To this end, the inquiry is whether or not the drug-induced protease mutations associated with coevolution of the substrate would corroborate this specific interaction by the re-orientation of fluctuations of hinge residues back to those of the wild type. The analysis of the orientational correlations (see Materials and methods) of the residue fluctuations among the mutant and the coevolved mutant structures will

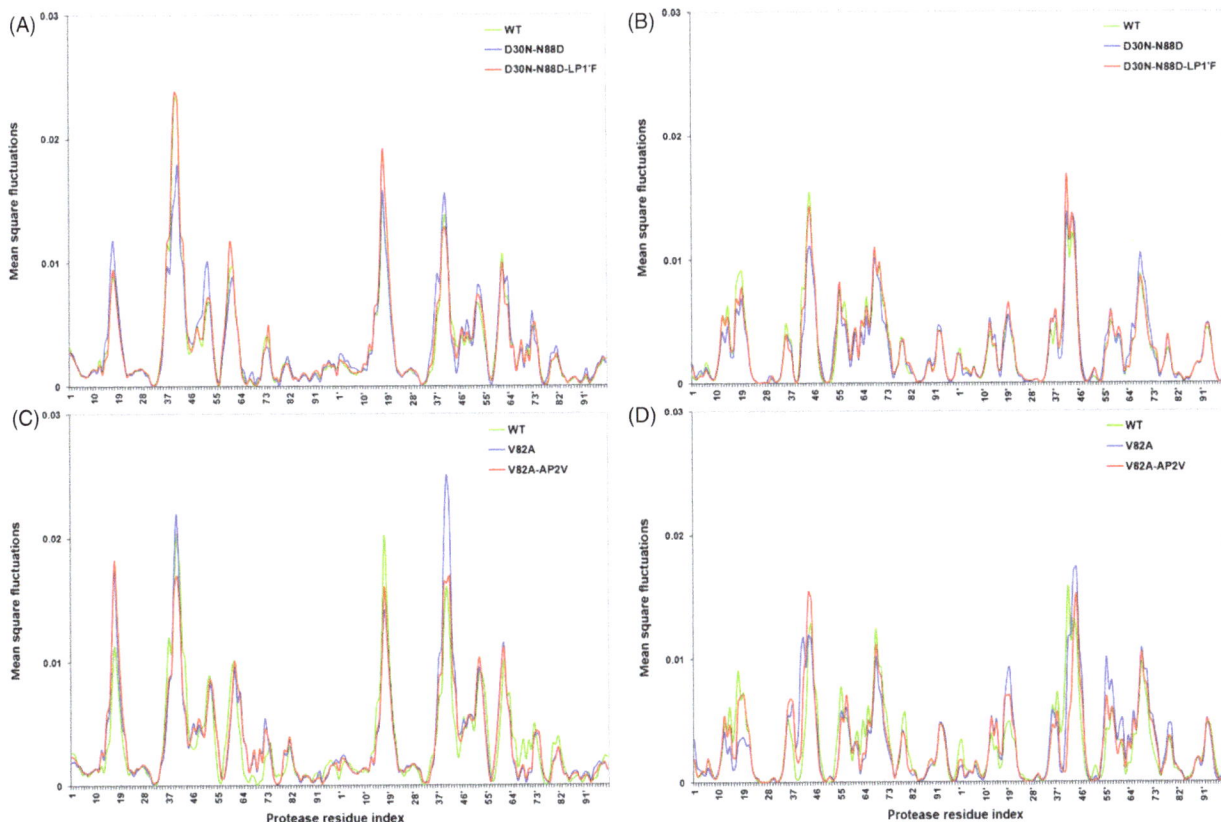

Figure 1 Mean-square fluctuations of the p1-p6 and NC-p1 bound protease complex structures in the first (A, C) and in the second (B, D) slowest Anisotropic Network Model (ANM) modes. The fluctuations are calculated for all the atoms in the structures, but for clearer representation, the fluctuations of C_α atoms are plotted, and the corresponding residue numbers are indicated on the x-axis.

provide a dynamic mechanistic perspective for the role of mutations.

Coevolution in p1-p6

The residue fluctuations in the slowest two modes of p1-p6$_{D30N}$, p1-p6$_{D30N/N88D}$, and $^{LP1'F}$p1-p6$_{D30N/N88D}$ variants are compared with those of p1-p6$_{WT}$ (Fig. 3). The correlation coefficient values between the fluctuation vectors of each residue in the two respective structures indicate the orientational (directional) variation in their fluctuations in the given mode. With the protease mutations, the largest orientational difference in the fluctuations is observed mostly at the proximity of residue 69 in the unprimed monomer and at residues 56, 78, 93 in both monomers in the slowest mode (Fig. 3A). In the second slowest mode, the fluctuations of residues 25, 26, 27, 39 in the unprimed monomer and residues 16, 49, 50, 51, 97 in both monomers display largest orientational difference (Fig. 3D). These residues lie at the hinge axes of rotational motion identified by the residue mobility profiles in Fig. 2.

The correlation coefficients of the residue fluctuation vectors between p1-p6$_{WT}$ and $^{LP1'F}$p1-p6$_{D30N/N88D}$ in both

modes are in general higher than they are between p1-p6$_{WT}$ and p1-p6$_{D30N/N88D}$. To quantify, when the residues with maximum variations in their orientational correlations that are below the lower standard deviation bound (average minus one standard deviation) are considered, the average correlation coefficient value which is 0.61 between p1-p6$_{WT}$ and p1-p6$_{D30N/N88D}$ increases to 0.90 between p1-p6$_{WT}$ and $^{LP1'F}$p1-p6$_{D30N/N88D}$ in the slowest mode. Similarly, in the second slowest mode, the average correlation coefficient value between p1-p6$_{WT}$ and p1-p6$_{D30N/N88D}$ is 0.63, whereas it is 0.69 between p1-p6$_{WT}$ and $^{LP1'F}$p1-p6$_{D30N/N88D}$. In both modes, the correlation of p1-p6$_{D30N}$ with p1-p6$_{WT}$ is even less than that of p1-p6$_{D30N/N88D}$ which has the signature mutations of nelfinavir resistance that occur in association with p1-p6 cleavage site mutations (Kolli et al. 2009).

The residues that display the maximum variations between the directions of fluctuations can also be visualized on the structures in Fig. 3, as color coded according to the change in the correlation values of p1-p6$_{D30N/N88D}$ (Fig. 3B,E) and $^{LP1'F}$p1-p6$_{D30N/N88D}$ (Fig. 3C,F) compared to p1-p6$_{WT}$. In the slowest mode, the higher correlation

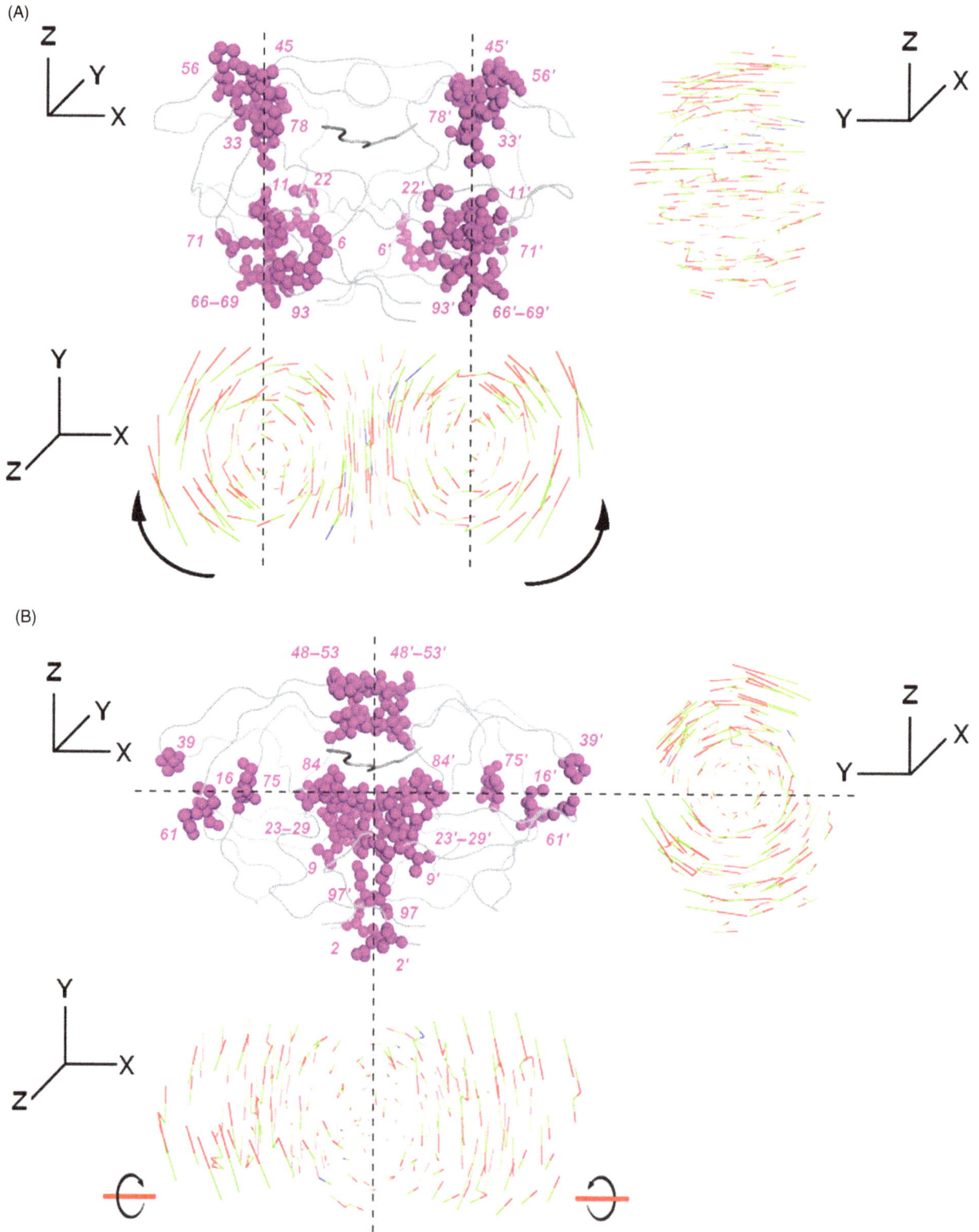

Figure 2 The regions of the orientational difference in the direction of fluctuations in the first (A) and second (B) slowest Anisotropic Network Model (ANM) modes. The least correlating residues in their fluctuations' directions between different complex structures of HIV-1 protease are displayed in magenta in the front view. Top and side views of the structure are shown, where the residue fluctuations in each mode are represented as moving between the conformations shown in green and red for the protease, and in green and blue for the peptide. The dashed lines indicate the hinge axes around which the monomers rotate. The coordinate system for the front, top, and side views is indicated next to the structures.

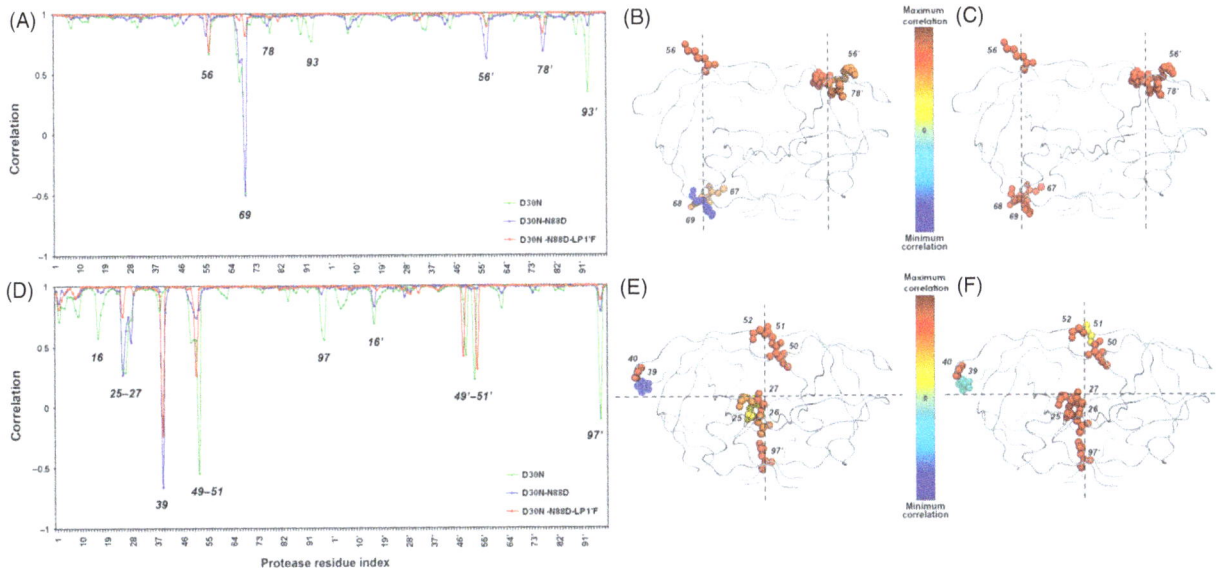

Figure 3 Orientational correlation of the protease residues' fluctuations of p1-p6$_{D30N}$, p1-p6$_{D30N/N88D}$, and $^{LP1'F}$p1-p6$_{D30N/N88D}$ to those of p1-p6$_{WT}$ in the first (A) and second (D) slowest modes. The residues that display the maximum variations between the directions of fluctuations are color coded according to the change in the correlations of p1-p6$_{D30N/N88D}$ and $^{LP1'F}$p1-p6$_{D30N/N88D}$ compared to p1-p6$_{WT}$; (B) and (C) in the first slowest mode, and (E) and (F) in the second slowest mode, respectively. The line plot connecting the points that indicate the correlation values is used to guide the eye.

values of residues 67, 68, 69, 56', and 78' between $^{LP1'F}$p1-p6$_{D30N/N88D}$ and p1-p6$_{WT}$ structures compared to those between p1-p6$_{D30N/N88D}$ and p1-p6$_{WT}$ structures are evident (Fig. 3B,C). On the other hand, the highest increase in the correlation values with p1-p6$_{WT}$ structure as a result of coevolution is observed for residues 25, 26, 27, and 39 in the second slowest mode (Fig. 3E,F).

The correlated mutations at residues 30 and 88 lead to a significant change in the orientation of the fluctuations of residue 69 in the slowest mode, which leads to the dramatic difference in the dynamic behavior with an additional asymmetry between the two monomers of the protease. The other significant changes in the fluctuation orientation in this mode are at residues 56' and 78'. Nevertheless, with the coevolving mutation at P1' site in the substrate, the re-orientation of the fluctuations of these residues close to their wild-type position is observed remarkably (Fig. 3A–C). Residues 56 and 78, being flap and substrate cleft residues with low mobility, are found at the hinge region that connects the 40's and 70's loops to the flaps, respectively. Residue 69 is found on the 70's loop which moves in a manner of a cantilever with the flaps, where flaps close as the cantilever moves up (Lebon and Ledecq 2000). Therefore, the motion of residues 56, 69, and 78 is coupled to the motion of the flaps, which is known to be the significant functional motion of the protease (Nicholson et al. 1995; Kurt et al. 2003; Hornak and Simmerling 2007). The importance of such protease regions which interact with

the flaps is implied in studies of developing allosteric inhibitors for HIV-1 protease that do not compete for the active site, where they are targeted as allosteric sites (Lebon and Ledecq 2000; Perryman et al. 2004; Hornak and Simmerling 2007; Yang et al. 2012). On the other hand, in the second slowest mode, protease mutations lead a fluctuation orientation difference particularly at residue 39 in the unprimed monomer together with the active site and dimerization interface residues, which is recovered by the coevolving mutation on the substrate (Fig. 3D–F). Residue 39 has high mobility in the slowest mode yet acts as a hinge in the second slowest mode, where it interacts and fluctuates in the opposite direction with the flaps of the same monomer (Ozer et al. 2010). Due to this anticorrelated behavior with the flap motion, 39 is also an important residue which has been considered as a potential allosteric inhibition site (Tozzini and McCammon 2005). By introducing mutations and specific cross-links at residues around 39 to restrict the hydrophobic core rearrangements, the essential role of core flexibility in modulating the activity of HIV-1 protease has been demonstrated (Mittal et al. 2012).

The most variation being in the orientation of the fluctuations around the hinge axes implies that the protease mutations perturb the mechanistically crucial sites that mainly coordinate the intrinsic dynamics of the protease in interaction with its substrate. The protease mutations with the example of p1-p6 substrate complex here show the deformation of the hinge axes, representing a global deformation with local perturbations. Nevertheless, the fluctuations in the coevolved structure have the highest correlation with that of the wild-type structure among other mutant structures. The deformation in the direction of fluctuations and the additional asymmetry of the protease monomers is recovered considerably by this coevolution. This provides a more comprehensive view with dynamics incorporated as an additional dimension, rather than just the structural information about the positions of the mutations.

The structures of p1-p6 bound protease studied here are the representative conformations of the largest of five, three, and four clusters in p1-p6$_{WT}$, p1-p6$_{D30N/N88D}$, and $^{LP1'F}$p1-p6$_{D30N/N88D}$ MD simulation trajectories, respectively (Table 1). Then, the ANM analyses of the representative members of the second largest clusters are also investigated. In the most cooperative modes of motion of these structures, the maximum orientational variations are observed in the same residues that lie along the hinge axes in p1-p6$_{WT}$ where the directions of their fluctuations are significantly distorted in p1-p6$_{D30N/N88D}$. On the other hand, these variations in the fluctuation directions of the hinge residues of $^{LP1'F}$p1-p6$_{D30N/N88D}$ largely remain within the variations observed in p1-p6$_{WT}$ compared to those observed in p1-p6$_{D30N/N88D}$.

Coevolution in NC-p1

Figure 4 displays the orientational correlation values of the fluctuation vectors of protease residues of NC-p1$_{V82A}$ and ^{AP2V}NC-p1$_{V82A}$ with respect to NC-p1$_{WT}$ for the slowest two modes. The residues of the protease mutant that display the largest orientational difference in their fluctuations with respect to the wild type are 56, 69, 78, 93 in both monomers in the slowest mode (Fig. 4A). In the second slowest mode, the fluctuations of residue 39 in the unprimed monomer and residues 25, 26, 27, 49, 50, 51, 97 in both monomers display the largest orientational difference as a result of the protease mutation (Fig. 4D). These residues with largest deviations in their orientations lie along the hinge axes of rotational motion in both modes (Fig. 2), and they are found at almost identical sites in the p1-p6 and NC-p1 bound structures, where they interact with functional regions such as flaps and cleft covering the active site.

The correlation coefficients of the residue fluctuation vectors between NC-p1$_{WT}$ and ^{AP2V}NC-p1$_{V82A}$ in both modes are higher than they are between NC-p1$_{WT}$ and NC-p1$_{V82A}$, yet the increases in the correlations are not as significant as in p1-p6. By considering the residues with maximum variations in their orientational correlations that are below the lower standard deviation bound (average minus one standard deviation), it is assessed that the

average correlation coefficient value which is 0.45 between NC-p1$_{WT}$ and NC-p1$_{V82A}$ increases to 0.58 between NC-p1$_{WT}$ and ^{AP2V}NC-p1$_{V82A}$ in the slowest mode. In the second slowest mode, the average coefficient value does not improve much (0.026–0.052, respectively) for the correlations of NC-p1$_{WT}$ with NC-p1$_{V82A}$ and ^{AP2V}NC-p1$_{V82A}$. The relatively low increase in the average correlation values is due to the opposing behavior observed in the residues as a result of coevolution; some correct their fluctuations while some fluctuate more diversely with respect to the wild type.

The structures in Fig. 4 display the residues that exhibit the maximum variations between the directions of fluctuations, as color coded according to the change in the correlation values of NC-p1$_{V82A}$ (Fig. 4B,E) and ^{AP2V}NC-p1$_{V82A}$ (Fig. 4C,F) compared to NC-p1$_{WT}$. In the slowest mode, the correlations of residues 66, 69, 89, 92, 93, 77', 89', and 93' between ^{AP2V}NC-p1$_{V82A}$ and NC-p1$_{WT}$ are higher than their correlations between NC-p1$_{V82A}$ and NC-p1$_{WT}$ (Fig. 4B,C). In the second slowest mode, an increase in the correlation values with NC-p1$_{WT}$ as a result of coevolution is observed for residues 25, 26, 27, 28, and 97 of both monomers, whereas the correlation values of residues 49, 50, and 51 on the flap regions decrease (Fig. 4E,F).

Additionally, the protease mutation at residue 82 causes the difference in the dynamic behavior with an additional asymmetry between the protease monomers, particularly by the change in the orientation of the fluctuation of residue 93 in the slowest mode and residue 97 in the second slowest mode. The deformed orientations of the residue fluctuations return closer to their behavior in the wild-type dynamics by the consequent coevolving mutation at P2 site in the NC-p1 substrate (Fig. 4). The importance of the dimerization interface regions was also emphasized in allosteric inhibition studies of HIV-1 protease: Besides the allosteric inhibition studies targeting the flap motion of HIV-1 protease, another class of allosteric inhibitors investigated are dimerization inhibitors that would prevent the formation of the active protease homodimer by binding to the dimerization interface (Hornak and Simmerling 2007; Yang et al. 2012).

The structures of NC-p1 bound protease studied here are the representative conformations of the largest of four, three, and five clusters in NC-p1$_{WT}$, NC-p1$_{V82A}$, and ^{AP2V}NC-p1$_{V82A}$ MD simulation trajectories, respectively (Table 1). As in p1-p6 bound structures, the representative members of the second largest clusters are also investigated by ANM. The residues that lie along the hinge axes affirm maximum orientational variations with significant distortions in the directions of their fluctuations in NC-p1$_{V82A}$ in the most cooperative modes of motion. Furthermore, the variations in the fluctuation directions of the hinge residues

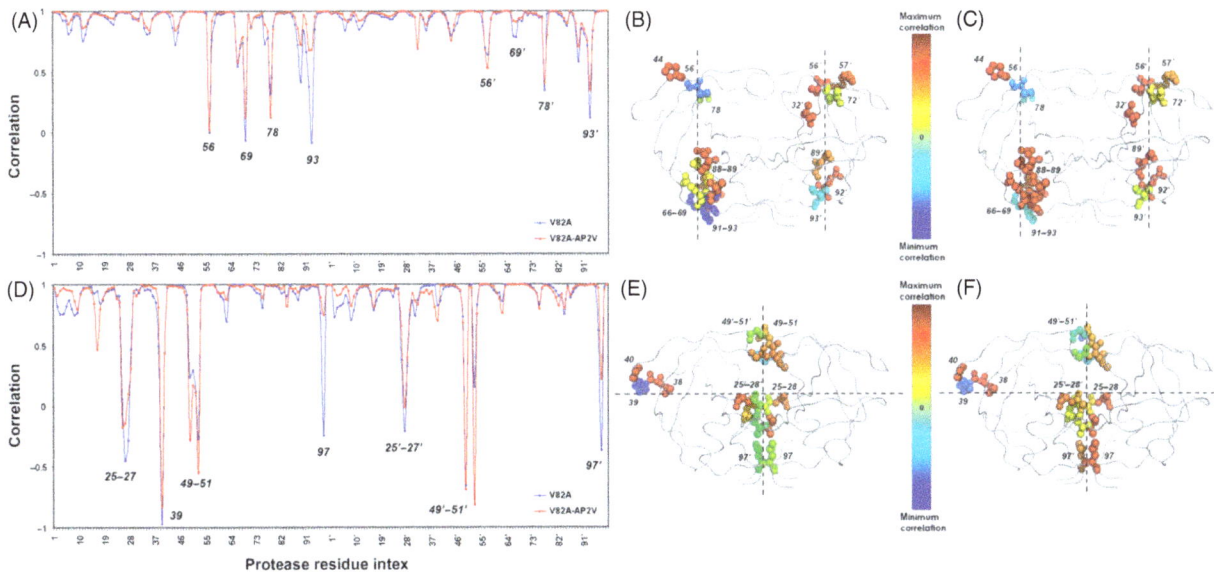

Figure 4 Orientational correlation of protease residues' fluctuations of NC-p1$_{V82A}$ and ^{AP2V}NC-p1$_{V82A}$ to those of NC-p1$_{WT}$ in the first (A) and second (D) slowest modes. The residues that display the maximum variations between the directions of fluctuations are color coded according to the change in the correlations of NC-p1$_{V82A}$ and ^{AP2V}NC-p1$_{V82A}$ compared to NC-p1$_{WT}$; (B) and (C) in the first slowest mode, and (E) and (F) in the second slowest mode, respectively. The line plot connecting the points that indicate the correlation values is used to guide the eye.

of ^{AP2V}NC-p1$_{V82A}$ remain within the variations observed in NC-p1$_{WT}$ compared to those observed in NC-p1$_{V82A}$.

Here, within the NC-p1 structures in the slowest two modes, higher correlation is observed in most of the hinge residues between wild-type and coevolved complex structures compared to that between wild-type and V82A mutant, resembling the orientational correlations within the p1-p6 structures. Yet, the lower correlations between the wild-type and the coevolved structures of NC-p1 compared to those in the p1-p6 structures, as well as the decreasing correlations for some of the hinge residues, suggest that the repossession of the structural dynamics as a result of the re-orientation by the coevolutionary mutation in the NC-p1 site is not as strong as that in the p1-p6 site.

Coevolved NC-p1 and other substrate complex structures

The dynamics of mutant NC-p1 complex structures are also studied with respect to that of the wild-type structures of HIV-1 protease bound to the natural substrates other than the NC-p1 itself, namely MA-CA, CA-p2, p2-NC, p1-p6, RT-RH, and RH-IN. The orientational correlations of the fluctuation vectors of protease residues of the NC-p1 complex structures with those of the wild-type complex structures of each of the other six natural substrates are calculated separately, and the average correlation per residue position over six correlation values is computed. The orientational correlation of protease residues of NC-p1$_{WT}$, NC-p1$_{V82A}$, and ^{AP2V}NC-p1$_{V82A}$ with respect to the average

of the wild-type structures of the other six natural substrates in the two most cooperative modes of motion is displayed in Fig. 5. The residues that display the maximum orientational difference in their fluctuations are observed at the similar hinge regions as in the previous cases, specifically at proximity of residues 56, 69, 78, 89, and 93 of both monomers in the slowest mode (Fig. 5A), and at residues 39 of the unprimed monomer, 97 of the primed monomer, and 25, 26, 27, 28, 52 of both monomers in the second slowest mode (Fig. 5D). The importance of these hinge regions in both function and allosteric communication, being either residues of the active site (25–28), the flaps (52, 56), the substrate cleft (78, 89), the dimerization interface (93, 97), or those exhibiting linked motion with the functionally important flaps (39, 69), should be noted (Lebon and Ledecq 2000; Perryman et al. 2004; Tozzini and McCammon 2005; del Sol et al. 2006; Hornak and Simmerling 2007).

In both modes, the correlations of the residue fluctuation vectors of the hinge regions in each of the natural substrate complexes other than NC-p1 with both of the mutant NC-p1 complex structures are higher compared to those with the wild-type NC-p1 complex structure. The correlation coefficients calculated over the least correlating residues (having correlation values below the lower standard deviation bound) are 0.64, 0.84, and 0.87 in the slowest mode and 0.32, 0.59, and 0.57 in the second slowest mode in turn for NC-p1$_{WT}$, NC-p1$_{V82A}$, and ^{AP2V}NC-p1$_{V82A}$ with respect to the average of the wild-type structures of the

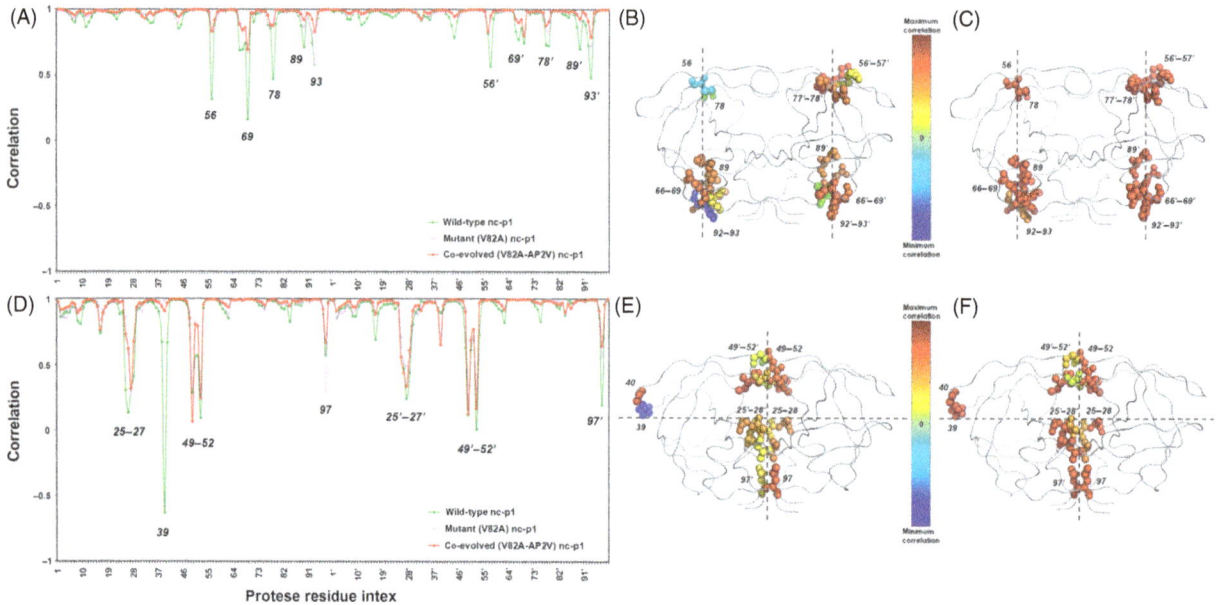

Figure 5 Orientational correlation of protease residues' fluctuations of NC-p1$_{WT}$, NC-p1$_{V82A}$, and ^{AP2V}NC-p1$_{V82A}$ to those of the averaged wild-type natural substrate complexes other than NC-p1 in the first (A) and second (D) slowest modes. The residues that display the maximum variations between the directions of fluctuations are color coded according to the change in the correlation of NC-p1$_{WT}$ and ^{AP2V}NC-p1$_{V82A}$ compared to the averaged wild-type natural substrate complexes other than NC-p1; (B) and (C) in the first slowest mode, and (E) and (F) in the second slowest mode, respectively. The line plot connecting the points that indicate the correlation values is used to guide the eye.

other six natural substrates. It appears that the correlation between the NC-p1 complex structure with the other natural substrate bound structures increases as a result of the mutation V82A in the protease. However, with the consecutive coevolving mutation AP2V in the substrate, this correlation does not increase significantly yet decreases slightly in the second slowest mode.

The residues with maximum orientational differences between their directions of fluctuations can also be observed on the structures in Fig. 5, as color coded according to the change in the correlation of the NC-p1$_{WT}$ (Fig. 5B,E) and ^{AP2V}NC-p1$_{V82A}$ (Fig. 5C,F) compared to the average of the wild-type protease structures bound to the other six natural substrates. As a result of the mutations in the protease and the NC-p1 substrate, the residues of the hinge regions explicitly mentioned above exhibit increased correlations with the rest of the natural substrate bound protease structures.

Here, it is interesting to note the higher number of protease residues in NC-p1$_{V82A}$ and ^{AP2V}NC-p1$_{V82A}$ possessing higher correlations with those in the wild-type complex structures bound to the rest of the natural substrates. This is consistent with the structural rationale for HIV-1 protease binding to the NC-p1 cleavage site given in Prabu-Jeyabalan's work (Prabu-Jeyabalan et al. 2004), where they solved the crystal structures of wild-type and V82A mutant proteases in complex with their respective wild-type and

AP2V mutant NC-p1 substrates. They observed that the AP2V mutant peptide bound the mutant protease more optimally than the wild-type NC-p1 peptide bound the wild-type protease. That is, the AP2V mutation on the peptide coevolving with the V82A mutation on the protease re-orients the peptide to a conformation which is more similar to those of the other natural substrate-protease complexes than the NC-p1 (Prabu-Jeyabalan et al. 2004).

In the analyses outlining the coupling between catalysis and conformational mechanics, there is growing evidence that enzymatic activity results from a delicate interplay between chemical kinetics and molecular motions (Yang and Bahar 2005). The catalytic sites are found at proximity of binding sites which enjoy flexibility to accommodate the ligand binding, and the accompanying large-scale conformational changes are connected to the hinge motion (Ferreiro et al. 2011). Overall, the variation in the residues at the hinge regions of HIV-1 protease in the functional modes is required for the protease to process different substrates, which results in specific cleavage rates for the proper functioning of the virus life cycle. Therefore, the interdependent nature of the substrate recognition allowing the protease to recognize various nonhomologous sequences as natural substrates may be partly due to this adaptability in the hinge regions which are the mechanistically crucial sites that mainly coordinate the intrinsic dynamics.

Conclusion

Structural dynamics analyses contribute largely to the understanding of functional and evolutionary properties of proteins, which suggest that the preservation of dynamic properties is critical for maintaining the biological function (Gerek et al. 2013). The global motions of the proteins are described by the most cooperative normal modes of the ANM. These modes have the highest contribution to the flexibility profiles, and they are predominantly defined by the proteins' architecture. Therefore, they are generally functionally relevant, evolutionarily conserved and are more robust with respect to mutational perturbations (Liu and Bahar 2012). Also, the conserved hinge regions identified in these conserved modes are shown to play decisive roles in conformational transitions induced by binding. Here, the examination of the structural and dynamic properties of the mutant and coevolved structures of p1-p6 and NC-p1 substrate complexes contributes to the understanding of the binding as well as the drug-resistant mechanism of HIV-1 protease. Overall, there seems to be interplay between the variation in the fluctuations of the important hinge regions and the variation in the substrate sites of the protease in regard to functionality. That is, a plausible complex dependence between the hinge behavior and the specific functionality of the substrate with respect to the rate of cleavage can be inferred. The mutation in the substrate allows the protease residues to re-orient and thus fluctuate as in the functional conformation, and justify the existence of this coevolutionary mutation for the conservation of, at least, the fluctuations and flexibility.

Understanding the determinants of ligand recognition and binding in sequence, structure, function, and dynamics paradigm is now a major challenge in drug discovery. The structural data for target proteins with different ligands display the contribution coming from both partners in selecting the bound forms. The intrinsic dynamics of proteins appears to be optimized by evolution for functional interactions, yet the ligand selects the one that best fits its structural and dynamic properties among the conformations accessible to the unbound protein. In binding, the variety of ligands with different compositions and shapes as well as affinity and selectivity can be explained by the conformational flexibility of receptors. Being related to dynamics and evolution, flexibility should also have impacts on structural divergence connected to the orientational difference in the fluctuation of the molecules. This implies that drug-resistant studies should go beyond the concept of inhibition of structure to the concept of inhibition of functional dynamics by focusing on flexibility as well.

Acknowledgements

We thank Dr. Nese Kurt Yilmaz for editorial assistance. This research is supported by NIH: AIDS-FIRCA RO3 TW006875-01. TH acknowledges Turkish State Planning Organization grant 2009K120520 and Betil Fund. CAS acknowledges R01 GM65347.

Literature cited

Atilgan, A. R., S. R. Durell, R. L. Jernigan, M. C. Demirel, O. Keskin, and I. Bahar 2001. Anisotropy of fluctuation dynamics of proteins with an elastic network model. Biophysical Journal **80**:505–515.

Bahar, I., A. R. Atilgan, and B. Erman 1997. Direct evaluation of thermal fluctuations in proteins using a single parameter harmonic potential. Folding and Design **2**:173–181.

Bahar, I., A. R. Atilgan, M. C. Demirel, and B. Erman 1998. Vibrational dynamics of folded proteins: significance of slow and fast motions in relation to function and stability. Physical Review Letters **80**:2733–2736.

Bahar, I., T. R. Lezon, L. Yang, and E. Eyal 2010. Global dynamics of proteins: bridging between structure and function. Annual Review of Biophysics **39**:23–42.

Bakan, A., and I. Bahar 2009. The intrinsic dynamics of enzymes plays a dominant role in determining the structural changes induced upon inhibitor binding. Proceedings of the National Academy of Sciences of the United States of America **106**:14349–14354.

Bally, F., R. Martinez, S. Peters, P. Sudre, and A. Telenti 2000. Polymorphism of HIV-1 type 1 Gag p7/p1 and p1/p6 cleavage sites: clinical significance and implications for resistance to protease inhibitors. AIDS Research and Human Retroviruses **16**:1209–1213.

Berkhout, B., and R. W. Sanders 2005. The evolvability of drug resistance: the HIV-1 case. Journal of Biomolecular Structure and Dynamics **22**:619–621.

Bernstein, E. E., T. F. Koetzle, G. J. B. Williams, J. E. F. Meyer, M. D. Brice, J. R. Rodgers, O. Kennard et al. 1977. The Protein Data Bank: a computer-based archival file for macromolecular structures. Journal of Molecular Biology **117**:535–542.

Chennubhotla, C., A. J. Rader, L. Yang, and I. Bahar 2005. Elastic network models for understanding biomolecular machinery: from enzymes to supramolecular assemblies. Physical Biology **2**:S173–S180.

Creavin, T. 2004. Evolvability: implications for drug design. Drug Discovery Today: Targets **3**:178.

Cui, Q., and I. Bahar 2005. Normal Mode Analysis: Theory and Applications to Biological and Chemical Systems. CRC Press, Boca Raton, FL.

Doyon, L., G. Croteau, D. Thibeault, F. Poulin, L. Pilote, and D. Lamarre 1996. Second locus involved in human immunodeficiency virus Type 1 resistance to protease inhibitors. Journal of Virology **78**:3763–3769.

Earl, D. J., and M. W. Deem 2004. Evolvability is a selectable trait. Proceedings of the National Academy of Sciences of the United States of America **101**:11531–11536.

Feher, A., I. T. Weber, P. Bagossi, P. Boross, B. Mahalingam, J. M. Louis, T. D. Copeland et al. 2002. Effect of sequence polymorphism and drug resistance on two HIV-1 Gag processing sites. European Journal of Biochemistry **269**:4114–4120.

Feig, M., J. Karanicolas, and C. L. Brooks 2004. MMTSB Tool Set: enhanced sampling and multiscale modeling methods for applications in structural biology. Journal of Molecular Graphics and Modelling **22**:377–395.

Ferreiro, D. U., J. A. Hegler, E. A. Komives, and P. G. Wolynes 2011. On the role of frustration in the energy landscapes of allosteric proteins. Proceedings of the National Academy of Sciences of the United States of America 108:3499–3503.

Friedland, G. D., N. A. Lakomek, C. Griesinger, J. Meiler, and T. Kortemme 2009. A correspondence between solution-state dynamics of an individual protein and the sequence and conformational diversity of its family. PLoS Computational Biology 5:1–16.

Gerek, Z. N., S. Kumar, and S. B. Ozkan 2013. Structural dynamics flexibility informs function and evolution at a proteome scale. Evolutionary Applications 6:423–433.

Gerstein, M., and N. Echols 2004. Exploring the range of protein flexibility, from a structural proteomics perspective. Current Opinion in Chemical Biology 8:14–19.

Gniewek, P., A. Kolinski, R. L. Jernigan, and A. Kloczkowski 2012. Elastic network normal modes provide a basis for protein structure refinement. The Journal of Chemical Physics 136:195101–195110.

Haliloglu, T., I. Bahar, and B. Erman 1997. Gaussian dynamics of folded proteins. Physical Review Letters 79:3090–3093.

Hamacher, K. 2010. Temperature dependence of fluctuations in HIV-1 protease. European Biophysics Journal: EBJ 39:1051–1056.

Hamacher, K., and J. A. McCammon 2006. Computing the amino acid specificity of fluctuations in biomolecular systems. Journal of Chemical Theory and Computation 2:873–878.

Hornak, V., and C. Simmerling 2007. Targeting structural flexibility in HIV-1 protease inhibitor binding. Drug Discovery Today 12:132–138.

Juan, D., F. Pazos, and A. Valencia 2008. Coevolution and co-adaptation in protein networks. FEBS Letters 582:1225–1230.

King, N., M. Prabu-Jeyabalan, E. A. Nalivaika, and C. A. Schiffer 2004. Combating susceptibility to drug resistance: lessons from HIV-1 protease. Chemistry and Biology 11:1333–1338.

Kolli, M., S. Lastere, and C. A. Schiffer 2006. Coevolution of nelfinavir-resistant HIV-1 protease and the p1-p6 substrate. Virology 347:405–409.

Kolli, M., E. Stawiski, C. Chappey, and C. A. Schiffer 2009. Human Immunodeficiency Virus type 1 protease-correlated cleavage site mutations enhance inhibitor resistance. Journal of Virology 83:11027–11042.

Kurt, N., W. R. Scott, C. A. Schiffer, and T. Haliloglu 2003. Cooperative fluctuations of unliganded and substrate-bound HIV-1 protease: a structure-based analysis on a variety of conformations from crystallography and molecular dynamics simulations. Proteins 51:409–422.

Lebon, F., and M. Ledecq 2000. Approaches to the design of effective HIV-1 protease inhibitors. Current Medicinal Chemistry 7:455–477.

Liu, Y., and I. Bahar 2012. Sequence evolution correlates with structural dynamics. Molecular Biology and Evolution 29:2253–2263.

Lovell, S. C., and D. L. Robertson 2010. An integrated view of molecular coevolution in protein-protein interactions. Molecular Biology and Evolution 27:2567–2575.

Lyman, E., and D. M. Zuckerman 2006. Ensemble-based convergence analysis of biomolecular trajectories. Biophysical Journal 91:164–172.

Ma, J. 2005. Usefulness and limitations of normal mode analysis in modeling dynamics of biomolecular complexes. Structure 13:373–380.

Maguid, S., S. Fernandez-Alberti, G. Parisi, and J. Echave 2008. Evolutionary conservation of protein vibrational dynamics. Gene 422:7–13.

Marianayagam, N. J., and S. E. Jackson 2005. Native-state dynamics of the ubiquitin family: implications for function and evolution. Journal of the Royal Society Interface 2:47–54.

Micheletti, C., P. Carloni, and A. Maritan 2004. Accurate and efficient description of protein vibrational dynamics: comparing molecular dynamics and Gaussian models. Proteins 55:635–645.

Mittal, S., Y. Cai, M. N. Nalam, D. N. A. Bolon, and C. A. Schiffer 2012. Hydrophobic core flexibility modulates enzyme activity in HIV-1 protease. Journal of the American Chemical Society 134:4163–4168.

Nalam, M. N., and C. A. Schiffer 2008. New approaches to HIV protease inhibitor drug design II: testing the substrate envelope hypothesis to avoid drug resistance and discover robust inhibitors. Current Opinion in HIV and AIDS 6:642–646.

Nicholson, L. K., T. Yamazaki, D. A. Torchia, S. Grzesiek, A. Bax, S. J. Stahl, J. D. Kaufman et al. 1995. Flexibility and function in HIV-1 protease. Nature Structural and Molecular Biology 2:274–280.

Nicolay, S., and Y. H. Sanejouand 2006. Functional modes of proteins are among the most robust. Physical Review Letters 96:078104.

Ozen, A., T. Haliloglu, and C. A. Schiffer 2011. Dynamics of preferential substrate recognition in HIV-1 protease: redefining the substrate envelope. Journal of Molecular Biology 410:726–744.

Ozen, A., T. Haliloglu, and C. A. Schiffer 2012. HIV-1 protease and substrate coevolution validates the substrate envelope as the substrate recognition pattern. Journal of Chemical Theory and Computation 8:703–714.

Ozer, N., C. A. Schiffer, and T. Haliloglu 2010. Rationale for more diverse inhibitors in competition with substrates in HIV-1 protease. Biophysical Journal 99:1650–1659.

Perryman, A. L., J. Lin, and J. A. McCammon 2004. HIV-1 protease molecular dynamics of a wild-type and of the V82F/I84V mutant: Possible contributions to drug resistance and a potential new target site for drugs. Protein Science 13:1108–1123.

Prabu-Jeyabalan, M., E. Nalivaika, and C. A. Schiffer 2002. Substrate shape determines specificity of recognition for HIV-1 protease: analysis of crystal structures of six substrate complexes. Structure 10:369–381.

Prabu-Jeyabalan, M., E. A. Nalivaika, N. M. King, and C. A. Schiffer 2004. Structural basis for coevolution of a human immunodeficiency virus type 1 nucleocapsid-p1 cleavage site with a V82A drug-resistant mutation in viral protease. Journal of Virology 78:12446–12454.

Ramanathan, A., and P. K. Agarwal 2011. Evolutionarily conserved linkage between enzyme fold, flexibility, and catalysis. PLoS Biology 9:1–17.

del Sol, A., H. Fujihashi, D. Amoros, and R. Nussinov 2006. Residues crucial for maintaining short paths in network communication mediate signaling in proteins. Molecular Systems Biology 2:1–12.

Teague, S. J. 2003. Implications of protein flexibility for drug discovery. Nature 2:527–541.

Thompson, J. N. 1994. The Coevolutionary Process. University of Chicago Press, Chicago, IL.

Tirion, M. M. 1996. Large amplitude elastic motions in proteins from a single-parameter, atomic analysis. Physical Review Letters 77:1905–1908.

Tomatis, P. E., S. M. Fabiane, F. Simona, P. Carloni, B. J. Sutton, and A. J. Vila 2008. Adaptive protein evolution grants organismal fitness by improving catalysis and flexibility. Proceedings of the National Academy of Sciences of the United States of America 105:20605–20610.

Tournier, A. L., and J. C. Smith 2003. Principal components of the protein dynamical transition. Physical Review Letters 91:208106.

Tözser, J., I. Blaha, T. D. Copeland, E. M. Wondrak, and S. Oroszlan 1991. Comparison of the HIV-1 and HIV-2 proteinases using oligopeptide substrates representing cleavage sites in Gag and Gag-Pol polyproteins. FEBS Letters 281:77–80.

Tozzini, V., and J. A. McCammon 2005. A coarse grained model for the dynamics of flap opening in HIV-1 protease. Chemical Physics Letters **413**:123–128.

Wlodawer, A., and J. W. Erickson 1993. Structure-based inhibitors of HIV-1 protease. Annual Review of Biochemistry **62**:543–585.

Xia, Y., and M. Levitt 2004. Simulating protein evolution in sequence and structure space. Current Opinion in Structural Biology **14**:202–207.

Yang, L., and I. Bahar 2005. Coupling between catalytic site and collective dynamics: a requirement for mechanochemical activity of enzymes. Structure **13**:893–904.

Yang, L., G. Song, A. Carriquiry, and R. L. Jernigan 2008. Close correspondence between the motions from principal component analysis of multiple HIV-1 protease structures and elastic network modes. Structure **16**:321–330.

Yang, H., J. Nkeze, and R. Y. Zhao 2012. Effects of HIV-1 protease on cellular functions and their potential applications in antiretroviral therapy. Cell & Bioscience **2**:32.

Zheng, W., and B. Brooks 2005. Identification of dynamical correlations within the myosin motor domain by the normal mode analysis of an elastic network model. Journal of Molecular Biology **346**:745–759.

Zoete, V., O. Michielin, and M. Karplus 2002. Relation between sequence and structure of HIV-1 protease inhibitor complexes: a model system for the analysis of protein flexibility. Journal of Molecular Biology **315**:21–52.

13

Impact of violated high-dose refuge assumptions on evolution of *Bt* resistance

Pascal Campagne,[1,2,3,4,5] Peter E. Smouse,[3] Rémy Pasquet,[1,2,4] Jean-François Silvain,[1,2] Bruno Le Ru[1,2,4] and Johnnie Van den Berg[6]

1 Laboratoire Évolution, Génome et Spéciation, CNRS UPR9034, Unité de Recherche IRD 072, Gif-sur-Yvette, France
2 Université Paris-Sud 11, Orsay, France
3 Department of Ecology, Evolution & Natural Resources, School of Environmental & Biological Sciences, Rutgers University, New Brunswick, NJ, USA
4 Noctuid Stem Borers Biodiversity in Africa Project, Environmental Health Division, International Centre for Insect Physiology & Ecology, Nairobi, Kenya
5 Institute of Integrative Biology, University of Liverpool, Liverpool, UK
6 School of Biological Sciences - Zoology, North-West University, Potchefstroom, South Africa

Keywords

fitness cost, high-dose, incomplete resistance, insecticide resistance, nonrandom mating, partial dominance, refuge strategy.

Correspondence

Pascal Campagne, Institute of Integrative Biology, Biosciences Building, University of Liverpool, Liverpool L69 7ZB, Liverpool, UK.

e-mail: Pascal.Campagne@liv.ac.uk

Abstract

Transgenic crops expressing *Bacillus thuringiensis* (*Bt*) toxins have been widely and successfully deployed for the control of target pests, while allowing a substantial reduction in insecticide use. The evolution of resistance (a heritable decrease in susceptibility to *Bt* toxins) can pose a threat to sustained control of target pests, but a high-dose refuge (HDR) management strategy has been key to delaying countervailing evolution of *Bt* resistance. The HDR strategy relies on the mating frequency between susceptible and resistant individuals, so either partial dominance of resistant alleles or nonrandom mating in the pest population itself could elevate the pace of resistance evolution. Using classic Wright-Fisher genetic models, we investigated the impact of deviations from standard refuge model assumptions on resistance evolution in the pest populations. We show that when *Bt* selection is strong, even deviations from random mating and/or strictly recessive resistance that are below the threshold of detection can yield dramatic increases in the pace of resistance evolution. Resistance evolution is hastened whenever the order of magnitude of model violations exceeds the initial frequency of resistant alleles. We also show that the existence of a fitness cost for resistant individuals on the refuge crop cannot easily overcome the effect of violated HDR assumptions. We propose a parametrically explicit framework that enables both comparison of various field situations and model inference. Using this model, we propose novel empiric estimators of the pace of resistance evolution (and time to loss of control), whose simple calculation relies on the observed change in resistance allele frequency.

Introduction

Genetically modified crops, expressing insecticidal toxins of *Bacillus thuringiensis* (*Bt*), were first introduced in 1995 and have now been adopted worldwide; by 2010, they had been planted on ~66 Mha of agricultural crop land (James 2011). While *Bt*-expressing crops have met with considerable success, resistance can arise whenever a pest population develops a genetically based decrease in susceptibility to the toxin (Tabashnik et al. 2009), which may lead in turn to drastic loss of *Bt* crop efficacy under field conditions (i.e., effective field resistance). While resistant mutations have been reported in many cases (Tabashnik et al. 2013), almost two decades after *Bt* crops were first deployed, clearly documented cases of effective field resistance have arisen in only four pests: *Busseola fusca* (South Africa, Van Rensburg 2007), *Spodoptera frugiperda* (Puerto Rico, Storer et al. 2010), *Pectinophora gossypiella* (India, Dhurua and Gujar 2011), and *Diabrotica virgifera virgifera* (USA, Gassmann et al. 2011).

Much attention has been devoted to the pace of resistance evolution (Tabashnik et al. 2013), as well as to developing operational strategies that can delay (Alstad and Andow 1995) or eventually reverse it (Carrière et al. 2010). Among them, the high-dose/refuge (henceforth, HDR) strategy, resulting in a lowered selection pressure on susceptible individuals (Carrière et al. 2010), has generally been effective (Huang et al. 2011), particularly in the USA, where its proper implementation has seldom led to loss of control (Tabashnik et al. 2013). This strategy amounts to planting nonresistant cultivars within or surrounding Bt-crop plantings, allowing the survival of some susceptible individuals in a Bt-dominated environment. If susceptible alleles (S) in the pest are dominant and rare resistant mutants (R) are completely recessive, then rare resistant individuals (RR) emerging from Bt plants will mate preferentially with susceptible individuals (SS) emerging from refuge plants. Crosses between (RR) and (SS) parents yield (RS) progeny, so if the dose of Bt toxin expressed is high enough to kill 100% of heterozygous (RS) larvae, the HDR strategy should strongly delay evolution of pest resistance to Bt toxins. Recommended refuge fractions for Bt crops have ranged from ~5% to 50% of crop acreage in the USA (Bates et al. 2005), depending notably on whether or not they were also sprayed with insecticide.

Theory shows that optimal efficiency of the HDR strategy is guaranteed when: (i) the genetic bases of resistance in natural populations and the dose of toxin expressed by the plant result in functionally recessive expression in the pest; (ii) mating is random among pest genotypes, with regard to Bt resistance; and (iii) the frequency of resistant mutants is low. The available data suggest that low background frequencies (q_0) of resistance alleles are associated with sustained susceptibility to Bt toxin (Tabashnik et al. 2013), so most modeling studies have explored cases where ($q_0 \leq 0.001$) (e.g., Tyutyunov et al. 2008).

·Success of the HDR strategy depends on the dominance level of the resistance allele ($1 > h > 0$), with $h = 0$ corresponding to a recessive trait and $h = 1$ to a dominant trait (Wright 1934). It also depends on the rate of nonrandom mating for resistant genotypes ($F > 0$), resulting in excesses of resistant homozygotes (RR), relative to panmictic expectation. Success also depends on the background frequency of (or rate of mutation to) resistant alleles ($q_0 > 0$), as well as to the proportion ($1 - \omega$) of the susceptible (refuge) crop that is planted.

The fraction of Bt crop planted in the landscape (ω) is expected to scale with the proportion of susceptible pest individuals killed by the toxin. A lack of refuge planting in India and China has apparently allowed rapid evolution of P. gossypiella resistance to Cry1Ac Bt cotton (Tabashnik et al. 2013). Similarly, low compliance among South African farmers in planting the recommended fraction of refuge Z. mays crop might have hastened the evolution of Bt resistance in the stem borer (B. fusca) (Kruger et al. 2012).

A review of documented cases of field monitoring has shown that rapid evolution of resistance occurs predominantly when the initial frequency of resistance allele (q_0) was above the threshold of detectability (Tabashnik et al. 2013). It has also been shown, however, that sustained susceptibility to Bt toxins can be achieved in the field, even when ($q_0 > 0.001$), when coupled with a high fraction ($1 - \omega$) > 40% of refuge acreage (Tabashnik et al. 2013).

Either failure to achieve a high-dose concentration of toxin in plant tissues and/or the presence of partially dominant ($h > 0$) resistance alleles yields a surviving fraction of heterozygous (RS) larvae on Bt plants, which compromises HDR success. Notwithstanding the potential problems, recessive inheritance has been supported by numerous studies of both laboratory-selected and field-evolved Bt resistance (e.g., Ferré and Van Rie 2002; Tabashnik et al. 2003). On the other hand, it is notoriously difficult to estimate dominance (h) levels reliably, under either field conditions (Moar et al. 2008; Tabashnik et al. 2008) or in the laboratory, largely attributable to concentration-dependent effects of the toxin (Gould et al. 1995; Tabashnik et al. 2002). There have also been more striking cases, for which (strong) partially dominant ($h > 0.5$) resistance has been observed, probably stemming from diverse inheritance or biochemical bases of resistance in a variety of different organisms (Zhang et al. 2012; Campagne et al. 2013; Jin et al. 2013).

Likewise, any elevated tendency ($F > 0$) for resistant individuals (emerging from the Bt crop) to mate with each other, rather than with susceptible individuals (emerging from the refuge crop), profoundly increasing the frequency of resistant (RR) homozygotes among the progeny [fr(RR progeny) = $(q^2 + Fpq)$] and compromising the efficacy of the HDR strategy. Promoting mating between resistant and susceptible individuals depends on both, the spatial structure of the Bt crop and refuge blocks and individual postemergence dispersal patterns (Alstad and Andow 1995). Many pest populations conform satisfactorily to Hardy–Weinberg expectations for selectively neutral markers (Han and Caprio 2002; Endersby et al. 2007; Krumm et al. 2008; Kim et al. 2009), suggesting a mating regime close to random, but whether that same condition obtains for genetic markers under strong and spatially structured Bt selection remains unclear. In spite of extensive genetic mixing and low inbreeding levels in the moth Ostrinia nubilalis (Bourguet et al. 2000), Dalecky et al. (2006) have demonstrated that this species would be prone to positive assortative mating in Bt-crop context. Indeed, mating between resistant individuals originating from a single Bt planting could reach a few percent, as a consequence of limited premating dispersal. The effects of the spatial structure of refuge plantings have been both contentious and extensive (Onstad et al. 2011). Some modeling studies have suggested that large block refuges could be more efficient in delaying resistance evolution than scattered refuges (Tyutyunov et al. 2008); others have suggested that seed blends (yielding a spatial mixture of *Bt* and non-*Bt* plants in the field) could provide at least as much HDR durability as block refuges (Pan et al. 2011). In practice, we still know very little about the empiric rates of nonrandom mating under field conditions for most pests.

Other crucial factors that might delay the evolution of resistance have been assessed in different pest species (Gassmann et al. 2009), among them: incomplete resistance, fitness cost, and the dominance of the fitness cost. Incomplete resistance denotes situations where the fitness of resistant individuals on *Bt* plants (V_{RR}) is lower than the fitness of susceptible individuals on non-*Bt* plants (U_{SS}), i.e., when ($V_{RR} < U_{SS}$), which reduces the selective advantage of resistant individuals in mixed plantings of *Bt* and non-*Bt* plants (Carrière et al. 2006). Fitness cost arises when a resistance allele reduces the fitness of homozygotes (RR) in environments that are toxin-free, so that ($U_{SS} - U_{RR} > 0$) (Tabashnik et al. 2014). Fitness cost may also exhibit a range of dominance levels ($0 \leq g = (U_{SS} - U_{RS})/(U_{SS} - U_{RR}) \leq 1$) as shown in Table 1. Available data suggest that a recessive ($g \approx 0$) fitness cost of 25% ($U_{SS} = 1$, $U_{RR} = 0.75$, $U_{SS} - U_{RR} = 0.25$) might be a reasonable average (Gassmann et al. 2009). Management accounting for fitness cost may strengthen the effects of the HDR strategy in delaying the evolution of resistance (e.g., Higginson et al. 2005).

Failure of standard HDR assumptions (Huang et al. 2011; Tabashnik et al. 2013) has led to occasional resistance development (Tabashnik et al. 2014), and the matter needs further exploration, both theoretically and empirically. Using Wright's (1942) classical genetic model, we here explore the sensitivity of resistance evolution to assumptions of strict randomness in mating and strictly recessive resistance alleles. This study is aimed at: (i) testing the robustness of the model when F and/or h might be slightly higher than 0; (ii) assessing the extent to which nonrecessive expression and nonrandom mating may balance the effects of fitness cost ($U_{SS} - U_{RR} > 0$ and $g > 0$) and incomplete resistance ($U_{SS} - V_{RR} > 0$); (iii) evaluating whether violations of model assumptions impact the expected time elapsed before buildup of resistance in the pest threatens the efficacy of the *Bt* crop itself.

Modeling evolution of *Bt* resistance

Resistance is considered to involve a single locus, with a susceptible allele (S), of frequency p, and a resistance allele (R), of frequency q. The survival probability of the genotypes RR, RS, and SS is denoted by (U_{RR}, U_{RS}, and U_{SS}) on refuge plants and (V_{RR}, V_{RS}, and V_{SS}) on *Bt* plants (Table 1). The proportion of *Bt* crop in the landscape (ω) determines the relative fitness of the three genotypes; for modeling purposes, the spatial distribution of *Bt* and non-*Bt* plants is considered continuous and random. The net relative fitness values of the three genotypes, emerging from a spatially randomized blend of *Bt* (ω) and refuge ($1 - \omega$) plants are as follows:

$$W_{SS} = (1 - \omega) \cdot U_{SS} + \omega \cdot V_{SS}$$
$$W_{RS} = (1 - \omega) \cdot U_{RS} + \omega \cdot V_{RS} \qquad (1)$$
$$W_{RR} = (1 - \omega) \cdot U_{RR} + \omega \cdot V_{RR}$$

Any tendency for preferential mating (to type), whether due to genetically programmed behavioral or spatially imposed dispersal patterns (to or from the refuge crop), will result in assortative mating ($F > 0$) among newly emerging individuals. Given these genotypic fitness values, the parental genotypic frequencies, and the value of (F) (Table 1), we define the average relative allelic fitness values on the refuge crop as:

Table 1. Summary of allelic fitness values, under the different parametric assumptions of the model.

Model parameters	SS	RS	RR	Planting fraction
Frequencies	$p^2 + pqF$	$2pq(1 - F)$	$q^2 + pqF$	
Bt fitness	$V_{SS} < V_{RR}$	$V_{RS} = V_{SS} + h(V_{RR} - V_{SS})$	V_{RR}	ω
Refuge fitness	U_{SS}	$U_{RS} = U_{RR} + g(U_{SS} - U_{RR})$	$U_{SS} > U_{RR}$	$(1 - \omega)$

Average allelic fitness values – *Bt* crop	
$\tilde{V}_S = [(p + qF) \cdot V_{SS} + q \cdot (1 - F) \cdot V_{RS}]$	$\tilde{V}_R = [(q + pF) \cdot V_{RR} + p \cdot (1 - F) \cdot V_{RS}]$

Average allelic fitness values – refuge crop	
$\tilde{U}_S = [(p + qF) \cdot U_{SS} + q \cdot (1 - F) \cdot U_{RS}]$	$\tilde{U}_R = [(q + pF) \cdot U_{RR} + p \cdot (1 - F) \cdot U_{RS}]$

Weighted average allelic fitness values – both crops	
$\tilde{W}_S = [(1 - \omega) \cdot \tilde{U}_S + \omega \cdot \tilde{V}_S]$	$\tilde{W}_R = [(1 - \omega) \cdot \tilde{U}_R + \omega \cdot \tilde{V}_R]$

$$\tilde{U}_S = [(p + qF) \cdot U_{SS} + q \cdot (1 - F) \cdot U_{RS}]$$
$$\tilde{U}_R = [(q + pF) \cdot U_{RR} + p \cdot (1 - F) \cdot U_{RS}] \quad (2a)$$

and on the *Bt* crop as:

$$\tilde{V}_S = [(p + qF) \cdot V_{SS} + q \cdot (1 - F) \cdot V_{RS}]$$
$$\tilde{V}_R = [(q + pF) \cdot V_{RR} + p \cdot (1 - F) \cdot V_{RS}] \quad (2b)$$

At landscape level, we can then define (see Table 1 and Appendix S1) weighted average allelic fitness values ($\tilde{W}_R$ and $\tilde{W}_S$ for the collective population (Table 1):

$$\tilde{W}_S = [(1 - \omega) \cdot \tilde{U}_S + \omega \cdot \tilde{V}_S] \text{ and}$$
$$\tilde{W}_R = [(1 - \omega) \cdot \tilde{U}_R + \omega \cdot \tilde{V}_R] \quad (3)$$

Standard theory (Wright 1942) shows that the change in the frequency of the (R) allele over a single discrete generation depends on the average fitness of the advantageous allele over the population average:

$$\Delta q = (q' - q) = q \cdot \left[\frac{\tilde{W}_R}{\overline{W}} - 1 \right],$$
$$\text{where } \overline{W} = [q \cdot \tilde{W}_R + p \cdot \tilde{W}_S] \quad (4)$$

It is convenient to define an equivalent form, using $y = q/(1 - q) = (q/p)$, so that (4) can be replaced with a more convenient analogue:

$$\Delta y = \left[\frac{\tilde{W}_R}{\tilde{W}_S} - 1 \right] \cdot y$$
$$= \left[\frac{\omega \cdot \tilde{V}_R + (1 - \omega) \cdot \tilde{U}_R}{\omega \cdot \tilde{V}_S + (1 - \omega) \cdot \tilde{U}_S} - 1 \right] \cdot y = \tilde{\Lambda} \cdot y \quad (5)$$

where $\tilde{\Lambda}$ accounts for all the parameters in the model, in its most general form (Table 1). In practice, q may either increase ($\Delta y > 0$, when $\tilde{W}_R > \tilde{W}_S$) or decrease ($\Delta y < 0$, when $\tilde{W}_R < \tilde{W}_S$), while the sets of parameters for which $\Delta y = 0$ ($0 < q < 1$) delineate two alternative trajectories of the resistance allele frequency. Equation (5) expresses a balance between the selective advantage of susceptible individuals on refuge and that of resistant individuals on *Bt* crop, balanced against the refuge crop fraction ($1 - \omega$). Comparing the values of $\tilde{W}_R$ and $\tilde{W}_S$ amounts to comparing $(\tilde{U}_S - \tilde{U}_R)/(\tilde{V}_R - \tilde{V}_S)$ with $\omega/(1 - \omega)$. If $(\tilde{U}_S - \tilde{U}_R)/(\tilde{V}_R - \tilde{V}_S) > \omega/(1 - \omega)$, the resistant allele (R) increases in frequency. Conversely, if $(\tilde{U}_S - \tilde{U}_R)/(\tilde{V}_R - \tilde{V}_S) < \omega/(1 - \omega)$, the resistant allele (R) decreases in frequency. In practice, the fitness of (RR) individuals on refuge plants may be lower than that of (SS) on refuge plants ($U_{SS} - U_{RR} \geq 0$), labeled a 'fitness cost' (e.g., Gassmann et al. 2009; Tabashnik et al. 2014). Moreover, heterozygote (RS) fitness on the refuge crop may also show partial dominance, yielding ($U_{SS} > U_{RS} > U_{RR}$) on the refuge crop, counterbalanced by ($V_{RR} > V_{RS} > V_{SS}$) on

the *Bt* crop. Finally, we must also consider incomplete resistance, cases where ($U_{SS} > V_{RR}$).

Time to loss of containment (passage time)

An adaptive resistance allele (R) will increase in frequency from very low to very high, in classic sigmoidal fashion. A convenient criterion used to assess the evolution of resistance is the number of generations (henceforth 'passage time') for which the frequency of the resistant (R) allele is lower than some critical frequency in the population (say, $q_k = 0.1$), as in Tyutyunov et al. (2008). If we denote the initial frequency of the resistant allele (R) as q_0 and that of the 'critical' value as q_k, then the passage time (T^k) for the allele frequency to increase from ($q_0 \rightarrow q_k$) may be obtained by iteration of equation (5). Discrete models do not yield closed form solutions for (T^k), but continuous approximations provide relatively simple (and parametrically explicit) approximations (see Felsenstein 2007). We constructed differential equations based on the difference equations, $\delta y/\delta t = \Delta y$, and used their solutions to derive an approximate formula for passage times (Appendix S2) for each of several models.

In the general form of equation (5), the increase in resistance allele frequency for a single generation can be calculated, based on the difference between y' and y. To solve the differential equation based on the difference equation (5), $\tilde{\Lambda}$ may be easily rewritten as a ratio of two linear functions of y, so the passage time (time to loss of containment) may be calculated for the general form of the model (Appendix S2). When y_0 is small, the expression for passage time can be further simplified (Appendix S2), and we achieve a relatively simple approximation of T^k_Λ for the case where $V_{SS} = 0$ and $U_{SS} = 1$ (the reference fitness values). Given initial (q_0) and critical (q_k) frequencies of the (R) allele, the approximate passage time can be written (in terms of $y = q/p$) as (see Appendix S2 for the full expression):

$$T^k_\Lambda \approx \frac{1 - \omega}{\omega \cdot \varepsilon \cdot V_{RR} - (1 - \omega) \cdot \chi \cdot (1 - U_{RR})} \cdot \ln \left(\frac{y_k}{y_0} \right) \quad (6)$$

where $\varepsilon = (F + h - F \cdot h)$ and $\chi = (F + g - F \cdot g)$ capture the essence of deviations from classic HDR assumptions on *Bt* and refuge crops, respectively.

We observe that mild deviations from HDR assumptions (e.g., $\varepsilon = 0.05$) dramatically shorten passage time, even when fitness cost and incomplete resistance are substantial (Fig. 1). Equation (6) suggests that the role of fitness cost, in terms of both ($1 - U_{RR}$) and (g), as well as incomplete resistance, denoted by ($V_{RR} < U_{SS}$), may have an impact for large fractions of refuge ($1 - \omega$) (Fig. 1). When most of the acreage is planted to the transgenic crop (ω is elevated), however, substantial levels of fitness cost and incomplete resistance are required to delay resistance evolution substantially. Transgenic crops will presumably be dominant in the landscape, so the sensitivity of passage time to the deviation parameter $\varepsilon = (F + h - F \cdot h)$ is greater than are the protective effects of fitness cost and incomplete resistance. By inflating the frequency of homozygotes (RR), F reinforces the role of fitness cost in the second part of the denominator of equation (6), so that the parameter h is expected to have a larger effect on T_Λ^k than will F, whenever fitness cost and incomplete resistance are sizeable.

We further explored the extent to which deviations from the idealized HDR assumptions ($\varepsilon = 0$) could be compensated for by increasing the fraction of refuge, fitness cost, and incomplete resistance. We can calculate the minimal fraction of refuge required to achieve a passage time greater than a given number of generations, based on equations (5) and (6) (see also Appendix S2). Insisting on a passage time of at least 40 generations, we assessed refuge requirements, based on our generalized model (Fig. 2). Refuge requirement appeared to depend more on ε than on any other feature of the model. While the amount of refuge $(1 - \omega)$ required to ensure that $(T_\Lambda^k > 40)$ generations was typically lower than the minimal requirement of 5% (unsprayed) refuge recommended for the classic model ($\varepsilon = 0$), a suitable refuge fraction was higher than 40% ($F = 0.05 = h$) when incomplete resistance and fitness cost were moderate ($V_{RR} = 0.9$ and $1 - U_{RR} = 0.1$), respectively (Fig. 2). A low refuge requirement $(1 - \omega < 0.10)$ was only appropriate for fairly incomplete resistance and high values of fitness cost, for example, ($q_0 = 10^{-4}$, $g = 0.4$, $\varepsilon \le 0.05$, $(1 - U_{RR}) > 0.4$, $V_{RR} < 0.65$). In overview, a sustained efficacy of Bt crops over a time horizon of 20 years appears attainable for most multivoltine species, but only with large fractions of refuge.

Along the same lines, some robust strategies might even be needed to ensure that $(\Delta y \le 0)$. In this case, the minimum fraction of refuge preventing an increase in resistance allele frequency, i.e. which guarantees $(\omega \cdot \varepsilon \cdot V_{RR}) < (1 - \omega) \cdot \chi \cdot (1 - U_{RR})$ when $(q_0 \to 0)$, would constitute an interesting benchmark [via equation (5)]:

$$(1 - \Omega) = \lim_{q \to 0}(1 - \omega)_{[\Delta y = 0]} = \frac{\varepsilon \cdot V_{RR}}{\varepsilon \cdot V_{RR} + \chi \cdot (1 - U_{RR})} \quad (7)$$

According to equation (7), $\Delta y < 0$ may be achieved only if $9 \times (\varepsilon \cdot V_{RR}) < \chi \cdot (1 - U_{RR})$, for a refuge fraction of 10%; or $4 \times (\varepsilon \cdot V_{RR}) < \chi \cdot (1 - U_{RR})$, for a refuge fraction of 20%, which clearly refers to cases where incomplete resistance ($V_{RR} \ll 1$), fitness cost ($1 - U_{RR} \gg 0$), and the dominance of this cost ($g \gg 0$) are considerable. Given that fitness cost might average at $(1 - U_{RR}) \approx 0.25$ and might be a rather recessive trait ($g < 0.25$), a decrease in resistance allele frequency might not be obtained for $(1 - \Omega) < 0.3$, in most cases.

Some simpler cases

While the general model illustrates the sensitivity of the pace of resistance evolution to even mild deviations from the ideal HDR assumptions, it is also useful to examine some special cases that elucidate particular features of the general problem, all involving relative fitness of SS pest genotypes ($U_{SS} = 1$) on refuge plants and ($V_{SS} = 0$) on Bt plants, where resistance is complete ($V_{RR} = 1$) and where there is no fitness cost ($U_{RR} = U_{RS} = U_{SS} = 1$).

Basic HDR model ($h = F = 0$)
In a strictly recessive model, the (RR) individuals are resistant to Bt ($V_{RR} = 1$), but both RS and SS individuals are fully susceptible ($V_{SS} = 0$, and $V_{RS} = h = 0$). Mating is assumed to be random, with respect to the genetic locus in question ($F = 0$). The proportion of Bt crop in the landscape (ω) alone determines the relative fitness of the three genotypes. Under such conditions, $\tilde{\Lambda}$ in equation (5) simplifies to:

$$\tilde{\Lambda} = \tilde{A} = \left(\frac{q \cdot \omega}{1 - \omega}\right) \quad (8)$$

The rate of resistance increase is determined by the ratio of (Bt/Refuge) crop fractions, $[\omega/(1 - \omega)]$. If y is initially low, the inflation due to the ratio ($\tilde{A}$) is moderate if the refuge fraction is above 10% [i.e., as long as ($\omega < 0.9$)]. There is very slow increase in the frequency of the (R) allele, at least until ($q^2 > 0.01$). We use (8) as the reference frame, against which to gauge the impact of violated HDR assumptions on the rate of resistance evolution.

Nonrandom mating ($F > 0$) and nonrecessive ($h > 0$) models
Next, we consider both the case of nonrecessive resistance and nonrandom mating, due to mating of relatives or to 'mating to type'. Nonrandom mating ($F > 0$) elevates the frequency of rare resistant homozygotes (RR), while $h > 0$ increases the fraction of heterozygotes (RS) surviving on Bt plants. Either nonrecessive resistance or nonrandom mating results in a dramatic increase in the rate of increase by the (R) allele, and a model with both yields an even more elevated rate of increase (see Appendix S1):

$$\Delta y_{\varepsilon > 0} = \tilde{\Lambda} \cdot y = \tilde{D} \cdot y,$$
$$\text{where } \tilde{D} \approx \left[1 + \left(\frac{p}{q}\right) \cdot (F + h - F \cdot h)\right] \cdot \tilde{A} \quad (9a)$$

with

$$\frac{\Delta y_{\varepsilon > 0}}{\Delta y_{\varepsilon = 0}} = \left(\frac{\tilde{D}}{\tilde{A}}\right) \approx \left[1 + \left(\frac{p}{q}\right) \cdot (F + h - F \cdot h)\right] \gg 1 \quad (9b)$$

The rate of Bt resistance evolution is profoundly elevated whenever either F and/or $h \gg q$. If both assumptions fail,

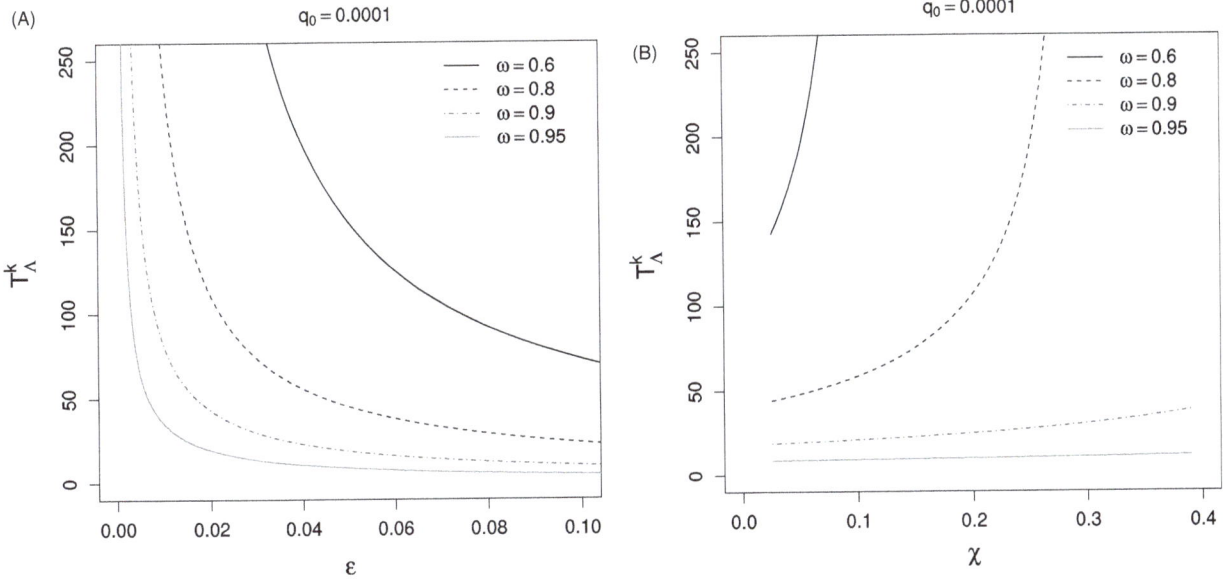

Figure 1 Passage time T_Λ^k (generations) from $q_0 = 10^{-4}$ to $q_k = 10^{-1}$, the critical allele frequency of the resistance allele (R), under two different scenarios. (A) Effects of deviations from the assumption $\varepsilon = 0$ ($F = 0 = h$) on passage time T_Λ^k with varied Bt-crop fraction: $0.6 \leq \omega \leq 0.95$. The parameters of the model were set as follow: $V_{RR} = 0.75$, $U_{RR} = 0.75$, $U_{SS} = 1$, $g = 0.05$. (B) Combined effects of F and g ($\chi = F + g - Fg$) on passage time T_Λ^k. Parameters of the model: $V_{RR} = 0.75$, $U_{RR} = 0.5$, $U_{SS} = 1$, $F = h = 0.025$ ($\varepsilon \approx 0.05$) and $0 < g < 0.375$ ($0 < \chi < 0.4$).

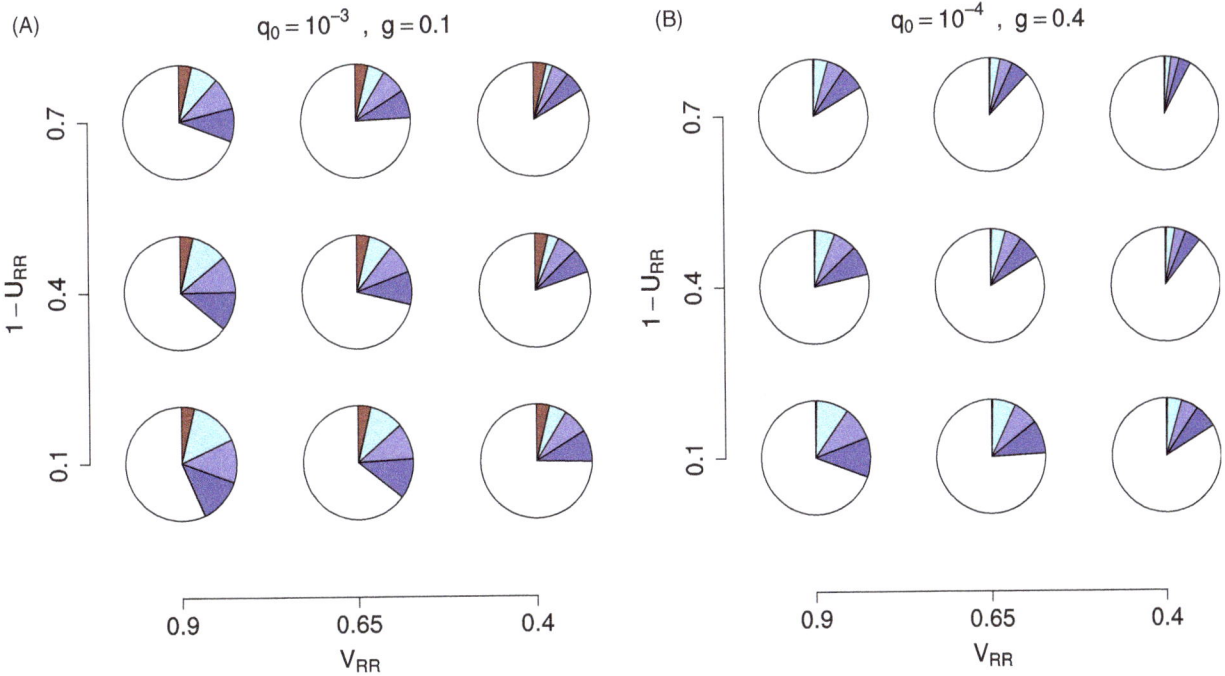

Figure 2 Additional proportion of refuge required ($1 - \omega$) to keep passage time T_Λ^k above 40 generations (blue slices) when the model deviates from strict recessivity and strict random-mating [$\varepsilon = 0$, equation (10) red slices]. Two scenarios were envisaged: (A) $q_0 = 10^{-3}$ and $g = 0.1$, and (B) $q_0 = 10^{-4}$ and $g = 0.4$. Additional refuge fractions, when deviation increases, were calculated based on equations (5) and (6): light blue slices represent $\varepsilon \approx 0.02$ ($F = 0.01 = h$); $\varepsilon \approx 0.05$ ($F = 0.025 = h$), middle blue; $\varepsilon \approx 0.10$ ($F = 0.05 = h$), darker blue. Various combinations of parameters were used for incomplete resistance ($0.4 \leq V_{RR} \leq 0.9$) and fitness cost ($0.1 \leq (1 - U_{RR}) \leq 0.7$) with $U_{SS} = 1$.

the effect on the pace of *Bt* resistance evolution is almost additive. As a consequence, the passage time expressions obtained for these two cases present striking differences. Solving the differential equation, for the basic HDR ($\varepsilon = 0$) case, yields:

$$T_{\text{A}}^k \equiv T_{\varepsilon=0}^k \approx \left(\frac{1-\omega}{\omega}\right) \cdot \left[\frac{1}{y_0} - \frac{1}{y_k} + \ln\left(\frac{y_k}{y_0}\right)\right] \quad (10)$$

Equation (10) typically yields long passage times, provided that ($y_0 < 0.01$). By contrast, for cases where ($h > 0$) and/or ($F > 0$), the ($1/y_0$) term disappears from the passage time equation, and:

$$T_{\text{D}}^k \equiv T_{\varepsilon>0}^k \approx \left(\frac{1-\omega}{\omega}\right) \cdot \frac{1}{\varepsilon} \cdot$$
$$\left[(\varepsilon - 1) \cdot \ln\left(\frac{\varepsilon + y_k}{\varepsilon + y_0}\right) + \ln\left(\frac{y_k}{y_0}\right)\right] + \alpha \quad (11a)$$

where

$$\alpha = \frac{\varepsilon^2 + \varepsilon - h}{\varepsilon} \cdot \ln\left(\frac{\varepsilon + y_k}{\varepsilon + y_0}\right) \quad (11b)$$

with ($\alpha < 0.10$), provided (ε and q_k) < (0.1), reducing (equation 11), relative to (10).

The shortening of passage time $\left[T_{\text{D}}^k - T_{\text{A}}^k\right]$ depends primarily on the product of $[(1-\omega)/\omega]$ and (p_0/q_0) = (y_0)$^{-1}$, and is dramatic when ($\varepsilon > 0$). For any value of ε and any starting value of y_0, reducing the refuge fraction ($1 - \omega$) shortens passage time. As an example, 5% refuge ($1 - \omega = 0.05$) shortens the passage time by a factor of five, relative to the rate for 20% refuge ($1 - \omega = 0.20$), everything else being equal. Similarly, for any given value of ω, the passage time is drastically reduced whenever ($\varepsilon \gg y_0$). For example, the set of parameters ($F = h = 0 \rightarrow \varepsilon = 0$, $\omega = 0.8$, $q_0 = 10^{-4}$, $q_k = 10^{-1}$) yields a passage time of $T_{\varepsilon=0}^k \approx 2500$ generations, which decreases for ($F = h = 0.025 \rightarrow \varepsilon \approx 0.05$, $\omega = 0.8$, $q_0 = 10^{-4}$, $q_k = 10^{-1}$) to a value of $T_{\varepsilon\approx0.05}^k \approx 30$ generations (see also Fig. 1A). In view of the fact that many pest species are multivoltine, empirical loss of containment can be anticipated within 10 years. Even low levels of dominance and/or nonrandom mating can compromise current HDR management protocols, even with high refuge fractions ($1 - \omega$).

Determining passage time from the evolutionary trajectory

Based on the approximation of passage time for the generalized model equation (6), we note that the ratio $T_{\text{A}}^k / \ln(y_k/y_0)$ is a logarithmic mean; i.e., a constant that reflects the pace of resistance evolution. Many monitoring surveys of resistance evolution provided data on observed change in resistance allele frequency ($q_0 \rightarrow q_j$) over an observed time lapse of T^j generations. Consistent with

equation (6), we can use T^{k*} and ξ^* as 1st approximations of passage time to q_k and the pace of resistance evolution, respectively:

$$\xi^* \approx \frac{\omega}{1-\omega} \cdot \varepsilon \cdot V_{\text{RR}} - \chi \cdot (1 - U_{\text{RR}}) = \frac{1}{T^j} \cdot \ln\left(\frac{y_j}{y_0}\right) \quad (12a)$$

equivalently

$$T^{k*} \approx \frac{1}{\xi^*} \cdot \ln\left(\frac{y_k}{y_0}\right) \quad (12b)$$

Because it relies on an approximation (equation 6) of the general model, T^{k*} is an upper-bound estimate of passage time ($T^{k*} > T_{\text{A}}^k$), but for ($T_{\text{A}}^k < 100$), it is a suitable estimate (see Appendix S2); i.e., ($T_{\text{A}}^{k*} \approx T_{\text{A}}^k$). The inverted logarithmic mean ξ^* defines the pace of resistance evolution; the higher the evolutionary rate, the shorter the passage time. The utility of such empirical estimates is that, while clearly related to equations (5) and (6), their calculation does not require detailed knowledge of the system, seldom understood well enough together to translate into precise values of the parameters (ω, U_{RR}, V_{RR}, g, h, and F).

Using published data reporting resistance evolution (Table 2), this exercise suggests a passage time (to $q_k = 0.1$) of about 10–15 years in the four cases for which suitable time-lapse data were available (i.e., $q_j < 0.1$, Table 2). These cases are acknowledged as situations where resistant mutations arose but for which control failure had not (yet) been observed (Tabashnik et al. 2013). In spite of some noticeable differences in terms of survey data, similar values of ξ^* have been observed in *H. armigera* in China and Australia, suggesting that this same approach may work reasonably well for similar examples of resistance evolution (Table 2).

Discussion

While iterative genetic simulations of resistance evolution have been used to compare theoretical expectations and empirical data (e.g., Tabashnik et al. 2008; Jin et al. 2015), we have here defined parametrically explicit predictions of the rate of evolution. We embedded our analyses in a general model, which should be useful for modeling a variety of single gene responses to selection in diploid pest organisms. Our approach is complementary to simulations of demogenetic and spatially explicit models, which may include additional levels of realism as well as increasing the number of parameters. Our model reveals contrasting outcomes that reflect the stringency of the HDR assumptions. Indeed, in the simplest cases, the structure of passage time equations differ drastically, depending on whether ε is assumed to be strictly '0' or not (see equations 10 and 11a).

The equations reveal the parameters of primary importance in the generalized model to be (q_0, ω, ε, and χ). By lowering the selective pressure on pest populations, the

Table 2. Empirical estimates of pace of resistance evolution ξ^* and passage time T^{k^*} (number of generations) from q_0 to $q_k = 0.1$, using survey data: q_0, the initial frequency of resistance alleles and q_j, the allele frequency measured T^j generations later. Are considered, 11 cases for which field-evolved resistance or field resistance has been reported (see Tabashnik et al. 2013).

Case summary					Survey data			Projections		
Pest species	Bt crop	Toxin	Country	Gener/Year	q_0	q_j	T^j	ξ^*	T^{k^*} Gener	Passage time (years)
Busseola fusca	Corn	Cry1Ab	South Africa	2	a	>0.1	<16	>0.336	NA	NA
Diatraea saccharalis	Corn	Cry1Ab	USA	4–5	0.0023	0.018	27	0.076	50.5	11.2
Helicoverpa armigera	Cotton	Cry1Ac	China	3–5	0.0058	0.075	36	0.069	40.5	10.5
Helicoverpa armigera	Cotton	Cry2Ab	Australia	3–5	0.0033	0.021	28	0.066	52.5	13.1
Helicoverpa punctigera	Cotton	Cry2Ab	Australia	3–5	0.0010	0.0091	28	0.093	54.5	13.6
Helicoverpa zea	Cotton	Cry1Ac	USA	3	0.0008	>0.1	<18	>0.273	NA	NA
Helicoverpa zea	Cotton	Cry2Ab	USA	3	0.0004	>0.1	<12	>0.471	NA	NA
Ostrina furnacalis	Corn	Cry1Ab	Philippines	6	a	>0.1	36	>0.130	NA	NA
Pectinophora gossypiella	Cotton	Cry1Ac	China	3	a	>0.1	39	>0.120	NA	NA
Pectinophora gossypiella	Cotton	Cry1Ac	India	4–6	a	>0.1	<30	>0.156	NA	NA
Spodoptera frugiperda	Corn	Cry1F	USA	10	a	>0.1	<30	>0.156	NA	NA

a, no empirical estimate of q_0 is available; in such cases, $q_0 < 0.001$ was assumed to provide an estimate of ξ^*. NA, cases for which $q > 0.1$ occurred within T^j, no projections of passage time were performed.

refuge strategy has been widely successful in delaying the evolution of *Bt* resistance in some major pest species since *Bt* crops were first deployed, 15 years ago (Huang et al. 2011). Notwithstanding the ensuing success, Tabashnik et al. (2013) have reported field-evolved resistance in 13 of 24 examined cases. Equations derived from a Wright–Fisher model show that passage time depends primarily on the (refuge/*Bt* crop) ratio $(1 - \omega)/\omega$, but also on the counterbalance between the benefits (ε) of resistant (R) alleles on *Bt* crops and those of susceptible (S) alleles on refuge crops (χ), highlighted in equation (12a). The utility of incorporating these countervailing adaptive payoffs in particular designs of the refuge strategy has been addressed by a number of studies (see Gassmann et al. 2009; Tabashnik et al. 2009), but wherever crops expressing insecticidal toxins dominate the landscape, the generalized version of the model is much more sensitive to (ε) than to (χ). Indeed, the effects of a recessive fitness cost of 25% ($U_{SS} - U_{RR} = 1 - 0.25$), which might be a reasonable average across species (Gassmann et al. 2009), appear limited whenever ($\varepsilon > 0.01$) and ($\omega > 0.7$). Given the sensitivity of the model to low values of ε, the question arises of how small deviations from classic HDR assumptions (ε) can be empirically detected, especially with respect to that of recessive resistance and random mating.

Partial dominance

Foremost, the difficulty of accurately estimating degrees of partial dominance under field conditions has been empha-

sized (Moar et al. 2008; Tabashnik et al. 2008). Although laboratory bioassays are indisputably useful for monitoring resistance evolution, the extent to which the dominance index (h), estimated under laboratory conditions, is an accurate indicator of field dominance is unclear (Bourguet et al. 1996). Indeed, both larval susceptibility to *Bt* toxin and the dominance level of any resistance are typically dosage dependent (Gould et al. 1995; Tabashnik et al. 2002). It follows that an estimate of dominance is highly context specific and its accuracy might be well below the standards that reliable predictions would require.

Assessing the partial dominance of R alleles at early stages of resistance evolution remains a challenge, since such alleles are rare, of potentially different mutational origins, and may catalyze divergent biological functions (Zhang et al. 2012; Jin et al. 2013). In addition, seasonal variation in toxin concentration within plant tissues may translate into temporal variation in functional dominance (Carrière et al. 2010). In a recent review study (Tabashnik et al. 2013), none of the 10 cases for which resistance had evolved to the point where more than 1% of individuals had become resistant could be considered 'high dose'. In addition, there have been a few published cases of newly emerging resistance alleles showing partial dominance under field conditions (Campagne et al. 2013; Jin et al. 2013). We may yet discover that *Bt* strategies based on a strictly recessive resistance assumption are overly vulnerable to the range of empirical evolutionary responses under field conditions.

Nonrandom mating

Secondly, the amount of nonrandom mating entrained by refuge structure and individual premating movement is not well understood. Generally, estimates of the randomness of mating often lack statistical power. The limited resolution of the genetic markers that have been routinely deployed in pest species (allozymes, AFLPs), or the frequent occurrence of null alleles in co-dominant genetic markers (microsatellites), has constrained our ability to detect small deviations from panmictic population structure, especially in Lepidopteran pests (Zhang 2004). For many population genetic studies of moth pests, the analytical power has been sufficient to detect only substantial deviations (F-values > 0.1) from Hardy–Weinberg frequencies (e.g., Bourguet et al. 2000; Han and Caprio 2002; Endersby et al. 2007; Kim et al. 2009). As a consequence, low levels of local nonrandom mating, crucial for HDR strategy, could not really be detected in pest species. We have here assumed an unstructured refuge/Bt-crop distribution and therefore dealt with effective fractions of refuge and Bt crop. The extent to which planted refuge within a field and landscape refuge (non-Bt farms) translate into comparable fractions of effective refuge is a pest-specific question that will need further clarification, particularly in terms of empirical data on actual pest species dispersal dynamics. As highlighted by Bourguet et al. (2000), high levels of gene flow within and among populations do not necessarily translate into a random-mating pattern, either in general or with regard to genotypes at Bt-relevant loci. It is noteworthy that assortative mating regimes may only be evident for loci closely linked to the chromosomal segments containing loci under selection for resistance. In the European Corn Borer (*O. nubilalis*), although no significant departure from Hardy–Weinberg equilibrium was initially identified (Bourguet et al. 2000), mating was found to take place at restricted spatial scales (within 50 m), effectively translating into an assortative mating rate of perhaps 5% (Dalecky et al. 2006; Bailey et al. 2007). Low premating movement is expected to increase the rate of assortative mating ($F > 0$) between individuals originating from the same block of Bt crop and has been suggested in few moths (see Cuong and Cohen 2003; Qureshi et al. 2006). In addition, some Bt-resistant pest strains evince slower larval development than Bt-susceptible conspecifics, potentially leading to emergence asynchrony of resistant and susceptible genotypes (c.f., Gryspeirt and Grégoire 2012), which could increase assortative mating (in general) but also an elevated rate of mating with resistant siblings. We clearly need better information on pest ecology, and in particular, information on dispersal behavior, with respect to the various contexts within which transgenic crops are grown.

Pace of resistance evolution and passage time

Both the pace of resistance evolution and the passage time can be described by simple combinations of model parameters. On the one hand, the expected rates of resistance evolution can be obtained by evaluating the (V_{RR}, U_{RR}, ε, χ, and ω) parameters when estimates of those parameters are attainable. On the other hand, the observed rates of evolution can be obtained by observing allele frequency changes under field conditions. That duality provides us with a simple framework for explicitly connecting empirical data and theory. While Δq, as a measure of resistance evolution, is completely dependent on the allele frequency q at any particular point on the trajectory, the rate ζ^* offers a standardized measure of the general pace of resistance evolution, even when precise estimates of (ω, U_{RR}, V_{RR}, g, h and F) are not available, provided that ($q < 0.1$) is below the 'loss of containment' threshold.

For the sake of illustration, we consider the case of *B. fusca* resistance in South Africa, for which no fitness cost has been observed (Kruger et al. 2014), and for which resistance seems complete and inherited as a dominant trait (Campagne et al. 2013). Moreover, the planted fraction of Bt maize averaged ($\omega < 0.30$) from 1998 to 2004 (~14 generations) in the area where this resistance evolved (Tabashnik et al. 2009). Assuming that initial allele frequency was low (i.e., $0.0001 < q_0 < 0.001$), the expected pace of resistance would then be $[\omega \cdot \varepsilon \cdot V_{RR}/(1 - \omega)] \approx [0.3 \cdot 1 \cdot 1/(1 - 0.3)] \approx 0.43$ (i.e., a passage time of ~11 to 16 generations, depending on q_0), roughly compatible with an empiric estimate of $\zeta^* \approx 0.34$ (i.e., ~14 generations), based on the rate of change in resistance frequency (Table 2).

Implications for resistance management and monitoring

The main option for delaying resistance evolution is to manipulate the fraction of refuge crop, either its proportion ($1 - \omega$), lowering selection pressure, or its spatial organization, reducing the impact of limited dispersal on $F > 0$. In a context where resistance evolution is not expected to follow the trajectory of a 'strictly recessive' allele (i.e., when $\varepsilon > 0$), and where the estimation of some important parameters might not be achievable, robust resistance management might have to involve substantial refuge fractions. Vacher et al. (2003) suggested refuge fractions of ~25% to minimize pest density while efficiently delaying resistance evolution. Similarly, our results show that ($1 - \omega$) < 0.20 are not likely to result in an effective expression of the fitness cost to increase passage time. Some strategies might even be needed to ensure that ($\Delta y \leq 0$). In this case, the minimum fraction of refuge preventing an increase in resistance allele frequency is expected to be

$(1 - \Omega) > 30\%$. Unsprayed refuge requirements as low as (5%) of the total planted with *Bt* crops (Bates et al. 2005) do not appear to be sufficient with respect to the statements above (see also Vacher et al. 2003). By contrast, in the Southern states of the USA, where cotton is grown, the decision was taken to establish more generous refuge fractions, $(1 - \omega) \approx \omega \approx 0.5$, in areas where other *Bt* crops were deployed.

The notion of high dose toxin, in the context of *Bt* crops, relies on a purely empirical criterion, a dose that kills 99.99% of susceptible individuals in the field 'to assure that 95% of heterozygotes would probably be killed' (USEPA 1998, see also Gould 1998), which translates as $(\varepsilon > h > 0.05)$. In this respect, our model results provide rationale to expect rapid evolution of resistance (for $h = 0.05$), typically requiring a high refuge fraction $(1 - \omega > 0.25)$ to achieve a passage time of $T^* \approx 40$ generations (with $q_0 = 0.001$, $q_k = 0.1$, $F = 0$, $V_{RR} = 1$, $U_{RR} = 0.75$, $g = 0.1$).

The model suggests that the definition of 'high-dose' should depend explicitly on (the unknown) q_0 [see equation (9b) and Appendix S1], since the variation in frequency of the resistance allele is inflated by a factor $[1 + (p/q)\varepsilon] \approx (1 + \varepsilon/q)$ whenever the system deviates from idealized HDR behavior. Assuming the parameters just above, if we set our 'dosage requirement' high enough to ensure that $\varepsilon = (F + h - F \cdot h) < q_0$, which would reduce $(\tilde{\Lambda} \rightarrow 2 \cdot \tilde{A})$ at most; we need a dose that kills a fraction p_0 of RS heterozygotes. Our model (with the same parameters as above) shows that we can attain a passage time of $T^* \approx 40$ with a refuge fraction of only $(1 - \omega) = 5\%$, but only if we can assure that $(\varepsilon < 0.007)$. Our findings suggest that, even with random mating, the current 'high-dose' requirement is inadequate for low refuge fractions. The 'dose' or the refuge fraction $(1 - \omega)$ needs to be increased.

An efficient insect resistance management strategy must be based on robust assumptions that ensure sustained toxicity of *Bt* crops under a variety of circumstances. Notably, insect survival on transgenic crops expressing at least two *Bt* toxins appeared to be higher than previously anticipated (Carrière et al. 2015). In this context, both, breeding programs and modeling studies may benefit from explicitly integrating other deviations from idealized situations in order to minimize the gap between theoretical expectations and empirical trends observed in the field. Better predictive models of resistance evolution may be a key for both designing sustainable strategies and anticipating eventual failures.

Acknowledgement

PC was supported by IRD and Natural Environment Research Council grant NE/J022993/1; PES was supported by USDA/NJAES-17160; RP, JFS and BLR were funded by IRD; JVdB was supported by Biosafety South Africa (Grant 08-001).

Competing interests

The authors declare no competing interests.

Literature cited

Alstad, D. N., and D. A. Andow 1995. Managing the evolution of insect resistance to transgenic plants. Science **268**:1894–1896.

Bailey, R. I., D. Bourguet, A. H. L. Pallec, and S. Ponsard 2007. Dispersal propensity and settling preferences of European corn borers in maize field borders. Journal of Applied Ecology **44**:385–394.

Bates, S. L., J. Z. Zhao, R. T. Roush, and A. M. Shelton 2005. Insect resistance management in GM crops: past present and future. Nature Biotechnology **23**:57–62.

Bourguet, D., M. Prout, and M. Raymond 1996. Dominance of insecticide resistance presents a plastic response. Genetics **143**:407–416.

Bourguet, D., M. T. Bethenod, N. Pasteur, and F. Viard 2000. Gene flow in the European corn borer *Ostrinia nubilalis*: implications for the sustainability of transgenic insecticidal maize. Proceedings of the Royal Society of London. Series B, Biological Sciences **267**:117–122.

Campagne, P., M. Kruger, R. Pasquet, B. Le Ru, and J. Van den Berg 2013. Dominant inheritance of field-evolved resistance to *Bt* corn in *Busseola fusca*. PLoS One **8**:e69675.

Carrière, Y., D. W. Crowder, and B. E. Tabashnik 2010. Evolutionary ecology of insect adaptation to *Bt* crops. Evolutionary Applications **3**:561–573.

Carrière, Y., C. Ellers-Kirk, R. W. Biggs, M. E. Nyboer, G. C. Unnithan, T. J. Dennehy, and B. E. Tabashnik 2006. Cadherin-based resistance to *Bacillus thuringiensis* cotton in hybrid strains of pink bollworm: fitness costs and incomplete resistance. Journal of Economic Entomology **99**:1925–1935.

Carrière, Y., N. Crickmore, and B. E. Tabashnik 2015. Optimizing pyramided transgenic crops for sustainable pest management. Nature Biotechnology **33**:161–168.

Cuong, N. L., and M. B. Cohen 2003. Mating and dispersal behaviour of *Scirpophaga incertulas* and *Chilo suppressalis* (Lepidoptera; Pyralidae) in relation to resistance management for rice transformed with *Bacillus thuringiensis* toxin genes. International Journal of Pest Management **49**:275–279.

Dalecky, A., S. Ponsard, R. I. Bailey, C. Pélissier, and D. Bourguet 2006. Resistance evolution to *Bt* crops: predispersal mating of European Corn Borers. PLoS Biology **4**:e181.

Dhurua, S., and G. T. Gujar 2011. Field-evolved resistance to *Bt* toxin Cry1Ac in the pink bollworm, *Pectinophora gossypiella* (Saunders) (Lepidoptera: Gelechiidae), from India. Pest Management Science **67**:898–903.

Endersby, N. M., A. A. Hoffman, S. W. McKechnie, and A. R. Weeks 2007. Is there genetic structure in populations of *Helicoverpa armigera* from Australia? Entomologia Experimentalis et Applicata **122**:253–263.

Felsenstein, J. 2007. Theoretical Evolutionary Genetics. University of Washington, Seattle. Available at http://evolution.genetics.washington.edu/pgbook/pgbook.html

Ferré, J., and J. Van Rie 2002. Biochemistry and genetics of insect resistance to *Bacillus thuringiensis*. Annual Review of Entomology **47**:501–533.

Gassmann, A. J., Y. Carrière, and B. E. Tabashnik 2009. Fitness cost of insect resistance to *Bacillus thuringiensis*. Annual Review of Entomology **54**:147–163.

Gassmann, A. J., J. L. Petzold-Maxwell, R. S. Keweshan, and M. W. Dunbar 2011. Field evolved resistance to *Bt* maize by western corn rootworm. PLoS One **6**:e22629.

Gould, F. 1998. Sustainability of transgenic insecticidal cultivars: integrating pest genetics and ecology. Annual Review of Entomology **43**:701–726.

Gould, F., A. Anderson, A. Reynolds, L. Bumgarner, and W. Moar 1995. Selection and genetic analysis of a *Heliothis virescens* (Lepidoptera: Noctuidae) strain with high levels of resistance to *Bacillus thuringiensis* toxins. Journal of Economic Entomology **88**: 1545–1559.

Gryspeirt, A., and J. C. Grégoire 2012. Lengthening of insect development on *Bt* zone results in adult emergence asynchrony: does it influence the effectiveness of the high dose/refuge zone strategy? Toxins **4**:1323–1342.

Han, Q., and M. A. Caprio 2002. Temporal and spatial patterns of allelic frequencies in Cotton Bollworm (Lepidoptera: Noctuidae). Environmental Entomology **31**:462–468.

Higginson, D. M., S. Morin, M. E. Nyboer, R. W. Biggs, B. E. Tabashnik, and Y. Carrière 2005. Evolutionary trade-offs of insect resistance to *Bacillus thuringiensis* crops: fitness cost affecting paternity. Evolution **59**:915–920.

Huang, F. N., D. A. Andow, and L. L. Buschman 2011. Success of the high-dose/refuge resistance management strategy after 15 years of *Bt* crop use in North America. Entomologia Experimentalis et Applicata **140**:1–16.

James, C. 2011. Global Status of Commercialized Biotech/GM Crops: 2011. ISAAA Brief No. 43. ISAAA, Ithaca, NY.

Jin, L., Y. Wei, L. Zhang, Y. Yang, B. E. Tabashnik, and Y. Wu 2013. Dominant resistance to *Bt* cotton and minor cross-resistance to Bt toxin Cry2Ab in cotton bollworm from China. Evolutionary Applications **6**:1222–1235.

Jin, L., H. Zhang, Y. Lu, Y. Yang, K. Wu, B. E. Tabashnik, and Y. Wu 2015. Large-scale test of the natural refuge strategy for delaying insect resistance to transgenic *Bt* crops. Nature Biotechnology **33**:169–174.

Kim, K. S., M. J. Bagley, B. S. Coates, R. L. Hellmich, and T. W. Sappington 2009. Spatial and temporal genetic analyses show high gene flow among European corn borer (Lepidoptera: Crambidae) populations across the central US corn belt. Environmental Entomology **38**:1312–1323.

Kruger, M., J. B. J. Van Rensburg, and J. Van den Berg 2012. Transgenic *Bt* maize: farmers' perceptions, refuge compliance and reports of stem borer resistance in South Africa. Journal of Applied Entomology **136**:38–50.

Kruger, M., J. B. J. Van Rensburg, and J. Van den Berg 2014. No fitness cost associated with resistance of *Busseola fusca* (Lepidoptera: Noctuidae) to genetically modified *Bt* maize. Crop Protection **55**:1–6.

Krumm, J. T., T. E. Hunt, S. R. Skoda, G. L. Hein, D. J. Lee, P. L. Clark, and J. E. Foster. 2008. Genetic variability of the European corn borer, *Ostrinia nubilalis*, suggests gene flow between populations in the Midwestern United States. Journal of Insect Science **8**:12.

Moar, W., R. Roush, A. Shelton, J. Ferré, S. MacIntosh, B. R. Leonard, and C. Abel 2008. Field-evolved resistance to *Bt* toxins. Nature Biotechnology **26**:1072–1074.

Onstad, D. W., P. D. Mitchell, T. M. Hurley, J. G. Lundgren, R. P. Porter, C. H. Krupke, J. L. Spencer et al. 2011. Seeds of change: corn seed mixtures for resistance management and integrated pest management. Journal of Economic Entomology **104**: 343–352.

Pan, Z., D. W. Onstad, T. M. Nowatzki, B. H. Stanley, L. J. Meinke, and J. L. Flexner 2011. Western corn rootworm (Coleoptera: Chrysolmelidae) dispersal and adaptation to single-toxin transgenic corn deployed with block or blended refuge. Environmental Entomology **40**:964–978.

Qureshi, J. A., L. L. Buschman, J. E. Throne, and S. B. Ramaswamy 2006. Dispersal of adult *Diatraea grandiosella* (Lepidoptera: Crambidae) and its implications for corn borer resistance management in *Bacillus thuringiensis* maize. Annals of the Entomological Society of America **99**:279–291.

Storer, N. P., J. M. Babcock, M. Schlenz, T. Meade, G. D. Thompson, J. W. Bing, and R. M. Huckaba 2010. Discovery and characterization of field resistance to Bt maize: *Spodoptera frugiperda* (Lepidoptera: Noctuidae) in Puerto Rico. Journal of Economic Entomology **103**:1031–1038.

Tabashnik, B. E., Y. B. Liu, T. J. Dennehy, S. A. Sims, M. S. Sisterson and Y. Carrière 2002. Inheritance of resistance to *Bt* toxin *Cry1Ac* in a field-derived strain of pink bollworm (Lepidoptera: Gelechiidae). Journal of Economic Entomology **95**:1018–1026.

Tabashnik, B. E., Y. Carrière, T. J. Dennehy, S. Morin, M. S. Sisterson, R. T. Roush, A. M. Shelton, et al. 2003. Insect resistance to transgenic *Bt* crops: lessons from the laboratory and field. Journal of Economic Entomology **96**:1031–1038.

Tabashnik, B. E., A. J. Gassmann, D. W. Crowder, and Y. Carriere 2008. Insect resistance to *Bt* crops: evidence versus theory. Nature Biotechnology **26**:199–202.

Tabashnik, B. E., J. B. J. Van Rensburg, and Y. Carriere 2009. Field-evolved insect resistance to *Bt* crops: definition, theory, and data. Journal of Economic Entomology **102**:2011–2025.

Tabashnik, B. E., T. Brévault, and Y. Carrière 2013. Insect resistance to Bt crops: lessons from the first billion acres. Nature Biotechnology **31**:510–521.

Tabashnik, B. E., D. Mota-Sanchez, M. E. Whalon, R. M. Hollingworth, and Y. Carrière 2014. Defining terms for proactive management of resistance to *Bt* crops and pesticides. Journal of Economic Entomology **107**:496–507.

Tyutyunov, Y., E. Zhadanovskaya, D. Bourguet, and R. Arditi 2008. Landscape refuges delay resistance of the European corn borer to Bt-maize: a demo-genetic dynamic model. Theoretical Population Biology **74**:138–146.

[USEPA], U.S. Environmental Protection Agency. 1998. Final report of the subpanel on *Bacillus thuringiensis* (*Bt*) plant-pesticides and resistance management.

Vacher, C., D. Bourguet, F. Rousset, C. Chevillon, and M. E. Hochberg 2003. Modelling the spatial configuration of refuges for a sustainable control of pests: a case study of *Bt* cotton. Journal of Evolutionary Biology **16**:378–387.

Van Rensburg, J. B. J. 2007. First report of field resistance by the stem borer, *Busseola fusca* (Fuller) to *Bt*-transgenic maize. South African Journal of Plant and Soil **24**:147–151.

Wright, S. 1934. Physiological and evolutionary theories of dominance. The American Naturalist **67**:24–53.

Wright, S. 1942. Statistical genetics and evolution. Bulletin of the American Mathematical Society **48**:223–246.

14

Will life find a way? Evolution of marine species under global change

Piero Calosi,[1] Pierre De Wit,[2] Peter Thor[3] and Sam Dupont[4]

[1] Département de Biologie Chimie et Géographie, Universitè du Québec à Rimouski, Rimouski, QC,Canada
[2] Department of Marine Sciences, University of Gothenburg, Strömstad, Sweden
[3] Norwegian Polar Institute, Fram Centre, Tromsø, Norway
[4] Department of Biological and Environmental Sciences, University of Gothenburg, Fiskebäckskil, Sweden

Keywords

marine evolution, ocean warming, ocean acidification, salinity, phenotypic plasticity, local adaptation, rapid adaptation, epigenetics, evolutionary modeling, transgenerational responses, DNA methylation, common garden experiment.

Correspondence

Piero Calosi, Département de Biologie, Chimie et Géographie, Université du Québec à Rimouski, 300 Allée des Ursuline, Rimouski, QC, Canada.

e-mail: piero_calosi@uqar.ca

Abstract

Projections of marine biodiversity and implementation of effective actions for its maintenance in the face of current rapid global environmental change are constrained by our limited understanding of species' adaptive responses, including transgenerational plasticity, epigenetics and natural selection. This special issue presents 13 novel studies, which employ experimental and modelling approaches to (i) investigate plastic and evolutionary responses of marine species to major global change drivers; (ii) ask relevant broad eco-evolutionary questions, implementing multiple species and populations studies; (iii) show the advantages of using advanced experimental designs and tools; (iv) construct novel model organisms for marine evolution; (v) help identifying future challenges for the field; and (vi) highlight the importance of incorporating existing evolutionary theory into management solutions for the marine realm. What emerges is that at least some populations of marine species have the ability to adapt to future global change conditions. However, marine organisms' capacity for adaptation appears finite, due to evolutionary trade-offs and possible rapid losses in genetic diversity. This further corroborates the idea that acquiring an evolutionary perspective on how marine life will respond to the selective pressure of future global changes will guide us in better identifying which conservation efforts will be most needed and most effective.

<<It is difficult to believe in the dreadful but quiet war lurking just below the serene facade of nature>> (Charles Darwin 1859)

The chemical and physical evidence for ongoing anthropogenic global change is now so prevalent that the conclusion that our climate is drastically changing is considered indisputable (IPCC 2013). On the other hand, and despite the tremendous effort by the Intergovernmental Panel on Climate Change (IPCC 2013) to synthesize our present understanding of the biological implications of global change, biological evidence corroborating the existence of ubiquitous mechanisms governing species' responses to future environmental challenges is somewhat lagging behind (Melzner et al. 2009; Dupont and Pörtner 2013; Kroeker et al. 2013; Wittmann and Pörtner 2013; Storch et al. 2014). This discrepancy has so far prevented us from producing more conclusive projections on the fate of living systems under global change. What appears to be certain is that we are on the brink of a global biodiversity crisis (Barnosky et al. 2011). It is thus unlikely that any extant species and ecosystem will be able to survive the ongoing planetary environmental changes without actually changing. In fact, whilst migration can temporarily help prevent a species' global extinction, ultimately it is only through evolutionary adaptation that populations and species can be rescued from local and global extinction (Gonzalez et al. 2013). Nonetheless, phenotypic plasticity may buy additional time for adaptation to occur (Godbold and Calosi 2013; Munday et al. 2013; Reusch 2014; Sunday et al. 2014) and also provide a mechanism for adaptation to occur rapidly (Pigliucci et al. 2006; Ghalambor et al. 2015). Finally, extant levels of adaptation to local conditions may mediate populations' sensitivity to future global change drivers (e.g. Lardies et al. 2014; Wood et al. 2016). For these reasons, the investigation of populations' and species' ability to mount plastic and adaptive responses to prevalent environmental changes is an absolute priority (e.g. Pespeni et al. 2013), if we are to identify which populations, species and assemblages will survive global change, and which are more likely to go extinct (Hall-Spencer et al. 2008; Calosi et al. 2013; Lucey et al. 2015). As current efforts have to a large extent focused on individual species' abilities to cope with short-term changes through plastic responses, in order to make critical predictions of long-term responses, it is essential to gain an understanding of the mechanisms behind the complex interactions between plasticity, evolution and nongenetic inheritance (epigenetics).

The study of evolution is one of the central themes of modern biology (Darwin 1859; Dobzanshky 1937, Huxley 1942; Dobzhansky 1973; Margulis 1999; Noble 2015). Species' capacity to mount evolutionary responses to fluctuations and changes in the environment has been investigated for decades both *via* comparative (see Somero and Hochachka 2002; Stillman and Paganini 2015) and correlative methods (see Colin and Dam 2002; Gaston et al. 2009; Bozinovic et al. 2011; Dam 2013). More recently, in order to overcome some of the limitations of these former methods, the implementation of experimental evolutionary methods has been favoured (e.g. Bennett et al. 1992; Garland and Rose 2009; Kellermann et al. 2009). Nonetheless, in the field of marine global change biology, the investigation of the capacity of biological systems to adapt to the ongoing rapid environmental change has been largely overlooked, at least until very recently (Godbold and Calosi 2013; Munday et al. 2013; Reusch 2014; Sunday et al. 2014). This situation may result from the historical lack of true marine model systems, particularly for multicellular organisms when compared to terrestrial systems (e.g. *Drosophila*, *Arabidopsis*). In part, this has been a consequence of the difficulty of working with long-lived species, in a poorly understood environment, as well as having to deal with maintaining desired environmental conditions in laboratory sea water. Nonetheless, as marine systems, just like other biological systems, are intrinsically plastic (Ghalambor et al. 2007) and have the ability to evolve (Darwin 1859), in some cases rapidly (e.g. Ghalambor et al. 2015; Thor and Dupont 2015), these features can no longer be ignored when trying to project the responses of marine populations, species and assemblages to rapid changes in multiple environmental drivers.

There is no doubt that the IPCC (2013) has generated an in-depth synthesis of the patterns through which marine species and ecosystems presently respond to the ongoing global change and may do so in the future (Pörtner et al.

2014). However, if we are to critically improve current predictions of the fate of global biodiversity under the current environmental change, advances in understanding of the drivers and mechanisms behind marine evolution are required. In this sense, the investigation of trans-generational plastic and evolutionary responses of fitness-related traits under global change scenarios, and the identification of the underpinning physiological genetic and nongenetic mechanisms, is central to advance our current understanding of how marine organisms will be able to cope with future environmental challenges. Using an evolutionary approach will help us avoid potential overestimations or underestimations of the biological implications of global change (Dam 2013).

Consequently, this special issue aims to collect novel, cutting-edge studies, which represent a further proof for the idea that the investigation of evolution within the context of marine global change is imperative, and can help guiding environmental management and conservation solutions under the ongoing rapid global change.

The specific objectives of this special issue are to (i) investigate plastic and evolutionary responses of ecologically important marine species to some of the major global change drivers (e.g. ocean warming, ocean acidification, salinity changes), both as single drivers but also combined (simultaneous and sequential); (ii) move towards asking relevant broad eco-evolutionary questions, implementing well-designed multiple species and populations studies; (iii) show the advantages of using advanced experimental designs and appropriate tools (from high-throughput DNA sequencing and novel methods for studying methylation patterns, to mathematical modelling); (iv) move beyond current limitations by constructing novel model organisms for evolution in the marine realm; (v) help identify some of the future challenges for the field of marine global change biology; and finally (vi) highlight the importance of incorporating existing evolutionary theory into management solutions for the marine realm.

This special issue consists of thirteen original manuscripts, focusing on unicellular organisms, macroalgae, invertebrates and vertebrates as study models. Most importantly, these works cover a broad range of approaches and topics relevant to the development of marine global change research. These include (i) the investigation of the significance of local adaptation in defining populations' responses; (ii) the importance of trans- and multigenerational responses to mediate species' plastic responses; (iii) the possibility for rapid evolution to occur; and (iv) the relevance of epigenetic mechanisms, as well as evolutionary trade-offs, in mediating species' responses. From these studies, a number of relevant messages and lessons have emerged and are briefly summarized below.

Local adaptation

A species' level of local adaptation, here defined as the process of evolution of a given population in response to the prevalent local environmental regimes (Williams 1966) in the face of gene flow from nearby populations, will be critical to define populations' responses to future environmental conditions, by either providing a buffer for future negative impacts, or increasing sensitivity levels (e.g. Sanford and Kelly 2011; Calosi et al. 2013; Dam 2013; Pespeni et al. 2013; Savolainen et al. 2013). Within this special issue, Padilla-Gamiño et al. (2016) have used multiple life stages of different species of coralline algae to test the hypothesis that populations living in habitats characterized by higher variability and elevated levels of seawater pCO_2 will be less affected by future ocean acidification, when compared to populations from habitats characterized by more stable and low levels of seawater pCO_2. They were able to show that spores are less sensitive to elevated pCO_2 than adults, and reported more marked impacts in populations found in habitats characterized by lower variability and lower levels of seawater pCO_2. These findings have important implications for the conservation of these important ecosystem engineers in the future ocean.

On the other hand, Lucey et al. (2016) carried out a reciprocal transplant on individuals of the sessile calcifying polychaete *Simplaria* sp. from a population inhabiting a naturally elevated pCO_2 volcanic vent area and a population from a nearby control area exposed to unaltered water chemistry conditions. Their results indicate that in this taxon neither local adaptation nor phenotypic plasticity may suffice to buffer the negative impacts of future ocean acidification. In more detail, Lucey et al. (2016) showed that regardless of their original environmental conditions, both populations showed low fitness levels, increased tube growth rates and similar plastic responses when exposed to elevated pCO_2 conditions, suggesting that local adaptation to a low pH environment had not occurred and that long-term exposure had not caused any substantial phenotypic changes.

Results from these two studies suggest that local adaptation to future conditions may not be a ubiquitous process in the marine environment. Large variability in evolutionary and plastic responses may exist, most likely resulting from differences in life-history strategies, population size, fecundity and gene flow. Understanding the relative contributions of these parameters to local adaptive capability will enable us to widen our knowledge on the importance of the process of adaptation to counter environmental change, and ultimately use it to promote the conservation of marine biodiversity. Indeed, the investigation of local adaptation must become a conservation and resource management priority (Lucey et al. 2016). Finally, Padilla-Gamiño et al. (2016) and Lucey et al. (2016) both show the value of comparing populations living under differing environmental regimes as an approach to study marine organisms potential for evolution under global change.

Trans-generational and multigenerational studies, and evidence for rapid selection

Trans-generational effects, defined as changes in offspring phenotype due to stress exposure of the parental generation, have the potential to buffer species against environmental changes (Sunday et al. 2014). In this special issue, Donelson et al. (2016) use a model coral reef fish (*Acanthochromis polyacanthus*) to investigate the impact of different heat exposure of parents on the next generation's reproductive output ability and the quality of offspring produced. Interestingly, they found that a gradual warming over two generations resulted in greater plasticity of the reproductive traits investigated, when compared to fish that experienced the same increase within one generation. Similarly, evidence for positive effect of trans-generational exposure in helping restabilizing reproductive output levels following a rapid change in pCO_2 is also provided by Rodriguez-Romero et al. (2015), using a laboratory strain of an emerging marine polychaete model (*Ophryotrocha labronica*). These studies (Donelson et al. 2016; Lucey et al. 2016) suggest that trans-generational plasticity can induce full restoration of fitness-related traits, which may not be observed with developmental plasticity alone. Furthermore, Rodriguez-Romero et al. (2015) also conducted a mutual transplant experiment, following seven generations of exposure to differing pCO_2 conditions, providing evidence for the possible occurrence of rapid adaptation in a marine organism to rapid environmental change. Rodriguez-Romero et al. (2015) show the importance of conducting multigenerational experiments in order to provide more realistic estimates for marine metazoans' responses to future environmental changes. However, they also highlight the limitations of interpreting the evolutionary significance of the outcome of transgenerational and multigenerational experiments, without the use of physiological and genetic tools, often not available for nonmodel organisms.

A number of studies in this special issue integrate novel physiological and genetic tools in the investigation of marine organisms' responses to global change drivers. For example, Shama et al. (2016) investigated differences in mitochondrial respiratory capacity and gene expression across three generations in marine sticklebacks (*Gasterosteus aculeatus*) exposed to heat stress, either in an acute fashion or throughout development, allowing for some acclimation to occur. They used an advanced cross-breeding experimental design and demonstrated that the mechanisms underlying trans-generational effects persist across multiple generations, leading to phenotypes for mitochondrial respiratory capacity and gene expression that depend on both the type of acclimation and the environmental mismatch between generations. In addition, De Wit et al. (2015) further corroborated the evolutionary significance of the mitochondrial function in underpinning species' transgenerational responses to global changes. In order to do this, they exposed specimens of the copepod *Pseudocalanus acuspes* to different pCO_2 conditions over two successive generations, followed by a reciprocal transplant experiment (Thor and Dupont 2015). After this, they used a physiological hypothesis-testing strategy to mine both gene expression and nucleotide sequence data showing that exposure to elevated pCO_2 appears to impose selection in copepods on both mitochondrial and ribosomal function, and that these changes might be related to changes in RNA transcription activity. The important consequence of this work is that De Wit et al. (2015) show that evolution of fitness-related traits can occur rapidly in marine metazoans exposed to future global change scenarios, especially in species with high standing genetic variation and large population sizes. This gives some hope that selection acting on exiting phenotypic and genetic diversity can promote the rescue of some marine metazoans within the context of future global change conditions (Munday et al. 2013; Reusch 2014; Sunday et al. 2014).

Genetic diversity could rapidly diminish in the face of rapid environmental changes, as shown by Lloyd et al. (2016) in the larvae of the purple sea urchin (*Strongylocentrotus purpuratus*) exposed to elevated pCO_2 conditions. Lloyd et al. (2016) showed a greater loss of nucleotide diversity under elevated pCO_2 conditions than in control settings, and the authors suggest that in wild populations, loss of genetic diversity could limit their capacity for further adaptation to future ocean acidification, or other drivers, in future generations. The authors concluded that whilst some natural populations may currently possess sufficient standing genetic variation to face future global changes, this latent ability of populations to deal with future environmental challenges may be rapidly dissipated by the ongoing environmental change.

Chakravarti et al. (2016), using an emerging marine polychaete model (*Ophryotrocha labronica*), exposed individuals to projected ocean warming and acidification conditions over successive generations, and showed that transgenerational exposure in the laboratory can improve offspring fitness under single driver exposure, but not across all traits measured, potentially due to genetic or physiological constraints or trade-offs. In addition, Chakravarti et al. (2016) found no significant effect of exposure to combined global change drivers. As a consequence, the utilisation of human-assisted acclimation may require an in depth

reflection before local and global proactive conservation plans are put into motion (Van Oppen et al. 2015).

The existence of trade-offs between tolerance traits to different stressors can limit both species' plastic and evolutionary responses (e.g. Hoffmann and Sgrò 2011; Dam 2013). In order to test this idea, Kelly et al. (2016) hybridized (here intended specifically as crossings) different populations of the intertidal copepod Tigriopus californicus, differing for both heat and salinity tolerance, and undertook a multigenerational selection experiment for tolerance to heat, hypo-osmotic and hyperosmotic conditions. They found that (i) heat-selected lines were more heat tolerant but showed lower fecundity, (ii) hyperosmotic-selected lines showed a reduction in tolerance to heat and (iii) lines selected for both heat and hypo-osmotic stress combined showed a reduced tolerance to heat, thus indicating, together with transcriptomic evidence, that energy trade-offs exists for these two tolerance traits.

Finally, in an impressively long-lasting selection experiment, Listmann et al. (2016) investigated changes in thermal reaction norms in the model calcifying coccolithophore *Emiliania huxleyi* in response to 2.5 years of experimental selection to two temperatures (1200 asexual generations). The different thermal selection regimes led to a marked divergence of thermal reaction norms for optimal growth and maximum persistence temperature to a range of temperatures and pCO_2. Altogether, Listmann et al. (2016) showed that thermal reaction norms in phytoplankton may evolve at a faster pace than that of predicted ocean warming, bringing some hope for the future of a key element of marine ecosystems.

Epigenetics responses

Among the mechanisms underlying both plastic and evolutionary processes, especially trans-generational effects, epigenetic mechanisms (e.g., DNA methylation or histone modification) have to date been understudied within the context of marine organisms' responses to global change (Bonduriaski et al. 2012). This may have been primarily caused by the lack of well-developed model organisms and tools for the marine realm, as well as the relatively recent discovery of the importance of these mechanisms. However, many current initiatives address this issue, with new technological advances making it possible to study epigenetic patterns even in less-than-fully developed model systems. Taking advantage of these recent advances, Putnam et al. (2016) tested whether scleractinian corals of the environmentally sensitive species *Pocillopora damicornis* and more environmentally robust species *Montipora capitata* exhibited differences in their phenotypic response that were associated with changes in DNA methylation levels following exposure to elevated pCO_2. Putnam et al. (2016)

showed that the more sensitive species exhibited a reduced calcification rate under elevated pCO_2, which was not seen in the more tolerant species. In addition, the sensitive species exhibited larger changes both in its metabolomic profile and DNA methylation pattern, when compared to the most robust species. This novel study highlights the relevance of investigating environmentally induced changes in DNA methylation, as mechanisms mediating the responses to major global change drivers of important ecosystem engineers, such as are corals, whilst asking relevant broad eco-evolutionary questions. This line of investigation could provide us with a tool to generate heritable plasticity, in support of future conservation actions, and to promote assisted evolution in marine organisms (Van Oppen et al. 2015). It will be critical to focus future work on the relationship between methylation patterns, gene expression and evolution in the generation of the observed phenotypic trans-generational effects that might provide a rescue mechanism for species facing global change.

The modelling approach

Experimental and field observational approaches have so far led the way in building our understanding of how future marine biotas will be shaped by ongoing environmental changes (Godbold and Calosi 2013; Munday et al. 2013; Reusch 2014; Sunday et al. 2014). Mathematical models may provide conceptual frameworks within which such experimental data can be placed in context. Further, models can be used as tools in order to design well-informed and well-designed experiments to produce much needed proof of concepts for key aspects of biological systems responses to the global change.

Using an individual-based model, Collins (2016) investigated the evolution of cell division rates in asexual populations of unicellular microbes maintained under chronic environmental nutrient enrichment over hundreds of generations. She found that after many generations, initially elevated growth rates appear to become limited by increases in cellular damage. This in turn causes the growth rates to decline to the ancestral state, which Collins (2016) calls the 'Prodigal Son dynamics,' in the absence of further evolution for increased tolerance to damages or decreasing in repair cost or decreasing in rate of damages. An implication from this work is that a continuous increase in growth rate, usually taken as a sign of increased fitness, might actually be detrimental to a population in the long run and that intermediate rates are more sustainable and are positively selected for. This theoretical approach is relevant to inform our understanding of how environmental enrichment can increase or control cell division rate in a sustainable fashion, these processes being central to important applications

such as biofuel reactors and controlling biofouling, respectively.

Finally, Marshall et al. (2016) used a heuristic model to explore how traits associated with complex life histories, often found in marine organisms, can alter a population's capacity to cope with environmental change. Marshall et al. (2016) found that an increase in life-history complexity decreases the potential for evolution of a species during environmental change. The authors go further, suggesting that levels of genetic correlations in stress tolerance between different life stages, genetic variance levels characterizing each life stage and the relative plasticity level found among different stages, all interact to determine the environmental change threshold any given species can tolerate before extinction occurs. Marshall et al. (2016) concluded based on their model that marine organisms possessing more complex life cycles are particularly sensitive to future global change drivers, but also warn us that for most species we still have to acquire experimental evidence for key traits.

A broader implementation of relatively simple models such as those developed by Collins (2016) and Marshall et al. (2016) could, if well employed and further parameterized with empirical data, rapidly improve our understanding on both specific trait responses and biodiversity responses to the global change.

Conclusion

This special issue collects a number of novel cutting-edge studies showcasing advanced experimental designs, approaches and tools to be used in the investigation of key aspects of marine organisms' evolutionary responses to ongoing and future rapid global changes. This new knowledge further demonstrates that 'Life may find a way', that is at least some populations of some marine species have the ability to adapt to future global change conditions, and illustrates, through transgenerational and epigenetic studies, some of the evolutionary pathways and mechanisms of adaptation that may occur over the next decades. At the same time, we have seen that the potential and capacity of marine organisms for adaptation are finite, due to the presence of evolutionary trade-offs among different traits, particularly when exposed to multiple global change drivers, and the possibility that extant genetic diversity, which enable populations to adapt to changing environments, is quickly reduced with ongoing environmental changes. Consequently, extinction caused by global change in some populations and species in the marine realm, particularly for metazoans, can be expected. This critical understanding of how marine organisms will change under the selective pressure of future global change drivers should be harnessed to help us better predict population-, community- and ecosystem-level responses. This is particularly relevant when considering the discrepancy between our current under-standing of the rate of change for environmental parameters versus the rate of change (through plasticity and adaptation) of biological systems, within the context of the ongoing global change (Dam 2013; Pörtner et al. 2014). Further, studies of species' local adaption, their capacity for trans- and multigenerational plasticity and rapid evolution, and the existence of epigenetic responses mediating species plasticity need to be increasingly incorporated into future models of evolution under global change. Such studies will provide powerful tools in our efforts to promote marine conservation and provide increasingly reliable projections on changes in marine biodiversity in the face of global change. At the same time, field observations aiming at detecting ongoing biological changes will be critical to assess whether plasticity and adaptation responses observed under laboratory conditions are actually observed in nature (Garland and Rose 2009), and are occurring at a rate which is fast enough to prevent local and global extinction. This integration will further support our ability to produce reliable projections on changes occurring from the species to the ecosystem level. Current evidence appears to suggest that plasticity and adaptation may not be fully effective in promoting evolutionary rescue (e.g. Pörtner et al. 2014), as past mass extinctions may also suggest, particularly considering the rapidity of the ongoing environmental change (Barnosky et al. 2011). Nonetheless, relevant evolutionary information will guide us in identifying which conservation efforts may be the most needed to prevent populations and species extinction and the most effective, i.e. epigenetic manipulation, laboratory transgenerational exposure, artificial selection (Van Oppen et al. 2015). Furthermore, this approach will help us identify what rate and magnitude of environmental change we can afford for life to be able to eventually adapt; in turn, which are the thresholds for the rate of environmental change beyond which evolution will not be effective in rescuing marine organisms? We hope that future efforts (including those by the IPCC) will increasingly incorporate our current, and rapidly increasing, knowledge on marine biological systems' evolutionary responses to rapid environmental changes.

Acknowledgements

We wish to thank Prof. Hans Dam and Prof. Hans-Otto Pörtner for their constructive criticisms to this manuscript. This work is largely the result of the presentations delivered at the 'Evolutionary Effects of Ocean Warming and Acidification' session at the 2015 Aquatic Sciences Meeting of the Association for the Sciences of Limnology and Oceanography, Granada (Spain), 22nd–27th of February 2015, and discussions that took place at the Marine Evolution under Climate Change advance course at the Sven Lovén Centre for Marine Sciences in Kristineberg funded by the Linnaeus Centre for Marine Evolutionary Biology (CeMEB). In addition, this work is a contribution to the work undertaken by CeMEB and the NERC UK Ocean Acidification Research Programme Benthic Consortium task 1.2 and 1.4.

Author contributions

All authors contributed to the rational for this work. PC, SD and PDW wrote the first draft of this manuscript, and all authors contributed to the final write-up.

Funding

PC was funded by the NERC UK Ocean Acidification Research Programme (NE/H017127/1), and he is supported by an NSERC Discovery Grant and a FRQNT New University Researchers Start-Up Program. PDW was supported by a postdoctoral grant from the Marcus and Amalia Wallenberg Foundation, as well as BONUS, funded jointly by the EU and the Swedish FORMAS. PT was funded through the 'Ocean Acidification and Ecosystem Effects in Northern Waters Flagship' from the High North Research Centre for Climate and the Environment (Fram Centre). SD was financially supported by the Linnaeus Centre for Marine Evolutionary Biology at the University of Gothenburg (http://www.cemeb.science.gu.se) and a Linnaeus grant from the Swedish Research Councils VR and Formas.

Literature cited

Barnosky, A. D., N. Matzke, S. Tomiya, G. O. U. Wogan, B. Swartz, T. B. Quental, C. Marshall et al. 2011. Has the Earth's sixth mass extinction already arrived? Nature **471**:51–57.

Bennett, A. F., R. E. Lenski, and J. E. Mittler 1992. Evolutionary adaptation to temperature. I. Fitness Responses of *Escherichia coli* to changes in thermal environment. Evolution **1**:16–30.

Bonduriaski, R., A. J. Crean, and T. Day 2012. The implications of non-genetic inheritance for evolution in changing environments. Evolutionary Applications **5**:192–201.

Bozinovic, F., P. Calosi, and J. I. Spicer 2011. Physiological correlates of geographic range in animals. Annual Review for Ecology, Evolution and Systematics **42**:155–179.

Calosi, P., S. P. S. Rastrick, C. Lombardi, H. J. De Guzman, L. Davidson, A. Giangrande, J. D. Hardege et al. 2013. Adaptation and acclimatization to ocean acidification in marine ectotherms: an *in situ* transplant with polychaetes at a shallow CO_2 vent system. Philosophical Transactions of the Royal Society of London B: Biological Sciences **368**:20120444.

Chakravarti, L. J., M. D. Jarrold, E. M. Gibbin, F. Christen, G. Massamba-N'Siala, P. U. Blier, and P. Calosi 2016. Can trans-generational experiments be used to enhance species resilience to ocean warming and acidification? Evolutionary Applications **9**:1133–1146.

Colin, S. P., and H. G. Dam 2002. Latitudinal differentiation in the effects of the toxic dinoflagellate *Alexandrium* spp. on the feeding and reproduction of populations of the copepod *Acartia hudsonica*. Harmful Algae 1:113–125.

Collins, S. 2016. Growth rate evolution in improved environments under Prodigal Son dynamics. Evolutionary Applications 9:1179–1188.

Dam, H. G. 2013. Evolutionary adaptation of marine zooplankton to global change. Annual Review of Marine Science 5:349–370.

Darwin, C. 1859. On the Origin of Species by Natural Selection. John Murray, London.

De Wit, P., S. Dupont, and P. Thor 2015. Selection on oxidative phosphorylation and ribosomal structure as a multigenerational response to ocean acidification in the common copepod *Pseudocalanus acuspes*. Evolutionary Applications 9:1112–1123.

Dobzhansky, T. 1937. Genetics and the Origin of Species. Columbia University Press, New York.

Dobzhansky, T. 1973. Nothing in biology makes sense except in the light of evolution. The American Biology Teacher 35:125–129.

Donelson, J. M., M. Wong, D. J. Booth, and P. L. Munday 2016. Transgenerational plasticity of reproduction depends on rate of warming across generations. Evolutionary Applications 9:1072–1081.

Dupont, S., and H.-O. Pörtner 2013. Marine science: get ready for ocean acidification. Nature 498:429.

Garland, T., and M. R. Rose 2009. Experimental Evolution: Concepts, Methods, and Applications of Selection Experiments. University of California Press, Berkeley.

Gaston, K. J., S. L. Chown, P. Calosi, J. Bernardo, D. T. Bilton, A. Clarke, S. Clusella-Trullas et al. 2009. Macrophysiology: a conceptual reunification. American Naturalist 174:595–612.

Ghalambor, C. K., J. K. Mckay, S. P. Carroll, and D. N. Reznick 2007. Adaptive versus non-adaptive phenotypic plasticity and the potential for contemporary adaptation in new environments. Functional Ecology 21:394–407.

Ghalambor, C. K., K. L. Hoke, E. W. Ruell, E. K. Fischer, D. N. Reznick, and K. A. Hughes 2015. Non-adaptive plasticity potentiates rapid adaptive evolution of gene expression in nature. Nature 525:372–375.

Godbold, J., and P. Calosi 2013. Ocean acidification and climate change: advances in ecology and evolution. Philosophical Transactions of the Royal Society of London B: Biological Sciences 368:20120448.

Gonzalez, A., O. Ronce, R. Ferriere, and M. E. Hochberg 2013. Evolutionary rescue: an emerging focus at the intersection between ecology and evolution. Philosophical Transactions of the Royal Society of London B: Biological Sciences 368:20120404.

Hall-Spencer, J. M., R. Rodolfo-Metalpa, S. Martin, E. Ransome, M. Fine, S. M. Turner, S. J. Rowley et al. 2008. Volcanic carbon dioxide vents show ecosystem effects of ocean acidification. Nature 454:96–99.

Hoffmann, A. A., and C. M. Sgrò 2011. Climate change and evolutionary adaptation. Nature 470:479–485.

Huxley, J. 1942. Evolution: The Modern Synthesis. Allen and Unwin, London.

IPCC 2013. Climate Change 2013: The Physical Science Basis: Contribution of Working Group I to the Fifth Assessment Report of the Intergovernmental Panel on Climate Change. Cambridge University Press, New York.

Kellermann, V., B. van Heerwaarden, C. M. Sgrò, and A. A. Hoffmann 2009. Fundamental evolutionary limits in ecological traits drive *Drosophila* species distributions. Science 325:1244–1246.

Kelly, M. W., M. B. DeBiasse, V. A. Villela, H. L. Roberts, and C. F. Cecola 2016. Adaptation to climate change: trade-offs among responses to multiple stressors in an intertidal crustacean. Evolutionary Applications 9:1147–1155.

Kroeker, K. J., R. L. Kordas, R. Crim, I. E. Hendriks, L. Ramajo, G. S. Singh, C. M. Duarte et al. 2013. Impacts of ocean acidification on marine organisms: quantifying sensitivities and interaction with warming. Global Change Biology 19:1884–1896.

Lardies, M. A., M. B. Arias, M. J. Poupin, P. H. Manríquez, R. Torres, C. A. Vargas, J. M. Navarro et al. 2014. Differential response to ocean acidification in physiological traits of *Concholepas concholepas* populations. Journal of Sea Research 90:127–134.

Listmann, L., M. LeRoch, L. Schlüter, M. K. Thomas, and T. B. H. Reusch 2016. Swift thermal reaction norm evolution in a key marine phytoplankton species. Evolutionary Applications 9:1156–1164.

Lloyd, M. M., A. D. Makukhov, and M. H. Pespeni 2016. Loss of genetic diversity as a consequence of selection in response to high pCO$_2$. Evolutionary Applications 9:1124–1132.

Lucey, N. M., C. Lombardi, L. DeMarchi, A. Schultze, M.-C. Gambi, and P. Calosi 2015. To brood or not to brood: are marine invertebrates that protect their offspring more resilient to ocean acidification? Scientific Reports 5:12009.

Lucey, N. M., C. Lombardi, M. Florio, L. DeMarchi, M. Nannini, S. Rundle, M. C. Gambi et al. 2016. An *in situ* assessment of local adaptation in a calcifying polychaete from a shallow CO$_2$ vent system. Evolutionary Applications 9:1054–1071.

Margulis, L. 1999. Symbiotic Planet: A New Look at Evolution. Weidenfeld & Nicolson, London.

Marshall, D. J., S. C. Burgess, and T. Connallon 2016. Global change, life-history complexity and the potential for evolutionary rescue. Evolutionary Applications 9:1189–1201.

Melzner, F., M. A. Gutowska, M. Langenbuch, S. Dupont, M. Lucassen, M. C. Thorndyke, M. Bleich et al. 2009. Physiological basis for high CO$_2$ tolerance in marine ectothermic animals: pre-adaptation through lifestyle and ontogeny? Biogeosciences 6:4693–4738.

Munday, P. L., R. R. Warner, K. Monro, J. M. Pandolfi, and D. J. Marshall 2013. Predicting evolutionary responses to climate change in the sea. Evolutionary Applications 16:1488–1500.

Noble, D. 2015. Evolution beyond neo-Darwinism: a new conceptual framework. Journal of Experimental Biology 218:7–13.

Padilla-Gamiño, J. L., J. D. Gaitán-Espitia, M. W. Kelly, and G. E. Hofmann 2016. Physiological plasticity and local adaptation to elevated pCO$_2$ in calcareous algae: an ontogenetic and geographic approach. Evolutionary Applications 9:1043–1053.

Pespeni, M. H., E. Sanford, B. Gaylord, T. M. Hill, J. D. Hosfelt, H. K. Jarisa, M. LaVigner et al. 2013. Evolutionary change during experimental ocean acidification. Proceedings of the National Academy of Sciences of the United States of America 110:6937–6942.

Pigliucci, M., C. J. Murren, and C. D. Schlichting 2006. Phenotypic plasticity and evolution by genetic assimilation. Journal of Experimental Biology 209:2362–2367.

Pörtner, H.-O., D. M. Karl, P. W. Boyd, W. Cheung, S. E. Lluch-Cota, Y. Nojiri, D. N. Schmidt et al. 2014. Ocean Systems, Chapter 6. Climate Change 2014: impacts, adaptation, and vulnerability. Part A: global and sectorial aspects. Contribution of working group II to the fifth assessment report of the intergovernmental panel on climate change. New York, NY: Cambridge University Press.

Putnam, H. M., J. M. Davidson, and R. D. Gates 2016. Ocean acidification influences host DNA methylation and phenotypic plasticity in environmentally susceptible corals. Evolutionary Applications 9:1165–1178.

Reusch, T. B. H. 2014. Climate change in the oceans: evolutionary *versus* phenotypically plastic responses of marine animals and plants. Evolutionary Applications 7:104–122.

Rodriguez-Romero, A., M. D. Jarrold, G. Massamba-N'Siala, J. I. Spicer, and P. Calosi 2015. Multi-generational responses of a marine polychaete to a rapid change in seawater pCO_2. Evolutionary Applications 9:1082–1095.

Sanford, E., and M. W. Kelly 2011. Local adaptation in marine invertebrates. Annual Review of Marine Science 3:509–535.

Savolainen, O., M. Lascoux, and J. Merilä 2013. Ecological genomics of local adaptation. Nature Reviews Genetics 14:807–820.

Shama, L. N. S., F. C. Mark, A. Strobel, A. Lokmer, U. John, and K. M. Wegner 2016. Transgenerational effects persist down the maternal line in marine sticklebacks: gene expression matches physiology in a warming ocean. Evolutionary Applications 9:1096–1111.

Somero, G. N., and P. W. Hochachka 2002. Biochemical Adaptation: Mechanism and Process in Physiological Evolution. Oxford University Press, New York.

Stillman, J. H., and A. W. Paganini 2015. Biochemical adaptation to ocean acidification. Journal of Experimental Biology 218:1946–1955.

Storch, D., L. Menzel, S. Frickenhaus, and H.-O. Pörtner 2014. Climate sensitivity across marine domains of life: limits to evolutionary adaptation shape species interactions. Global Change Biology 20:3059–3067.

Sunday, J. M., P. Calosi, S. Dupont, P. L. Munday, J. H. Stillman, and T. B. H. Reusch 2014. Evolution in an acidifying ocean. Trends in Ecology and Evolution 29:117–125.

Thor, P., and S. Dupont 2015. Transgenerational effects alleviate severe fecundity loss during ocean acidification in a ubiquitous planktonic copepod. Global Change Biology 21:2261–2271.

Van Oppen, M. J. H., J. K. Oliver, H. M. Putnam, and R. D. Gates 2015. Building coral reef resilience through assisted evolution. Proceedings of the National Academy of Science s of the United States of America 112:2307–2313.

Williams, G. C. 1966. Adaptation and Natural Selection. Princeton University Press, Princeton, NJ.

Wittmann, A. C., and H.-O. Pörtner 2013. Sensitivities of extant animal taxa to ocean acidification. Nature Climate Change 3:995–1001.

Wood, H. L., K. Sundell, B. C. Almroth, H. N. Sköld, and S. P. Eriksson 2016. Population-dependent effects of ocean acidification. Proceedings of the Royal Society of London B 283:20160163.

15

Virulence evolution of a generalist plant virus in a heterogeneous host system

Mónica Betancourt,[1,3] Fernando Escriu,[2] Aurora Fraile[1] and Fernando García-Arenal[1]

1 Centro de Biotecnología y Genómica de Plantas UPM-INIA and E.T.S.I. Agrónomos, Universidad Politécnica de Madrid, Campus de Montegancedo Madrid, Spain
2 Centro de Investigación y Tecnología Agroalimentaria de Aragón, Unidad de Sanidad Vegetal Zaragoza, Spain
3 Present address: Facultad de Ciencias Agrícolas, Universidad Santa Rosa de Cabal, UNISARC, Campus el Jazmín Santa Rosa, Colombia

Keywords
Cucumber mosaic virus, multihost parasites, virulence evolution, virus emergence

Correspondence
Fernando García-Arenal,
Centro de Biotecnología y Genómica de Plantas UPM-INIA and E.T.S.I. Agrónomos, Universidad Politécnica de Madrid, Campus de Montegancedo, 28223 Pozuelo de Alarcón, Madrid, Spain.

e-mail: fernando.garciaarenal@upm.es

Abstract

Modelling virulence evolution of multihost parasites in heterogeneous host systems requires knowledge of the parasite biology over its various hosts. We modelled the evolution of virulence of a generalist plant virus, *Cucumber mosaic virus* (CMV) over two hosts, in which CMV genotypes differ for within-host multiplication and virulence. According to knowledge on CMV biology over different hosts, the model allows for inoculum flows between hosts and for host co-infection by competing virus genotypes, competition affecting transmission rates to new hosts. Parameters of within-host multiplication, within-host competition, virulence and transmission were determined experimentally for different CMV genotypes in each host. Emergence of highly virulent genotypes was predicted to occur as mixed infections, favoured by high vector densities. For most simulated conditions, evolution to high virulence in the more competent Host 1 was little dependent on inoculum flow from Host 2, while in Host 2, it depended on transmission from Host 1. Virulence evolution bifurcated in each host at low, but not at high, vector densities. There was no evidence of between-host trade-offs in CMV life-history traits, at odds with most theoretical assumptions. Predictions agreed with field observations and are relevant for designing control strategies for multihost plant viruses.

Introduction

A major topic of evolutionary biology is the study of infectious diseases, and the evolution of virulence, defined as the negative effect of infection on host fitness (Read 1994), has been extensively modelled. Models assume trade-offs between parasite life-history traits, mostly between transmission and virulence, by considering different factors and, have identified the selective forces acting on parasites (Bull 1994; Frank 1996; Lipsitch and Moxon 1997; Ebert and Bull 2003; Gandon and Day 2003; Day and Proulx 2004; Alizon et al. 2009). Most work has focused on single-host, obligate parasites (Gandon 2004; Brown et al. 2012; Williams 2012). This is in spite that a large fraction of pathogens of humans, other animals and plants are generalists or multihost parasites, that is, they are able to infect different hosts belonging to different taxa (Woolhouse et al. 2001), and that generalists may be, or

behave as, opportunists for a focal host (Woolhouse et al. 2001; Haydon et al. 2002; Brown et al. 2012). Analyses considering multiple hosts identify among-host heterogeneity in resistance and virulence, costs of infecting different hosts and differences in within-host and between-host transmission rates, as major factors driving the evolution of generalist parasites, and mostly predict that virulence will evolve to levels below the optima for each host (Ebert and Hamilton 1996; Regoes et al. 2000; Gandon et al. 2002; Dobson 2004; Gandon 2004; Williams 2012; but see Ganusov et al. 2002). As is the case for single-host analyses, there is a general paucity of experimental data on the values of key parameters in models, and empirical tests of theoretical predictions have not been frequent (Pfennig 2001; Gandon 2004; van den Bosch et al. 2006; Jeger et al. 2006). This is particularly so for plant pathogens, for which even the basic assumption of a trade-off between parasite virulence and transmission has been evaluated in few instances (Jarosz and Davelos 1995; Sacristán and García-Arenal 2008).

The purpose of this work is to analyse the factors that drive the evolution of virulence of a generalist plant virus, *Cucumber mosaic virus* (CMV, family *Bromoviridae*). CMV has a single-stranded, messenger-sense RNA genome built of three segments that are separately encapsidated in isometric particles. CMV has the broadest host range described for a plant virus, infecting more than 1200 species in more than 100 plant families. CMV is transmitted by more than 80 species of aphids (Hemiptera: Aphididae). Transmission is nonpersistent, that is, the virus does not infect the insect vector, but is retained in its mouth parts, and the aphid is able to transmit the virus for a short time (<2 h) after acquisition from an infected plant. CMV is also transmitted through the seed, with efficiency varying largely according to the plant species. Seed transmission may be epidemiologically relevant in weed reservoirs that, together with other crops, are inoculum sources for epidemics in crops (for a review on CMV, see Jacquemond 2012). CMV is the helper virus for a satellite RNA (satRNA), which is a small, noncoding, single-stranded RNA, not infectious by itself but depends on CMV for its replication, encapsidation and transmission. The presence of a satRNA results in a depression of CMV accumulation in the infected plant, so that it behaves as a molecular parasite of CMV. CMV-satRNA may modulate the pathogenicity of CMV in a way that depends on the strains of CMV and satRNA and on the species of host plant. While most satRNA variants do not modify or attenuate CMV symptoms in most plant species, in tomato, two main phenotypes can be distinguished, those that attenuate CMV symptoms (A-satRNAs) and those that aggravate them to a systemic necrosis (N-satRNAs). Most described CMV isolates do not support a satRNA, and CMV-satRNAs occur with low frequency in the field; high satRNA prevalence has been mostly associated with epidemics of tomato necrosis (for reviews on CMV-satRNA, see García-Arenal and Palukaitis 1999; Palukaitis and García-Arenal 2003).

From 1986 to 1992, one such epidemics of systemic necrosis occurred in tomato crops in eastern Spain, caused by CMV plus satRNAs (Jordá et al. 1992; Escriu et al. 2000a). CMV isolates collected during this epidemic caused three different symptoms in tomato plants: a systemic necrosis (N isolates), a stunting of the plant and curling of the leaves (A isolates) and a stunting of the plant with extreme reduction in the leaf lamina (Y isolates). N and A isolates were associated with satRNA-variants necrogenic and non-necrogenic (i.e. attenuative of CMV symptoms), respectively, while Y isolates were not associated with satRNAs (Jordá et al. 1992). The symptoms caused by N and A isolates were determined solely by the presence and nature of the associated satRNA and not by the interaction between satRNA variant and CMV variant (Escriu et al.

2000a). It should be noted that satRNAs associated with CMV isolates during this epidemic showed high genetic variation due to mutation accumulation and recombination but had only two phenotypes on tomato plants, necrogenic and attenuative as described above; attenuative and necrogenic satRNAs belonged to two clearly different evolutionary lineages (Aranda et al. 1993, 1997; Escriu et al. 2000a). In other host species, isolates Y, N and A did not obviously differ in symptoms, but a deeper analysis showed that in melon plants, in spite that Y, N and A isolates all caused a similar leaf mosaic, A and N isolates reduced plant growth similarly and more severely than Y isolates (Betancourt et al. 2011). Thus, CMV virulence in different host plant species is genetically determined, as it is modulated by the presence of satRNAs that can be considered as a fourth nonessential component of the genome of CMV.

Some years ago, we analysed the factors leading to the emergence of the tomato necrosis syndrome, that is, the factors that determined the invasion of the CMV population by N isolates. For this, model parameters for within-host multiplication, competition in mixed infections, virulence and transmission were determined experimentally for N, A and Y isolates (Escriu et al. 2000a,b). A model that allowed co-infection of a single host by different isolate types, and competition between types with an effect on transmission explained satisfactorily the invasion of the CMV population by N isolates at the beginning of the tomato necrosis epidemic, and its predictions also agreed with the long-term evolution of the CMV population according to field data. Important conclusions from this analysis were that the invasion of the CMV population by N isolates occurred in co-infection with A isolates and required high densities of the aphid vector's population (Escriu et al. 2003). In that analysis, the role that other CMV hosts in which N and A isolates would not have a specific phenotype (i.e. the large majority of CMV hosts) could play in N isolates emergence was not considered. This is the goal of the present work.

Here, we extend the analysis of CMV virulence evolution to a system considering two host species, among which isolates of N, A and Y genotypes will differ in within-host multiplication, competition in mixed infections, virulence and transmission. In addition to tomato, the focal host in which the necrosis epidemic emerged, melon was chosen as the second host. As satRNA variants responsible for the N and A CMV types in tomato do not differ in phenotype in melon plants (Betancourt et al. 2011), melon can be considered as representative of the large majority of CMV host plant species in this respect. Also, melon shares with most vegetables and weeds the trait of being a poorer host of CMV-satRNA than tomato and other species from the *Solanaceae* (García-Arenal and Palukaitis 1999; Betancourt et al. 2011). Last, melon is the most important CMV host crop sharing a geographical area, and overlapping in time, with tomato crops in Mediterranean Spain, where the epidemic of tomato necrosis occurred. Results indicate that the rate of transmission, determined by the density of the aphid vector population, is the key factor in CMV virulence evolution. Results also show that between-host and within-host transmission rate variation determines the possibility of emergence of highly virulent isolates in either hosts, but has different effects on the dynamics of CMV infection in each host.

Models

Models description

We have used SIR-like models allowing for co-infection of a single host, with within-host competition among co-infecting isolates, which will influence transmission rates. These models were derived from that initially proposed by Mosquera and Adler (1998). An important difference is that recovery of infected plants is not considered, as CMV causes systemic persistent infections so that plants, once CMV-infected, remain so until the end of their life cycle. We used epidemiological models in which mutations having an effect in virulence were not considered, as our previous results indicated that conversion of A-satRNAs into N-satRNAs, or *vice versa*, by mutation or recombination would be extremely rare events (Aranda et al. 1997; Escriu et al. 2000a). For a single host, the dynamics of the model is described by the equations (Escriu et al. 2003):

$$\frac{dS}{dt} = \theta - \sum_{J\in\{Y,A,N,M\}} \beta_J SJ - bS \qquad (1a)$$

$$\frac{dY}{dt} = \beta_Y SY - \sum_{J\in\{A,N,M\}} \beta_J YJ - (b+\alpha_Y)Y \qquad (1b)$$

$$\frac{dA}{dt} = \sum_{J\in\{S,Y\}} \beta_A JA - p_N \sum_{J\in\{N,M\}} \beta_J AJ + \gamma_N M \\ - (b+\alpha_A)A \qquad (1c)$$

$$\frac{dN}{dt} = \sum_{J\in\{S,Y\}} \beta_N JN - p_A \sum_{J\in\{A,M\}} \beta_J NJ + \gamma_A M \\ - (b+\alpha_N)N \qquad (1d)$$

$$\frac{dM}{dt} = \sum_{J\in\{S,Y\}} \beta_M JM + p_N \sum_{J\in\{N,M\}} \beta_J AJ \\ + p_A \sum_{J\in\{A,M\}} \beta_J NJ \\ - (\gamma_A + \gamma_N)M - (b+\alpha_M)M \qquad (1e)$$

Equations represent the variation with time (days) of density (plants/m²) of susceptible noninfected plants (S) or plants infected by isolate J (J being isolates Y, A, N and M,

M indicating mixed infection by A and N isolates, J in capitals for populations of infected plants or as subscripts for model parameters). Parameter α_J indicates the virulence of isolate J, expressed as the increase in *per capita* host mortality rate due to infection. The transmission rate β_J represents the number of virus transmissions per unit time per infected host per available susceptible host. CMV isolates Y, A and N can infect healthy susceptible plants (S) resulting in Y-, A- and N-infected plants. Besides, A and N isolates can infect Y plants, resulting in A and N plants, because acquisition of satRNAs from A or N isolates by Y isolates will convert these into A and N isolates, respectively. Last, A or N isolates can infect N or A plants, respectively, resulting in a new population of A+N mixed-infected plants (M). Parameter p_J (J = A, N) represents the frequency of success of parasite J in competing with an established parasite (i.e. N, A) when infecting an already infected plant, resulting in flows into the M plant class. γ_A and γ_N represent the rate per plant and unit time at which A isolates are displaced by N isolates, or *vice versa*, respectively, when A and N isolates compete within M plants, resulting in flows from M class plants. Parameters p_J and γ_J are related through the Lotka–Volterra competition model, as further explained below (*Estimation of competition parameters*). Note that the flow from A, N or M to Y plants is not considered. This simplification was introduced as it was experimentally shown that the fraction of transmissions from N or A plants resulting in Y plants was negligible in tomato and much lower than the fraction resulting in N or A plants in melon (Escriu et al. 2000b; Betancourt et al. 2011). Also, it was considered that transmission of A+N isolates from M plants is much more probable than that of A or N isolates alone: if the proportion of isolates A and N in M plants is f_A and f_N, and a transmission event involves k virus particles, A and N isolates will be transmitted with probabilities f_A^k and f_N^k, and the probability of transmission of A+N isolates will be $1 - f_A^k - f_N^k$. Note that during nonpersistent aphid transmission, it is assumed that virus particles are sampled at random from the source virus population (Betancourt et al. 2008). Although the effective number of particles transmitted by a single aphid is small (Betancourt et al. 2008), as soon as more than one aphid is involved in a transmission event $1 - f_A^k - f_N^k$ will be much bigger than f_A^k and f_N^k.

We considered a monomolecular growth of the population of susceptible plants. For any host H, $\theta_H = r(K_H - T)$, where T is the total plant population, K is the maximum size of the population and r is its rate of growth. In both crops, plant populations stay constant during a growing season, so we set r = 1 to get a constant value of T = K, that is, the crop does not change with time. According with the crop conditions, K was of 4 plants/m² for tomato (Host 1) and of 0.8 plants/m² for melon (Host 2). Parameters are described in Table 1, and a full description of this model is in Escriu et al.'s study (2003).

This model was extended to two hosts, and its dynamics for Host 1 are described by the set of equations:

$$\frac{dS_1}{dt} = \theta_1 - \sum_{J\in\{Y,A,N,M\}} \sum_{H=1}^{2} \beta_{H1J} S_1 J_H - b_1 S_1 \quad (2a)$$

$$\frac{dY_1}{dt} = \sum_{H=1}^{2} \beta_{H1Y} S_1 Y_{H1} - \sum_{J\in\{A,N,M\}} \sum_{H=1}^{2} \beta_{H1J} Y_1 J_H$$
$$- (b_1 + \alpha_{1Y}) Y_1 \quad (2b)$$

$$\frac{dA_1}{dt} = \sum_{J\in\{S,Y\}} \sum_{H=1}^{2} \beta_{H1A} J_1 A_H$$
$$- p_{1N} \sum_{J\in\{N,M\}} \sum_{H=1}^{2} \beta_{H1J} A_1 J_H + \gamma_{1N} M_1 \quad (2c)$$
$$- (b_1 + \alpha_{1A}) A_1$$

$$\frac{dN_1}{dt} = \sum_{J\in\{S,Y\}} \sum_{H=1}^{2} \beta_{H1N} J_1 N_H$$
$$- p_{1A} \sum_{J\in\{A,M\}} \sum_{H=1}^{2} \beta_{H1J} N_1 J_H + \gamma_{1A} M_1 \quad (2d)$$
$$- (b_1 + \alpha_{1N}) N_1$$

$$\frac{dM_1}{dt} = \sum_{J\in\{S,Y\}} \sum_{H=1}^{2} \beta_{H1M} J_1 M_H$$
$$+ p_{1N} \sum_{J\in\{N,M\}} \sum_{H=1}^{2} \beta_{H1J} A_1 J_H$$
$$+ p_{1A} \sum_{J\in\{A,M\}} \sum_{H=1}^{2} \beta_{H1J} N_1 J_H$$
$$- (\gamma_{1N} + \gamma_{1A}) M_1 - (b_1 + \alpha_{1M}) M_1 \quad (2e)$$

Subscripts 1 and 2 denote the host, and the model differs from the single host one (Eqn. 1) in that it allows for infection of Host 1 from Host 2 (parameters β_{21J}) in addition to the infection of Host 1 from Host 1 (parameters β_{11J}). A second difference is the parameter H, denoting the host plant species that may be $H = 1$ for Host 1 and $H = 2$ for Host 2. The flow diagram for the model is shown in Figure S1. A similar set of equations describes the dynamics for Host 2 (not shown).

Estimation of model parameters

The values of the parameters in the models above had been estimated experimentally for tomato, and the experimental procedures and values have been reported previously (Escriu et al. 2003). The same methodology was used for the estimation of parameters for melon, based on previously published results on the interaction of CMV and satRNAs with this host plant (Betancourt et al. 2011). Melon plants (*Cucumis melo* L) cv. Piel de Sapo were used in all experiments. As is the case for all melon cultivars grown in Spain, Piel de Sapo is fully susceptible to CMV. For all experiments, CMV strain Fny (Fny-CMV, Owen and Palukaitis 1988) was used alone (Y isolate) or as a helper virus for ten satRNA genetic variants with a necrogenic phenotype in tomato (N isolates) and ten satRNA genetic variants with a non-necrogenic (i.e. attenuative) phenotype in tomato (A isolates). These satRNAs were randomly chosen from a collection of satRNA isolates from the field and were the same used previously to estimate model parameters for tomato (Escriu et al. 2000a). Both CMV and satR-NAs were derived from infectious RNA transcripts of full-length cDNA clones (Rizzo and Palukaitis 1990; Escriu et al. 2000a) to minimize mutation accumulation and selection during experimentation in different host plants.

Estimation of virulence

As is the case for most plant viruses, CMV infection is not lethal, with the exception of N isolates in tomato. Hence, it is difficult to quantify virulence as the instantaneous mortality rate and, following Day (2002), virulence was quantified as the reduction in the host expected lifespan by infection. Instantaneous mortality rates relate to lifespan by $b = 1/D_S$ for noninfected plants or by $(b + \alpha_J) = 1/D_J$ for plants infected by isolate J, being D_S and D_J the lifespan of healthy and J-infected melon plants, respectively. D_s was estimated as of 100 days according to the agricultural practices in Spain (Alonso-Prados et al. 2003), hence, $b = 0.01/$ day. D_J was estimated experimentally in the form $D_J = d_J \cdot D_S$, where d_J represents the survival of J-infected plants relative to healthy ones.

As reported for tomato (Escriu et al. 2003), both for infected and mock-inoculated melon plants, a linear regression was found between the lifespan of each leaf (leaf survival, LS) and the square root of its biomass plus that of all previously senesced leaves of the same plant (senescent biomass, SB), that is, the lifespan of each leaf was dependent on the previous growth of the plant. The slope of the linear regression of LS on the square root of SB was significantly different between mock-inoculated and CMV-infected plants, but did not differ among Y-, N- and A-infected plants (Betancourt et al. 2011). The infection of melon plants by CMV had the effect of significantly reducing plant growth as compared to mock-inoculated controls. Growth was more severely reduced by N and A isolates than by Y isolates, but there were no significant differences between the growth of plants infected by N and A isolates or by A+N isolates in mixed infections (Betancourt et al. 2011). With these data, values for d_J and α_J were calculated for

Table 1. Estimates of parameters of virulence, transmission and competition for *Cucumber mosaic virus* (CMV) genotypes in two hosts, tomato and melon.

Parameters	Description	Tomato* (Host 1)	Melon† (Host 2)
b	*Per capita* mortality rate of uninfected plants	0.00952	0.01000
α_Y	*Per capita* plant mortality rate increase due to infection by isolates Y	0.00142	0.00150
α_A	*Per capita* plant mortality rate increase due to infection by isolates A	0.00004	0.01413
α_N	*Per capita* plant mortality rate increase due to infection by isolates N	0.01120	0.01311
α_M	*Per capita* plant mortality rate increase due to infection by isolates M (= A+N)	0.01120	0.01220
β_{pY} $(i = 1)$	Probability of transmission of isolates Y for each aphid-mediated contact	0.52795	0.32703
β_{pA} $(i = 1)$	Probability of transmission of isolates A for each aphid-mediated contact	0.28771	0.25088
β_{pN} $(i = 1)$	Probability of transmission of isolates N for each aphid-mediated contact	0.19979	0.17965
β_{pM} $(i = 1)$	Probability of transmission of isolates M (= A+N) for each aphid-mediated contact	0.19979	0.17965
p_A	Frequency of success of isolates A at infecting a plant already infected by isolates N	0.83	1
p_N	Frequency of success of isolates N at infecting a plant already infected by isolates A	1	1
γ_A	*Per capita* rate at which isolates A are displaced by isolates N through within-plant competition	0.0113	0
γ_N	*Per capita* rate at which isolates N are displaced by isolates A through within-plant competition	0	0

*Data from Escriu et al. 2003.
†Derived from data in Betancourt et al. 2011 as explained in main text.

each Y, N and A isolate and for mixed infections between N and A isolates (Table S1). Mean values of α_J for each type of isolate are shown in Table 1, showing that N and A isolates were similarly virulent on melon and were more virulent than Y isolates; virulence in mixed infections did not differ significantly from virulence of A and N isolates in single infection. Because differences in plant growth or virulence between N, A and M isolates were nonsignificant, mean virulence values could have been used in simulations, as well as those in Table 1. The use of mean virulence values did not affect any of the reported results (not shown). Hence, virulence of the different types of CMV isolates was not the same for both hosts, because for tomato, N isolates were the most virulent, Y isolates had an intermediate virulence, and A isolates had very low virulence; virulence of mixed infections was as that of N isolates (Table 1).

Estimation of transmission rates
The transmission rate for each CMV isolate, β_J, was considered as the product of two terms $\beta_J = \beta_e(i) \cdot \beta_{pJ}(i)$ as in Escriu et al.'s study (2003). The first term represents the number of aphid-mediated contacts between plants, that is, the number of events per unit time and plant in which one (or several) aphid(s) leaves an infected plant and feeds in another one; the second term is the probability of virus transmission of isolate J for each of these events (Day 2001). Both β_e and β_{pJ} may vary with the number of aphids per plant, i.

In both melon and tomato, the frequency of transmission by *Aphis gossypii* for one single aphid $(i = 1)$ was shown to be determined by virus accumulation levels in the source leaf, although the relationship between both variables differed for each host (Escriu et al. 2000b; Betancourt et al. 2011). Similarly to tomato, accumulation levels of Y,

N and A isolates differed significantly in melon plants, being smaller for N isolates, intermediate for A isolates and highest for Y isolates (Betancourt et al. 2011). With these data, values of the probability of transmission of each isolate by a single aphid β_{pJ} $(i = 1)$ were calculated and are shown in Table S2, and average values for each type of isolate are shown in Table 1. Note that β_{pJ} $(i = 1)$ values ranged similarly for the different types of isolates in both hosts (i.e. Y>A>N), but because CMV multiplication is more efficient in tomato than in melon (see below), absolute values are higher in this host (Table 1) (Escriu et al. 2000a; Betancourt et al. 2011). As values of β_{pJ} $(i = 1)$ depended on the virus accumulation levels in the source leaf, for between-host transmissions, we assumed that β_{pJ} from melon to tomato was as β_{pJ} for melon and that β_{pJ} from tomato to melon was as β_{pJ} for tomato.

$\beta_{pJ}(i)$ was calculated for other i values according to the expression proposed by Gibbs and Gower (1960): $\beta_{pJ}(i) = 1 - [1 - \beta_{pJ} (i = 1)]^i$. Note that as more aphids participate in each transmission event, that is, the higher the i value, the smaller the difference in $\beta_p(i)$ values for the three CMV genotypes (Figure S2). We were unable to estimate experimentally the rate of transmission events β_e for any value of i and have used arbitrary values of $\beta_e(i)$ varying between 0.0001 and 0.1/days; these values may be realistic as epidemiological studies of CMV in different regions of Spain for different years indicate transmission rates of 0.008–0.122/days (Alonso-Prados et al. 2003).

Estimation of competition parameters
Parameters p_A, p_N, γ_A and γ_N depend on competition between A and N isolates in M plants. Dynamics of competition was simulated by the logistic equations of Lotka–Volterra model (Bulmer 1994). Our previous results had

shown that in melon plants, the accumulation in single infection of N-satRNAs was more efficient than the accumulation of A-satRNA; indeed, four of ten assayed A-satRNA did not accumulate to detectable levels in systemically infected melon leaves (Betancourt et al. 2011; see also Tables S1 and S2). In mixed infections, the accumulation of N-satRNAs was significantly depressed as compared to single infections (0.35 ± 0.01 μg satRNA per g of leaf in mixed vs 0.46 ± 0.03 μg/g in single infections), while accumulation of A-satRNAs was unaffected by the presence of N-satRNAs (0.16 ± 0.02 μg/g in mixed vs 0.14 ± 0.01 μg/g in single infection; Table 1 in Betancourt et al. 2011). These data were used to estimate the competition parameters c_{ij} (inhibitory effect of parasite j on parasite i) in the model of Lotka–Volterra, giving the values $c_{AN} = 0$ and $c_{NA} = 0.733$. These values were used in 100 simulations of the competition model, letting them vary at 10% (close to the standard error of the original data on accumulation). The resulting competition dynamics for A and N isolates in M plants was given as follows: frequency of co-existence of genotypes A and N when infecting a plant already infected by the N or A genotype, $p_A = p_N = 1$, and frequency of displacement of A by N and of N by A $\gamma_A = \gamma_N = 0$ (Table 1). Thus, although A isolates inhibited N-isolate multiplication in mixed infections, this effect was not so strong than N isolates were displaced; Lotka–Volterra frequencies at equilibrium being 0.301 ± 0.0004 and 0.699 ± 0.0004 for A and N isolates, respectively, in good agreement with experimentally determined values.

In summary, the behaviour of N and A types in both hosts is broadly different: in tomato, N- and A-satRNAs accumulated to similar levels in single infection, but N-satRNAs successfully outcompeted A-satRNA in M plants. In melon, N-satRNAs accumulated to higher levels than A-satRNA in single infection, but suffered the effect of competition of A-satRNAs in mixed infection (Table 1). Note also that the multiplication of any type of sat RNA in melon was much lower than in tomato, about 75-fold lower for N-satRNAs and about 200-fold lower for A-satRNAs (Escriu et al. 2000a; Betancourt et al. 2011).

Results

Evolution of CMV virulence in the melon crop

We analysed first the evolution of CMV virulence in the melon crop by itself. Isolates J (J = Y, N and A) differed in their basic reproductive value, $R_0 = \beta_J T/(b+\alpha_J)$ (Frank 1996). Both in tomato and melon, R_0 values ranked Y>A>N at low or moderate aphid densities. As the aphid density, i, increased, R_0 values increased and differences between Y, N and A isolates decreased, so that at high i values, R_0 for Y and A isolates in tomato, and for A and N isolates in melon, did not differ (Table 2). Maximization of

Table 2. Basic reproductive value, R_0, for Y, A and N isolates and different aphid vector densities*.

Genotype	$i = 1$	$i = 5$	$i = 10$
Tomato			
Y	9.6658	17.8565	18.2717
A	5.9452	16.7959	19.8660
N	1.8784	6.2921	8.3515
Melon			
Y	1.1375	2.9982	3.4120
A	0.4251	1.2904	1.5946
N	0.3191	1.1058	1.5129

*R_0 was calculated for different aphid densities and for $\beta_e = 0.05$. For A and N isolates, values are mean for at least five isolates. i = Number of aphids per plant.

R_0, however, did not explain the invasion of the CMV population by N isolates. As previously shown for tomato, the invasion of the CMV population by N isolates was only predicted using the co-infection model represented by eqn (1). Simulations of this model were done for β_e values between 0.01 and 0.15 and for i values of 0.5–30 aphids per plant, and with initial conditions of $S = 0.77$, $Y = N = A = 0.01$, $M = 0$ plants/m^2. These simulations yielded data on the density of plants infected by isolates Y, A, N and M (Y, A, N and M plants), on which the relative frequency of Y, A, N and M isolates in the virus population could be determined and hence the average virulence of the population. Relative isolate frequencies and average virulence varied with time (i.e. evolved) until reaching an equilibrium that differed under different scenarios (Fig. 1A). The model predicted that invasion of the CMV population by N or A isolates occurred mostly in mixed infections (M plants). Also, as previously shown for tomato, the major factor determining the invasion of the CMV population by N isolates was the density of the aphid population. For a rate of transmission events $\beta_e = 0.03$, Y-CMV isolates become the most prevalent in the melon population when the density of aphids exceeded 1 per plant, while for N and A isolates to become the most prevalent ones in mixed infections (M isolates), aphid densities of more than 5 per plant were required and much higher transmission events ($\beta_e = 0.06$). This is an important difference respective to tomato, in which M isolates were the most prevalent in the population at aphid densities above 3 per plant for $\beta_e = 0.03$ (Fig. 1A). These results reflect that melon is a poorer host for CMV multiplication and transmission than tomato. Another important difference between hosts is that average virulence of the virus population steadily increased with i and β_e in melon, while it showed a relative minimum in tomato for low β_e values and moderate aphid densities (Fig. 1A). Variation of the initial conditions did not change the outcome of the simulations.

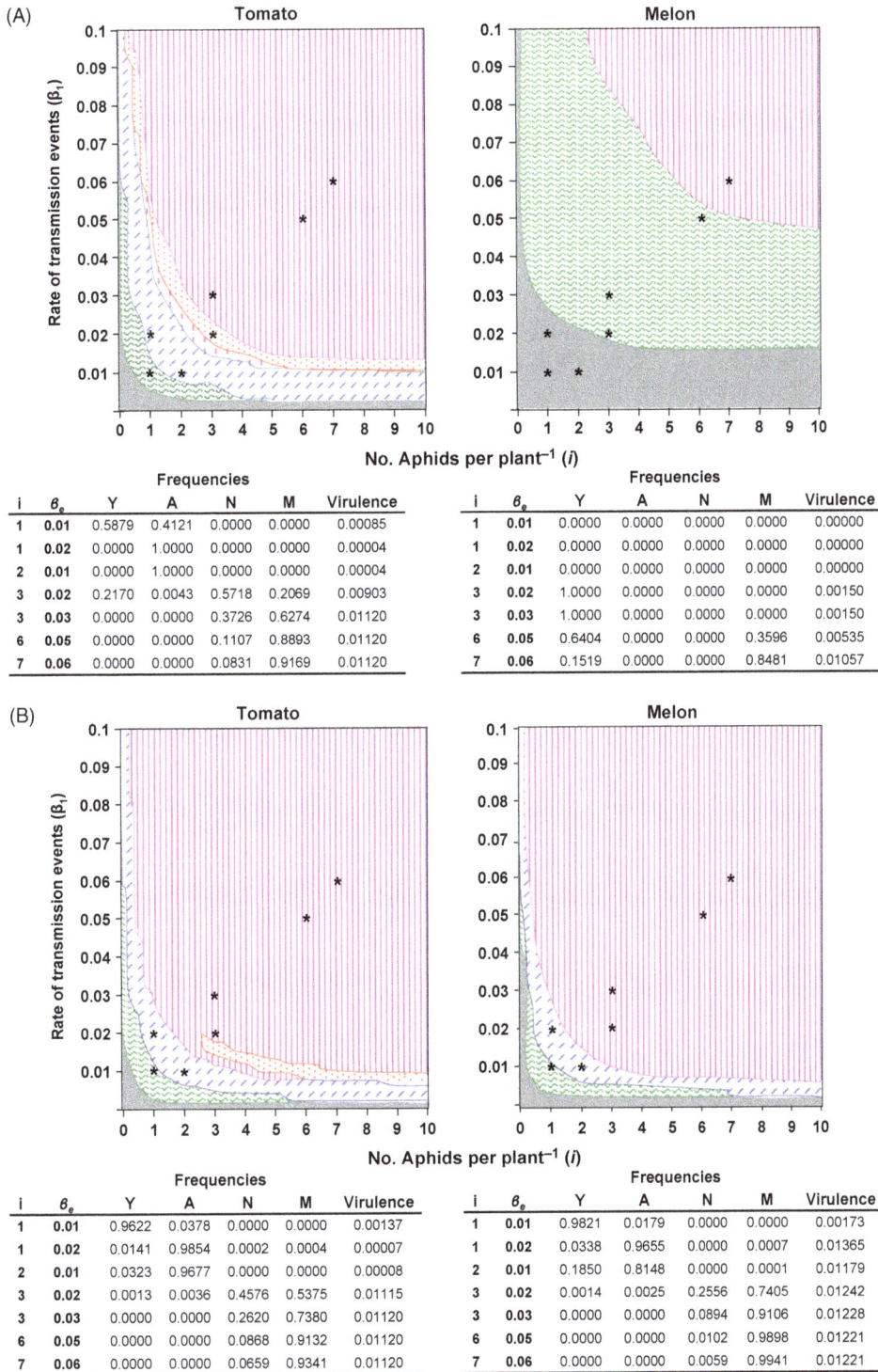

(A)

i	β_e	Frequencies				Virulence
		Y	A	N	M	
1	0.01	0.5879	0.4121	0.0000	0.0000	0.00085
1	0.02	0.0000	1.0000	0.0000	0.0000	0.00004
2	0.01	0.0000	1.0000	0.0000	0.0000	0.00004
3	0.02	0.2170	0.0043	0.5718	0.2069	0.00903
3	0.03	0.0000	0.0000	0.3726	0.6274	0.01120
6	0.05	0.0000	0.0000	0.1107	0.8893	0.01120
7	0.06	0.0000	0.0000	0.0831	0.9169	0.01120

i	β_e	Frequencies				Virulence
		Y	A	N	M	
1	0.01	0.0000	0.0000	0.0000	0.0000	0.00000
1	0.02	0.0000	0.0000	0.0000	0.0000	0.00000
2	0.01	0.0000	0.0000	0.0000	0.0000	0.00000
3	0.02	1.0000	0.0000	0.0000	0.0000	0.00150
3	0.03	1.0000	0.0000	0.0000	0.0000	0.00150
6	0.05	0.6404	0.0000	0.0000	0.3596	0.00535
7	0.06	0.1519	0.0000	0.0000	0.8481	0.01057

(B)

i	β_e	Frequencies				Virulence
		Y	A	N	M	
1	0.01	0.9622	0.0378	0.0000	0.0000	0.00137
1	0.02	0.0141	0.9854	0.0002	0.0004	0.00007
2	0.01	0.0323	0.9677	0.0000	0.0000	0.00008
3	0.02	0.0013	0.0036	0.4576	0.5375	0.01115
3	0.03	0.0000	0.0000	0.2620	0.7380	0.01120
6	0.05	0.0000	0.0000	0.0868	0.9132	0.01120
7	0.06	0.0000	0.0000	0.0659	0.9341	0.01120

i	β_e	Frequencies				Virulence
		Y	A	N	M	
1	0.01	0.9821	0.0179	0.0000	0.0000	0.00173
1	0.02	0.0338	0.9655	0.0000	0.0007	0.01365
2	0.01	0.1850	0.8148	0.0000	0.0001	0.01179
3	0.02	0.0014	0.0025	0.2556	0.7405	0.01242
3	0.03	0.0000	0.0000	0.0894	0.9106	0.01228
6	0.05	0.0000	0.0000	0.0102	0.9898	0.01221
7	0.06	0.0000	0.0000	0.0059	0.9941	0.01221

Figure 1 Predictions of co-infection models for one host (A) or for two hosts (B) for virulence evolution of CMV in Host 1 (tomato) and Host 2 (melon), as a function of the number of aphids per plant (i) and of the rate of transmission events (β_e). Graphs indicate areas in which there is no infection (S) or where Y isolates (Y), A isolates (A), N isolates (N) or A+N isolates (M) are the most prevalent in the virus populations. Figure 1A for tomato was redrawn from the study by Escriu et al. (2003). For a series of i and β_e values, indicated by asterisks in the figure, the relative equilibrium frequency of the different virus genotypes in, and the average virulence of, the virus population is indicated.

Evolution of CMV virulence in two hosts growing synchronically

In the analysis of CMV virulence evolution in a two-host system, we considered first the situation in which both hosts, that is, tomato (Host 1) and melon (Host 2), grow during the same period within the year. This is a realistic condition that could represent the case of synchronous crops, but also the case of weeds (i.e. Host 2) growing within the crop (i.e. Host 1). Because either host may be a source of inoculum for the other, we considered the same or different rates of transmission events (β_e) within and between hosts. Equal values of β_e within and between hosts imply a close spatial proximity between hosts and no vector preference for one host, while different β_e values within and between hosts might be due to spatial partition of host distribution and/or vector host preference. Both are realistic assumptions.

We considered first equal rates of transmission events within and between hosts. The model was simulated for β_e values between 0.01 and 0.15 and for i values of 0.5–30 aphids per plant. Initial conditions were $S_2 = 0.77$, $Y_2 = N_2 = A_2 = 0.01$, $M_2 = 0$ plants/m^2 and $S_1 = 4.0$, $Y_1 = N_1 = A_1 = M_1 = 0$ plants/m^2, that is, Host 2 was the inoculum source for Host 1. As before, changes in the genetic composition and in average virulence of the virus population at equilibrium could be determined (Fig. 1B). Under the above assumptions, inoculum flows between hosts resulted in very similar dynamics of Y, N, A and M isolates for both of them. For $\beta_e < 0.01$ and $i < 1$, Y isolates were the most prevalent in both populations whenever there was infection, as i increased from 1 to 3, A isolates became more prevalent than Y, and for $i \geq 2$ and $\beta_e \geq 0.02$, M isolates were the most prevalent in the populations. A summary of these results is shown in Fig. 1B. Note that virulence evolution showed different trajectories in each host, with average virulence having a relative minimum in tomato, and a relative maximum in melon, pending on i and β_e values. The comparison with Fig. 1A clearly shows the effect of inoculum flows from Host 1 to Host 2 in the dynamics of infection in Host 2. Varying the initial conditions or making Host 1, the initial inoculum source for Host 2 did not change these results (not shown). For all initial conditions, equilibrium densities of S, Y, A, N and M plants were reached faster the higher the transmission rates (not shown).

In nature, it might be more frequent that rates of transmission events are different within hosts than between hosts and, specifically, that they are higher within than between hosts. However, simulations were done exploring all possibilities, so that β_e varied within and between hosts in the range 0.0001–0.1; and for the different β_e values, i varied between 0.5 and 10 aphids per plant. Results on the predicted densities of S, Y, A, N and M plants, and average

virulences, are summarized in Fig. 2 for the extreme values of within- and between-host rates of transmission events and for initial conditions in which Host 1 was the infected host ($S_1 = 3.97$, $Y_1 = N_1 = A_1 = 0.1$, $M_1 = 0$ plants/m^2; $S_2 = 0.8$, $Y_2 = N_2 = A_2 = M_2 = 0$ plants/m^2). The reduction in the between-host values of β_e had a higher impact on Host 2 than on Host 1: when between host $\beta_e = 0.0001$, it was required that $i \geq 2$ for A, N or M CMV isolates to have any frequency, and $i \geq 8$ for mixed infections of A and N (M isolates) to have a frequency $\geq 50\%$ in Host 2 (Fig. 2A). On the other hand, reducing within-host β_e had a bigger effect on Host 1, as it could reduce the frequency of infected plants (all types) below 25% (Fig. 2B). Increasing between-host β_e resulted in higher frequency of M plants in Host 2 (Fig. 2B). Note that variation in within- and between-host β_e values had a limited effect on the average virulence of the virus population in Host 1, in spite of its dramatic effects on infection frequency, while the reduction in between-host β_e values resulted in a reduction in both infection frequency and average virulence in Host 2, particularly noticeable at $i < 5$ (Fig. 2). Thus, virulence evolved to different values in each host according to transmission rates and vector densities, bifurcating at the lower vector densities.

Last, we considered the situation in which β_e values differing within and between hosts also represent asymmetric inoculum flows between both hosts. Figure 3 summarizes the results for the extreme situation in which inoculum flows only occurred from Host 2 to Host 1 (Fig. 3A) or vice versa (Fig. 3B). If inoculum flow occurred from Host 2 to Host 1, it sufficed that $i \geq 1$ for Host 1 to become infected with genotypes Y and N and for mixed infections of N+A (M plants) reaching a high frequency. At these aphid densities, though, Host 2 was only infected by Y isolates, and much higher aphid densities ($i \geq 7$) were required for N isolates to occur in mixed infection with A isolates and for M plants to be the most frequent infected plants; for all i values, S plants were the most prevalent (Fig. 3A). If transmission occurred from Host 1 to Host 2, the dynamics of infection in Host 1 was little affected, while in Host 2, high frequency of mixed infections occurred at much lower aphid densities ($i \geq 2$, Fig. 3B). Thus, both hosts showed different sensitivity to variation of inoculum flow from the other one: in Host 1, the dynamics of infection was quite independent of inoculum flows from Host 2, while in Host 2, it was highly dependent on transmission from Host 1. In other words, in Host 1 within-host transmission is more relevant than between-host transmission, while for Host 2, transmission from Host 1 is more relevant than within-host transmission. This conclusion was also reached in simulations in which within-host transmission was not allowed (not shown). However, Host 2 was not irrelevant for infection dynamics in Host 1, as it could be a highly efficient inoculum source for Host 1. Note that when direction of

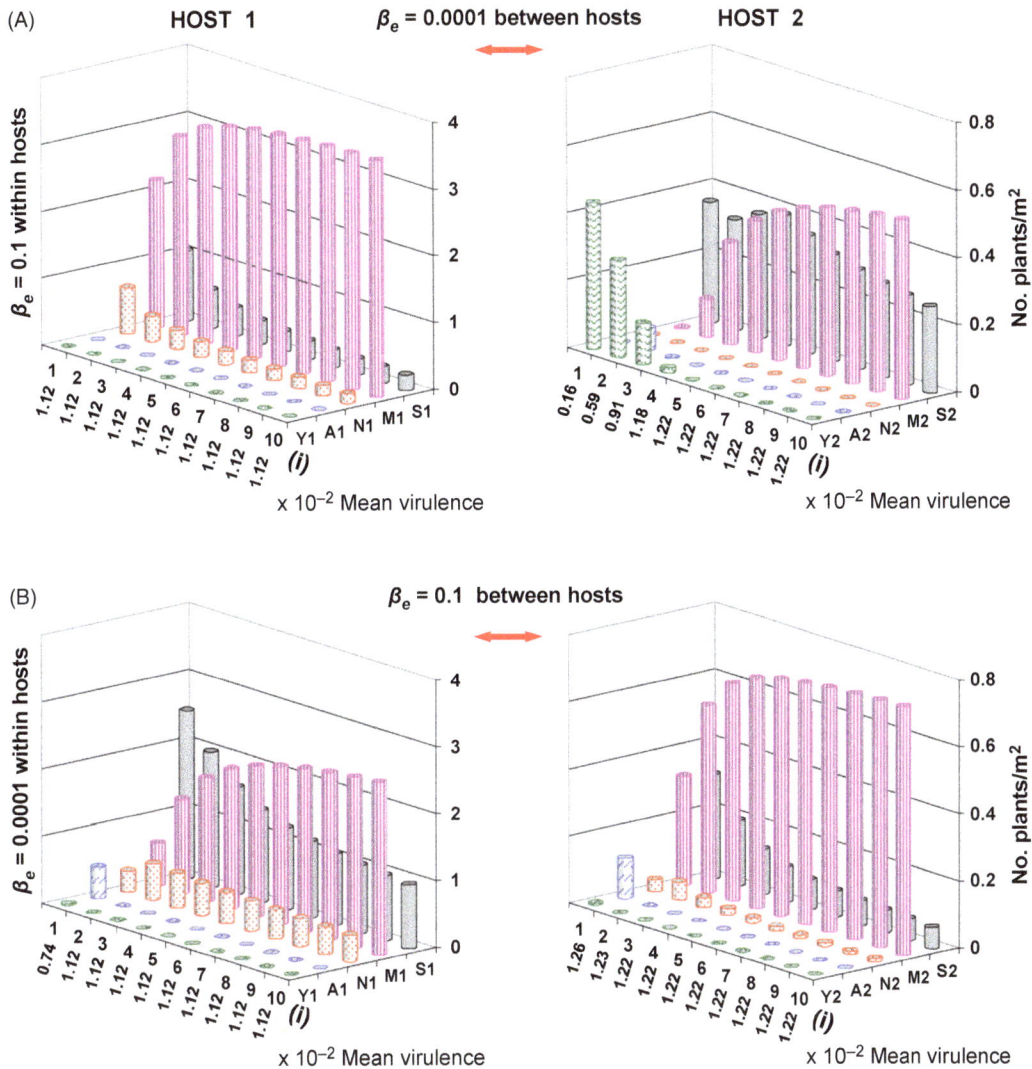

Figure 2 Equilibrium density of susceptible noninfected plants, S, and of plants infected by CMV isolates Y, A, N or mixed infected by CMV isolates A+N (M) according to a co-infection model for two hosts, when within-host and between-host rates of transmission events (β_e) differ. Presented results are for within-host $\beta_e = 0.1$ and between-host $\beta_e = 0.0001$ (A) or when within-host $\beta_e = 0.0001$ and between-host $\beta_e = 0.1$ (B). Number of aphids per plant, i, varied from 1 to 10. Initial conditions were as follows: $S_1 = 3.97$, $Y_1 = A_1 = N_1 = 0.01$, $M_1 = 0$ and $S_2 = 0.8$, $Y_2 = A_2 = N_2 = M_2 = 0$ plants/m². Bars represent plant density for different i values for noninfected plants (S_1; S_2 ▨) or for plants infected by CMV isolates Y_1, Y_2 (▨), A_1, A_2 (▨), N_1, N_2 (▨) or mixed infected with A+N isolates (M_1, M_2 ▨).

inoculums transmission between hosts changed, the average virulence did not vary in Host 1 in parallel with infection frequency, while it dramatically changed in Host 2 (compare virulence in both hosts for $i \leq 7$, Fig. 3A,B). Thus, bifurcation occurred between hosts pending on between-host transmission values.

Evolution of CMV virulence in two hosts that rotate in time

In agroecosystems, it frequently occurs that different hosts of the same pathogen (either crops or weeds) have different growing cycles along the year (i.e. rotate temporally) so that they are alternatively inoculum reservoirs for the other hosts. This situation was simulated by making Host 1 and Host 2 rotate in time, with different time overlaps between their biological cycles and, again, considering the same or different β_e values within and between hosts. The results largely agree with the conclusion from the previous section in that the dynamics of infection in Host 1 was largely independent of transmission from Host 2, once infection had started, while the dynamics of infection in Host 2 was largely determined by the continuous transmission from Host 1. For simplicity, we shall present only the

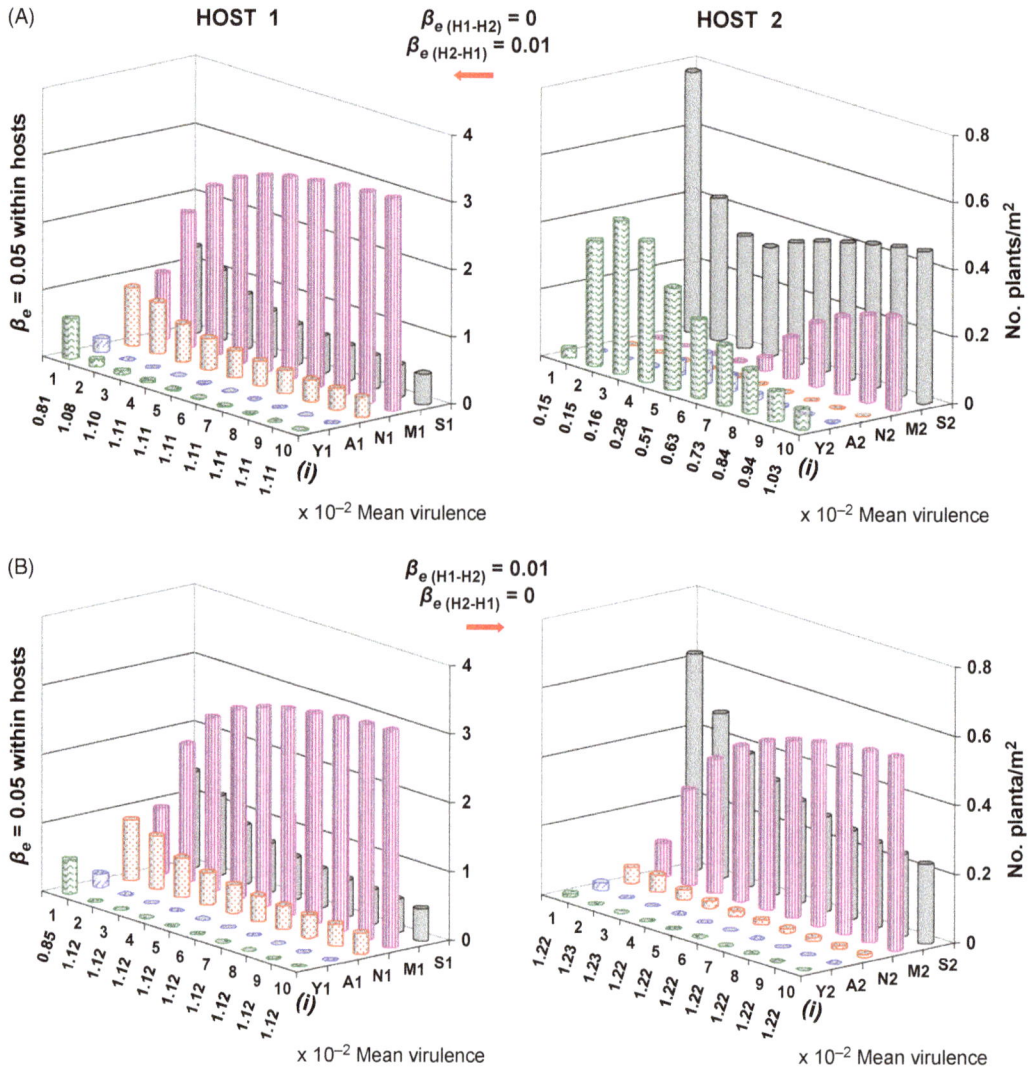

Figure 3 Equilibrium density of susceptible noninfected plants, S, and of plants infected by CMV isolates Y, A, N or mixed infected by CMV isolates A+N (M) according to a co-infection model for two hosts when within-host and between-host rates of transmission events (β_e) differ. Presented results are for within-host $\beta_e = 0.05$ and transmission events from Host 1 to Host 2 $\beta_{e(H1-H2)} = 0$ and from Host 2 to Host 1 $\beta_{e(H2-H1)} = 0.01$ (A) or when within-host $\beta_e = 0.05$ and transmission events from Host 1 to Host 2 $\beta_{e(H1-H2)} = 0.01$ and from Host 2 to Host 1 $\beta_{e(H2-H1)} = 0$ (B). Number of aphids per plant, i, varied from 1 to 10. Initial conditions were as follows: $S_1 = 3.97$, $Y_1 = A_1 = N_1 = 0.01$, $M_1 = 0$ and $S_2 = 0.8$, $Y_2 = A_2 = N_2 = M_2 = 0$ plants/m^2. Bars represent plant density for different i values for noninfected plants (S_1; S_2 ▨) or for plants infected by CMV isolates Y_1, Y_2 (▨), A_1, A_2 (▨), N_1, N_2 (▨) or mixed infected with A+N isolates (M_1, M_2 ▨).

simulations in which the rate of transmission events, β_e, was different within and between hosts.

Simulations were done for within- and between-host β_e values varying from 0.0001 to 0.1, for each β_e value i varying between 0.5 and 30 aphids per plant, and for conditions in which Host 1 initiates the rotation and is thus the inoculum source for Host 2, and *vice versa*. Figure 4 presents the results for the extreme β_e values and for initial conditions: $S_1 = 3.97$, $Y_1 = N_1 = A_1 = 0.1$, $M_1 = 0$ plants/m^2; $S_2 = 0.8$, $Y_2 = N_2 = A_2 = M_2 = 0$ plants/m^2. When the rate of transmission events between hosts was very low,

$\beta_e = 0.0001$, the prevalence of infection in Host 2 was below 5%. The virus genotypes that were transmitted between hosts depended on their prevalence at the end of the overlapping period between hosts, for instance in Fig. 4A, N+A in mixed infection were the most prevalent in Host 1 at the end of its growth period and were those transmitted to Host 2. However, N or A genotypes cannot be maintained in Host 2 for aphid densities of i $\leq$ 2. Conversely, if intrahost β_e values are very low (Fig. 4B), the prevalence of infected plants in Host 1 will be very low until the temporal overlap with Host 2. Thus, in this

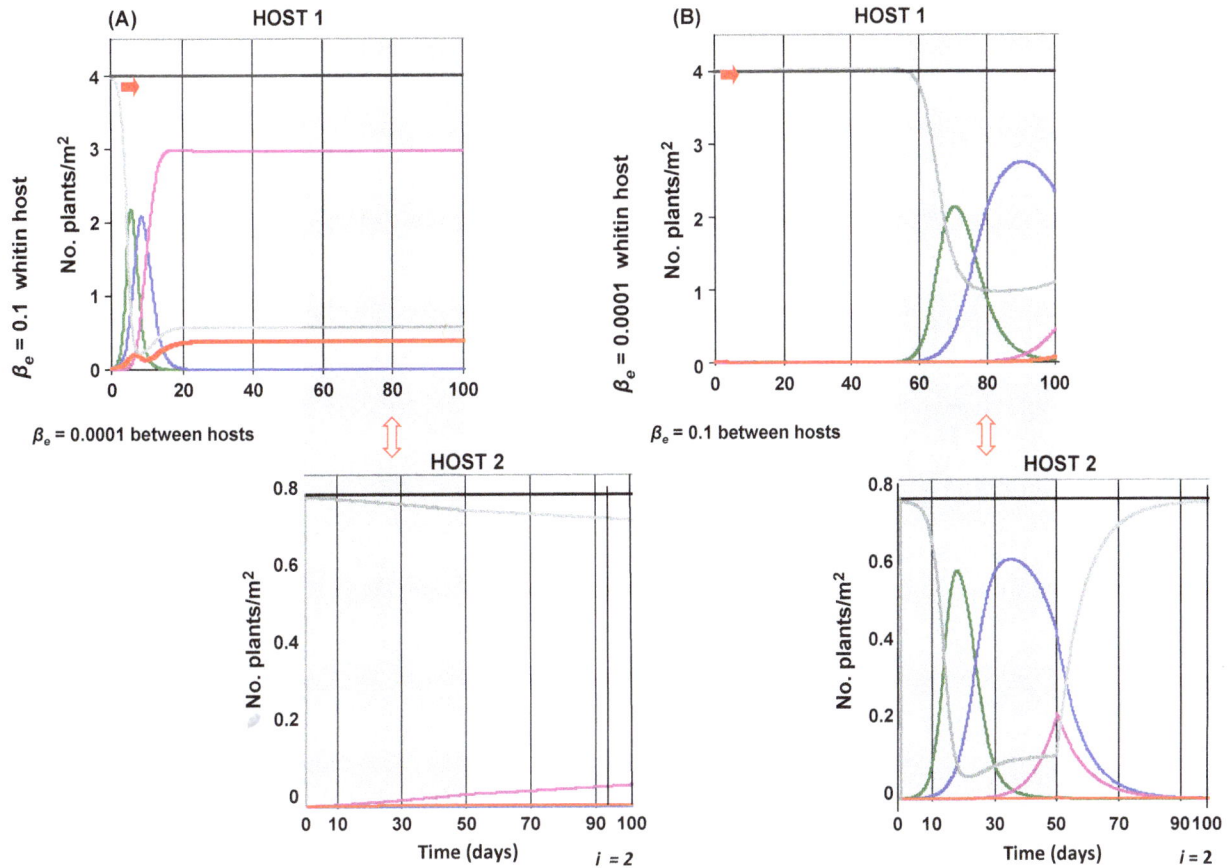

Figure 4 Dynamics of the populations of susceptible noninfected plants, *S*, and of plants infected by CMV isolates Y, A, N or mixed infected by CMV isolates A+N (M) according to a co-infection model for two hosts that rotate in time. Presented results are for a temporal overlap of hosts equivalent to half their life cycle (i.e. for 50 days) and for within-host $\beta_e = 0.1$ and between-host $\beta_e = 0.0001$ (A) or within-host $\beta_e = 0.0001$ and between-host $\beta_e = 0.1$ (B), for *i* = 0.5 and *i* = 2. Initial conditions were as follows: $S_1 = 3.97$, $Y_1 = A_1 = N_1 = 0.01$, $M_1 = 0$ and $S_2 = 0.8$, $Y_2 = A_2 = N_2 = M_2 = 0$ plants/m^2. Curves represent the variation in time of the density of noninfected plants (S ——) or plants infected by Y isolates (Y ——), by A isolates (A ——), by N isolates (N ——) or by A+N isolates (M ——). The shadow indicates the overlapping of the life cycles of Host 1 and Host 2, and the arrow indicates the host that initiates the rotation.

extreme situation, the dynamics of infection in Host 1 depends on transmission from Host 2.

Figure 5 summarizes the results of simulations in which the rotation was initiated by Host 2 (initial conditions: $S_2 = 0.77$, $Y_2 = N_2 = A_2 = 0.1$, $M_2 = 0$ plants/m^2; $S_1 = 4$, $Y_1 = N_1 = A_1 = M_1 = 0$ plants/m^2). Figure 5A shows that if within-host β_e values are high, prevalence of Y, A or M types in Host 1 can be high, even at low between-host β_e values. When within-host β_e values were low (Fig. 5B), prevalence of infection in Host 2 was always low and in Host 1 only increased during the overlapping period, thus depending on inoculum flows from Host 2.

Modifying the overlapping period between hosts did not affect the above conclusions (not shown). It was shown than an overlap period of 10 days was sufficient for infection of the noninfected host. Under this low overlapping period, conditions for infection of Host 2 from Host 1 were

between-host $\beta_e \geq 0.01$, $i \geq 1$, and condition for infection of Host 1 from Host 2 were between-host $\beta_e \geq 0.02$, $i \geq 5$, again underlining that Host 2 is a poorer host and, hence, a poorer inoculum source than Host 1.

Discussion

In this work, we analyse the conditions that may determine the invasion of the population of a generalist plant virus, CMV, by genotypes highly pathogenic for a focal host, resulting in the emergence of a new disease syndrome. For this, we consider the evolution of CMV virulence in the focal host, tomato (Host 1), and in other hosts, exemplified by melon (Host 2). The within-host multiplication and the virulence of the different CMV genotypes differ in both hosts (Escriu et al. 2003; Betancourt et al. 2011). For this analysis, we have used a model that allows for co-infection

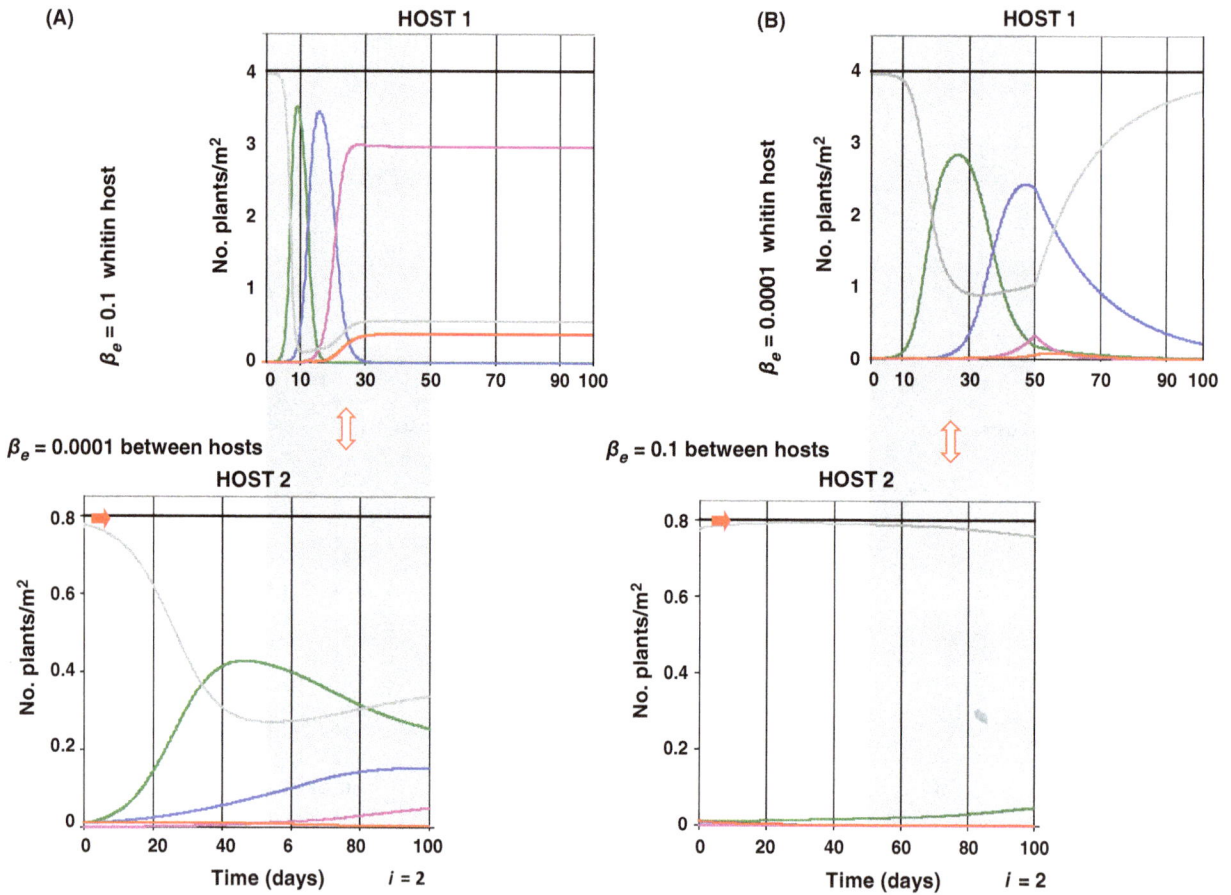

Figure 5 Dynamics of the populations of susceptible noninfected plants, S, and of plants infected by CMV isolates Y, A, N or mixed infected by CMV isolates A+N (M) according to a co-infection model for two hosts that rotate in time. Presented results are for a temporal overlap of hosts equivalent to half their life cycle (i.e. for 50 days) and for within-host $\beta_e = 0.1$ and between-host $\beta_e = 0.0001$ (A) or within-host $\beta_e = 0.0001$ and between-host $\beta_e = 0.1$ (B), for $i = 0.5$ and $i = 2$. Initial conditions were as follows: $S_1 = 4$, $Y_1 = A_1 = N_1 = M_1 = 0$ and $S_2 = 0.77$, $Y_2 = A_2 = N_2 = 0.01$, $M_2 = 0$ plants/m². Curves represent the variation in time of the density of noninfected plants (S ⎯) or plants infected by Y isolates (Y ⎯), by A isolates (A ⎯), by N isolates (N ⎯) or by A+N isolates (M ⎯). The shadow indicates the overlapping of the life cycles of Host 2 and Host 1, and the arrow indicates the host that initiates the rotation.

of a single host by different genotypes which compete, the outcome of the competition affecting the transmission rates to new hosts. This model had been developed for a single-host system (Escriu et al. 2003) and was extended now to include two hosts and to allow for inoculum flows between them. Modelling virulence evolution of multihost parasites in heterogeneous host systems may be limited by a poor knowledge of the parasite's life cycle over its various hosts, what may hinder the development of models with realistic assumptions (Day 2002; Galvani 2003). The model used in this work rests on detailed epidemiological and genetic analyses of CMV and CMV-satRNA in different hosts in Spain, demonstrating that different CMV hosts may be inoculum sources with varying effectiveness for each other, that individual hosts are often infected by different CMV genotypes that compete in mixed-infected hosts and that

CMV-satRNA spreads as a molecular parasite on the CMV population, converting pre-existing CMV genotypes (i.e. with no satRNA, Y isolates in this work) into new genotypes (N, A or M isolates in this work) (Jordá et al. 1992; Aranda et al. 1993; Alonso-Prados et al. 1998, 2003; Sacristán et al. 2004; Bonnet et al. 2005). These traits of CMV biology were all made explicit in the model described by eqn (2). Moreover, as our and other's work indicated that different CMV genotypes may broadly differ in phenotype according to host (García-Arenal and Palukaitis 1999; Palukaitis and García-Arenal 2003), key evolutionary parameters of within-host multiplication, within-host competition, between-host transmission and virulence were experimentally estimated (Escriu et al. 2000a,b, 2003; Betancourt et al. 2011, this work) so that model simulations could approach realistic situations.

A first conclusion of this work is that in both tomato and melon, N isolates highly virulent for tomato can invade the CMV population, but only in co-infection with A isolates, which do not differ in virulence with N isolates in melon. This conclusion agrees with field data (Alonso-Prados et al. 1998; Escriu et al. 2003). In either host, invasion of N and A isolates depended on the density of aphid vector populations, invasion being favoured by higher vector densities. The dynamics of CMV virulence in melon and tomato when considered as single-host systems followed similar patterns, with the important difference that invasion of N and A isolates in melon required much higher aphid densities than in tomato. This is the consequence of the highly relevant fact that melon-like hosts are less competent hosts for the multiplication and transmission of CMV and, specifically, of satRNAs than tomato-like hosts (Escriu et al. 2000a; Betancourt et al. 2011). On both hosts, the most virulent virus genotypes are favoured when transmission is less limiting, that is, at higher vector densities. Note, however, that at no vector density, between-host trade-offs occur in the analysed two-host system, as both within-host multiplication and the probability of transmission per contact event (β_p) ranged similarly for the three CMV genotypes in both hosts (Y>A>N, Escriu et al. 2000a; Betancourt et al. 2011; Table 1), in spite that the transmission rate of each virus genotype varied with vector density in a host-specific way. Note also that between-host trade-offs, which have been identified as central determinants of virulence evolution in heterogeneous host systems, have been estimated seldom (Ganusov et al. 2002; Osnas and Dobson 2011), and it is uncertain how generally they occur in multihost parasites.

The second key factor for virulence evolution in heterogeneous host systems, effectiveness of between-host transmission (Gandon 2004; Osnas and Dobson 2011; Williams 2012), was made to vary in simulations of the model within ranges compatible with apparent infection rates of CMV disease progress curves (Alonso-Prados et al. 2003). When inoculum flows were allowed between Host 1 and Host 2, the model predicted the evolution of the CMV population to high virulence levels in both hosts, again as mixed N+A infections, and depend on the rate of between-host contacts (β_e), thus again on aphid population densities. Interestingly, in both hosts, the average virulence of mixed infections was that of the more virulent genotype (N for Host 1, N and A for Host 2), a condition that according to some analyses should prevent genotype co-existence in mixed infections (Alizon 2008). Within most of the range of simulated between-host and within-host rates of transmission events, the dynamics of CMV evolution in the species that is a more competent host for CMV and, particularly, for CMV-satRNA, that is, Host 1, was quite independent of transmission from the other host, while for the less

competent Host 2, dependency on transmission from Host 1 was central. Thus, between-host transmission had the effect of reducing the vector density required for the invasion of Host 2 by highly virulent genotypes, due to flows from Host 1. Under these conditions, the dynamics of genotype CMV evolution in both hosts was similar: at high vector density, N and A genotypes invaded the CMV population in mixed infection, and at low vector densities, Y genotypes predominated, again in agreement with field observations (Alonso-Prados et al. 1998; Escriu et al. 2003). However, because the three CMV genotypes do not range similarly for virulence in Host 1 and Host 2 (see Table 1), virulence evolution differs in both hosts: at high vector density, the highly virulent genotypes N (for Host 1) or N and A (for Host 2) prevail, while at low vector densities, virulence drops to intermediate (Host 1) or to the lowest (Host 2) levels. Thus, the differential effect of vector densities on virulence evolution over hosts results in a situation that is more complex than that predicted in theoretical analyses that give as an outcome of host heterogeneity either lower virulence over hosts (Ebert and Hamilton 1996; Regoes et al. 2000; Gandon et al. 2002; Dobson 2004; Gandon 2004; Williams 2012) or different evolutionary pathways in each host (Dobson 2004; Gandon 2004). Our results show that evolutionary pathways differ between hosts, that is, virulence bifurcates, when aphid densities are lower, at odds with other predictions (Gandon 2004). This difference between our results and theoretical predictions may derive from the fact that in our system, the genotypes with different virulence are not differentially adapted to each host, that is, there is not a trade-off across hosts, as pointed out above. Our results agree better with the predictions of Ganusov et al. (2002), which do not assume such a trade-off. More generally, in our system, there is not a trade-off between virulence and transmission (see Table 1) as assumed by most theory. An interesting outcome of our analyses is that the effects of vector density on average virulence at equilibrium may differ broadly from its effects on infection frequency, again underscoring the complexity of the system.

The results of the present work may also be relevant for the control of diseases caused by generalist plant viruses. A first conclusion derives from the asymmetrical role of low- and high-competent virus hosts as inoculum source for each other. Because the different hosts of a generalist pathogen may differ in their efficiency for within-host multiplication and between-host transmission and, hence, as inoculum sources for each other, selective elimination of specific host species may be efficient for disease control. In natural ecosystems, it has been shown that highly competent plant host species may determine the ecology of virus infection in less competent hosts, in which infection proceeds mostly by 'spill-over' from the most competent host

(Power and Mitchell 2004; Cronin et al. 2010). If the virus is more virulent in the less competent hosts, this asymmetry may have deep consequences in ecosystem composition and dynamics (Power and Mitchell 2004; Malmstrom et al. 2005; Power et al. 2011). Differences in host competence have mostly not been considered in control strategies of viral diseases in agroecosystems: for generalist plant viruses, it has mostly been assumed that crop rotations or host elimination would be not efficient control strategies (Zitter and Simons 1980). Our results show that this may not be always the case. Thus, elimination of Host 1-like species, or avoiding their overlapping in the rotation, may result in efficient virus control in Host 2-like crops, with a cumulative effect over rotations.

The second conclusion relates to the effects of reducing the density of vector populations, an important control strategy for plant viruses, which results in a decrease in infection rates and prevalence. Theoretical analyses of the effect of virus transmission mechanisms on epidemics have shown that the reduction in vector density has a lesser effect on the prevalence of nonpersistently transmitted viruses, such as CMV, than on semi-persistent or persistent/circulative viruses (Madden et al. 2000) in agreement with empirical evidence (Perring et al. 1999). These analyses also showed the high sensitivity of nonpersistent virus epidemics to variation in the number of plants visited per day by an insect vector (Madden et al. 2000). This parameter may be approximated to the rate of transmission events, β_e, in our model, which arrives to similar conclusions from different approaches. Hence, the interest to analyse experimentally the relationship between the rate of transmission events and vector density which to our knowledge is an unexplored subject. Our present and previous (Escriu et al. 2003) results also show that reducing the density of virus vectors may have the additional benefit of preventing the invasion of the virus population by highly virulent genotypes. Note that the effect of vector density reduction on virulence evolution would be independent on the virus transmission mechanism, as this factor was not considered in our analyses. However, the specific relationships between transmission mechanisms, vector density and virulence would require further analyses. Our results also show that the effect of vector control on virulence would be more effective in less competent hosts (Host 2), thus reducing their efficiency as reservoirs for highly competent hosts (Host 1). In addition with selective host rotation or elimination, the effects of reducing vector population density, particularly over periods of vector migration between hosts, on virus prevalence and virulence would be enhanced.

In conclusion, a model for the evolution of the virulence of a multihost plant virus, based on a detailed knowledge of the virus biology in its different hosts, was able to explain satisfactorily the emergence of highly virulent genotypes for a focal host and the long-term evolution of virulence over the different hosts. Moreover, predictions of this model under situations common in agroecosystems revealed the value of control measures that traditionally have been considered impractical. These results may help developing long-term strategies for the control of virus diseases. Interestingly, assumptions of trade-offs between different life-history traits of parasites that are central to most theory on virulence evolution in heterogeneous host systems did not hold for the system analysed here. This underscores the need to evaluate how generally do these trade-offs occur and to couple theoretical analyses with empirical and experimental knowledge on host–parasite systems.

Acknowledgements

This work was in part funded by grant AGL2008-02458 from Ministerio de Ciencia e Innovación, Spain, to FGA. MB was under a commission for studies from Universidad Rosa de Cabal, Colombia.

Literature cited

Alizon, S. 2008. Decreased overall virulence in coinfected hosts leads to the persistence of virulent parasites. American Naturalist **172**:E67–E79.

Alizon, S., A. Hurford, N. Mideo, and M. Van Baalen 2009. Virulence evolution and the trade-off hypothesis: history, current state of affairs and the future. Journal of Evolutionary Biology **22**:245–259.

Alonso-Prados, J. L., M. A. Aranda, J. M. Malpica, F. García-Arenal, and A. Fraile 1998. Satellite RNA of *Cucumber mosaic* cucumovirus spreads epidemically in natural populations of its helper virus. Phytopathology **88**:520–524.

Alonso-Prados, J. L., M. Luis-Arteaga, J. M. Alvarez, E. Moriones, A. Batle, F. García-Arenal, and A. Fraile 2003. Epidemics of aphid transmitted viruses in melon crops in Spain. European Journal of Plant Pathology **109**:129–138.

Aranda, M. A., A. Fraile, and F. García-Arenal 1993. Genetic variability and evolution of the satellite RNA of *cucumber mosaic virus* during natural epidemics. Journal of Virology **67**:5896–5901.

Aranda, M. A., A. Fraile, J. Dopazo, J. M. Malipca, and F. García-Arenal 1997. Contribution of mutation and RNA recombination to the evolution of a plant pathogenic RNA. Journal of Molecular Evolution **44**:81–88.

Betancourt, M., A. Fereres, A. Fraile, and F. García-Arenal 2008. Estimation of the effective number of founders that initiate en infection after aphid transmission of a multipartite plant virus. Journal of Virology **82**:12416–12421.

Betancourt, M., A. Fraile, and F. García-Arenal 2011. Cucumber mosaic virus satellite RNAs that induce similar symptoms in melon plants show large differences in fitness. Journal of General Virology **92**: 1930–1938.

Bonnet, J., A. Fraile, S. Sacristán, J. M. Malpica, and F. García-Arenal 2005. Role of recombination in the evolution of natural populations of *Cucumber mosaic virus*, a tripartite RNA plant virus. Virology **332**:359–368.

van den Bosch, F., G. Akudibilah, S. Seal, and M. Jeger 2006. Host resistance and the evolutionary response of plant viruses. Journal of Applied Ecology **43**:506–516.

Brown, S. P., D. M. Cornforth, and N. Mideo 2012. Evolution of virulence in opportunistic pathogens: generalism, plasticity, and control. Trends in Microbiology **20**:336–342.

Bull, J. J. 1994. Perspective virulence. Evolution **48**:1423–1437.

Bulmer, M. 1994. Theoretical Evolutionary Ecology. Sinauer Associates, Sunderland, MA.

Cronin, J. P., M. E. Welsh, M. G. Dekkers, S. T. Abercrombie, and C. E. Mitchell 2010. Host physiological phenotype explains pathogen reservoir potential. Ecology Letters **13**:1221–1232.

Day, T. 2001. Parasite transmission modes and the evolution of virulence. Evolution **55**:2389–2400.

Day, T. 2002. On the evolution of virulence and the relationship between various measures of mortality. Proceedings of the Royal Society of London B **269**:1317–1323.

Day, T., and S. R. Proulx 2004. A general theory for the evolutionary dynamics of virulence. American Naturalist **163**:E40–E63.

Dobson, A. 2004. Population dynamics of pathogens with multiple host species. American Naturalist **164**:S64–S78.

Ebert, D., and J. J. Bull 2003. Challenging the trade-off model for the evolution of virulence: is virulence management feasible? Trends in Microbiology **11**:15–20.

Ebert, D., and W. D. Hamilton 1996. Sex against virulence: the coevolution of parasitic diseases. Trends in Ecology & Evolution **11**:79–82.

Escriu, F., A. Fraile, and F. García-Arenal 2000a. Evolution of virulence in natural populations of the satellite RNA of *Cucumber mosaic virus*. Phytopathology **90**:480–485.

Escriu, F., K. Perry, and F. García-Arenal 2000b. Transmissibility of *Cucumber mosaic virus* by *Aphis gossypii* correlates with viral accumulation and is affected by the presence of its satellite RNA. Phytopathology **90**:1068–1072.

Escriu, F., A. Fraile, and F. García-Arenal 2003. The evolution of virulence in a plant virus. Evolution **57**:755–765.

Frank, S. A. 1996. Models of parasite virulence. The Quarterly Review of Biology **71**:37–78.

Galvani, A. P. 2003. Epidemiology meets evolutionary ecology. Trends in Ecology & Evolution **18**:132–139.

Gandon, S. 2004. Evolution of multihost parasites. Evolution **58**:455–469.

Gandon, S., and T. Day 2003. Understanding and managing pathogen virulence: a way forward. Trends in Microbiology **11**:206–207.

Gandon, S., M. Van Baalen, and V. A. Jansen 2002. The evolution of parasite virulence, superinfection, and host resistance. American Naturalist **159**:658–669.

Ganusov, V. V., C. T. Bergstrom, and R. Antia 2002. Within-host population dynamics and the evolution of microparasites in a heterogeneous host population. Evolution **56**:213–223.

García-Arenal, F., and P. Palukaitis. 1999. Structure and functional relationships of satellite RNAs of *cucumber mosaic virus*. In P. K. Vogt, and A. O. Jackson, eds. Current Topics in Microbiology and Immunology. 239: Satellites and Defective Viral RNAs, pp. 37–63. Springer-Verlag, Berlin.

Gibbs, A. J., and J. C. Gower 1960. The use of a multiple-transfer method in plant virus transmission studies – some statistical points arising in the analysis of results. Annals of Applied Biology **48**:75–83.

Haydon, D. T., S. Cleveland, L. H. Taylor, and M. K. Laurenson 2002. Identifying reservoirs of infection: a conceptual and practical challenge. Emerging Infectious Diseases **8**:1468–1473.

Jacquemond, M. 2012. Cucumber mosaic virus. Advances in Virus Research **84**:439–504.

Jarosz, A. M., and A. I. Davelos 1995. Effects of disease in wild plant populations and the evolution of pathogen aggressiveness. New Phytologist **129**:371–387.

Jeger, M. J., S. E. Seal, and F. van den Bosch 2006. Evolutionary epidemiology of plant virus disease. Advances in Virus Research **67**:163–203.

Jordá, C., A. Alfaro, M. A. Aranda, E. Moriones, and F. García-Arenal 1992. Epidemic of *Cucumber mosaic virus* plus satellite RNA in tomatoes in eastern Spain. Plant Disease **76**:363–366.

Lipsitch, M., and R. Moxon 1997. Virulence and transmissibility of pathogens: what is the relationship? Trends in Microbiology **5**: 31–37.

Madden, L. V., M. J. Jeger, and F. van den Bosch 2000. A theoretical assessment of the effects of vector-virus transmission mechanism on plant virus disease epidemics. Phytopathology **90**:576–594.

Malmstrom, C. M., C. C. Hughes, L. A. Newton, and C. J. Stoner 2005. Virus infection in remnant native bunchgrasses from invaded California grasslands. New Phytologist **168**:217–230.

Mosquera, L. J., and F. R. Adler 1998. Evolution of virulence: a unified framework for coinfection and superinfection. Journal of Theoretical Biology **195**:293–313.

Osnas, E. E., and A. F. Dobson 2011. Evolution of virulence in heterogeneous host communities under multiple trade-offs. Evolution **66**: 391–401.

Owen, J., and P. Palukaitis 1988. Characterization of *Cucumber mosaic virus* I. Molecular heterogeneity mapping of RNA 3 in eight CMV strains. Virology **166**:495–502.

Palukaitis, P., and F. García-Arenal 2003. Cucumoviruses. Advances in Virus Research **62**:241–323.

Perring, T. M., N. M. Gruenhagen, and C. A. Farrar 1999. Management of plant viral diseases through chemical control of insect vectors. Annual Review of Entomology **44**:457–481.

Pfennig, K. S. 2001. Evolution of pathogen virulence: the role of variation in host phenotype. Proceedings of the Royal Society London B **268**:755–760.

Power, A. G., and C. E. Mitchell 2004. Pathogen spillover in disease epidemics. American Naturalist **164**:S79–S89.

Power, A. G., E. T. Borer, P. Hosseini, C. E. Mitchell, and E. W. Seabloom 2011. The community ecology of barley/cereal yellow dwarf viruses in Western US grasslands. Virus Research **159**:95–100.

Read, A. F. 1994. The evolution of virulence. Trends in Microbiology **2**:73–76.

Regoes, R. R., M. A. Nowak, and S. Bonhoeffer 2000. Evolution of virulence in a heterogeneous host population. Evolution **54**:64–71.

Rizzo, T. M., and P. Palukaitis 1990. Construction of full-length cDNA of *Cucumber mosaic virus* RNAs 1, 2 and 3: generation of infectious RNA transcripts. Molecular and General Genetics **222**:249–256.

Sacristán, S., and F. García-Arenal 2008. The evolution of virulence and pathogenicity in plant pathogen populations. Molecular Plant Pathology **9**:369–384.

Sacristán, S., A. Fraile, and F. García-Arenal 2004. Population dynamics of *Cucumber mosaic virus* in melon crops and in weeds in central Spain. Phytopathology **94**:992–998.

Williams, P. D. 2012. New insights into virulence evolution in multigroup hosts. American Naturalist **179**:228–239.

Woolhouse, M. E. J., L. H. Taylor, and D. T. Haydon 2001. Population biology of multi-host pathogens. Science **292**:1109–1112.

16

Directional selection for flowering time leads to adaptive evolution in *Raphanus raphanistrum* (Wild radish)

Michael B. Ashworth,[1,2] Michael J. Walsh,[1,3] Ken C. Flower,[3] Martin M. Vila-Aiub[1,4] and Stephen B. Powles[1,3]

1 Australian Herbicide Resistance Initiative, School of Plant Biology, The University of Western Australia, Crawley, WA, Australia
2 Department of Agriculture and Environment, School of Science, Curtin University, Bentley, WA, Australia
3 School of Plant Biology, The University of Western Australia, Crawley, WA, Australia
4 IFEVA-CONICET, Facultad de Agronomía, Universidad de Buenos Aires, Buenos Aires, Argentina

Keywords
biomass, evolution, flowering height, flowering time, phenotypic resistance, wild radish.

Correspondence
Stephen B. Powles, Australian Herbicide Resistance Initiative, School of Plant Biology, The University of Western Australia, Crawley, WA 6009, Australia.

e-mail: stephen.powles@uwa.edu.au

Abstract

Herbicides have been the primary tool for controlling large populations of yield depleting weeds from agro-ecosystems, resulting in the evolution of widespread herbicide resistance. In response, nonherbicidal techniques have been developed which intercept weed seeds at harvest before they enter the soil seed bank. However, the efficiency of these techniques allows an intense selection for any trait that enables weeds to evade collection, with early-flowering ecotypes considered likely to result in early seed shedding. Using a field-collected wild radish population, five recurrent generations were selected for early maturity and three generations for late maturity. Phenology associated with flowering time and growth traits were measured. Our results demonstrate the adaptive capacity of wild radish to halve its time to flowering following five generations of early-flowering selection. Early-maturing phenotypes had reduced height and biomass at maturity, leading to less competitive, more prostrate growth forms. Following three generations of late-flowering selection, wild radish doubled its time to flowering time leading to increased biomass and flowering height at maturity. This study demonstrates the potential for the rapid evolution in growth traits in response to highly effective seed collection techniques that imposed a selection on weed populations within agro-ecosystems at harvest.

Introduction

Agro-ecosystems are productive environments placed under intense disturbance (Grime 1977). Despite this disturbance, genetically diverse weed species exhibit ruderal strategies that enable them to colonize, establish and successfully persist despite efforts to eradicate them (Grime 1977; Harper 1977). Recurrent use of chemical (herbicides), physical (cultivation) and cultural (agronomy) techniques allows intense selection on the life history, phenological and growth traits of plants (Mortimer 1997). For example, herbicide selection often results in the evolution herbicide resistance (Powles and Yu 2010) and adaptive changes in the timing of seed germination and seedling emergence (Kleemann and Gill 2013; Owen et al. 2014).

Herbicides for weed control are the dominant and most intensive selective force used in modern agriculture, result-

ing in the widespread evolution of herbicide resistance in 246 weed species worldwide (Heap 2015). However, with few new herbicide modes of action (Duke 2012) and the loss of available herbicides through regulation or the evolution of herbicide resistance, it has become necessary to develop new nonherbicidal weed control strategies (Murphy et al. 1998; Madafiglio et al. 2006; Walsh and Powles 2007). The most prominent of these are a range of techniques which intercept weed seeds at harvest before they re-enter the soil seed bank (techniques collectively termed harvest weed seed control). These agricultural techniques have been reviewed (Walsh et al. 2012, 2013; Walsh and Powles 2014a,b) and are now being employed over large areas in Australia as well as being investigated for use in other grain-growing nations.

Seed dispersal and seed return to the soil seed bank are key factors in the persistence of weed populations

(Fernandez-Quintanilla 1988). The flowering time of many weed species is synchronized with crop flowering (Tremblay and Colasanti 2007), so weeds often mature concurrently with crops. Consequently, grain harvesting techniques effectively intercept and redistribute weed seeds back onto the soil surface, replenishing the soil weed seed bank (Walsh et al. 2013). Intercepting and destroying weed seeds of annual weed species is a new technique to manage weeds in agro-ecosystems (Walsh et al. 2013).

It is likely that weeds can adapt to any selective force (Jordan and Jannink 1997). Harvest weed seed control is a selective force favouring any mechanism that will enable plants to evade harvest interception. The efficacy of harvest weed seed control is contingent upon weed seeds being collected during the harvesting process, which is dependent upon the amount of weed seed retained on standing plants at crop harvest (Walsh and Powles 2014a,b). More prostrate forms (Ferris 2007) and/or earlier-seed shedding phenotypes may evade harvest collection (Baker 1974). The selection of earlier-flowering ecotypes is likely to increase the risk of seed/fruit abscission prior to harvest, resulting in harvest weed seed control evasion (Panetta et al. 1988).

Raphanus raphanistrum (wild radish) is among the worst weeds in global agriculture (Snow and Campbell 2005). In Australia, wild radish is considered to be the most problematic dicotyledonous weed species (Alemseged et al. 2001), causing significant yield losses in grain and horticultural crops (Code and Donaldson 1996; Blackshaw et al. 2002). Wild radish exhibits sufficient standing genetic variation (Conner et al. 2003; Madhou et al. 2005) to enable adaptive resistance evolution to multiple herbicide chemical classes (Hashem et al. 2001; Walsh et al. 2004a,b; Ashworth et al. 2014).

As approximately 95% of wild radish seed production is retained on the parent plant at harvest, harvest weed seed collection is ideal for controlling wild radish populations (Walsh and Powles 2014a,b). Currently, wild radish flowering time is synchronized with dryland field crops in Mediterranean climates. However, with the significant phenotypic variability in flowering time evident both within and between wild radish populations (Kercher and Conner 1996; Conner et al. 2003; Madhou et al. 2005), it is speculated that persistent collection of wild radish seed collection at crop harvest could impose a selection for early flowering time. The selection of earlier-flowering phenotypes would likely result in the evolution of wild radish populations that display a shorter life cycle, allowing plants to set and shed seed prior to crop harvest (i.e. crop maturity). This study investigated the potential for recurrent directional selection to result in heritable changes in flowering time and fitness traits in wild radish.

Materials and methods

Plant material

This selection study was conducted using a wild radish population (WARR7, referred hereafter as G0), originally collected in 1999 from Yuna, Western Australia (WA) (28.34°S, 115.01°E). This population has never been exposed to selection by herbicides or agronomic practices such as weed seed collection at harvest (Walsh et al. 2004a). Since collection, seed stocks of this herbicide susceptible population have been maintained and multiplied, ensuring no cross-pollination with other populations to maintain its susceptibility. Commencing with this population, (G0), five successive generations of recurrent early-flowering time (FT) selection was conducted in October 2011 (EF1), December 2011 (EF2), March 2012 (EF3), July 2012 (EF4) and December 2012 (EF5). Concurrently, three generations of late-FT selection were conducted in September 2012 (LF1), January 2013 (LF2) and April 2013 (LF3). During each selection, control populations were maintained without selection in the same experimental conditions, except for the absence of FT selection (CE1–CE5; CL1–CL3) (Fig. 1). At all times, plants were well watered with optimum fertilization.

Initial flowering date selection procedure

The initial FT selection (EF1) was made from a starting G0 population of 1300 plants (Table 1). Wild radish seeds (G0) greater than 2.2 mm in diameter were pregerminated on agar-solidified water (0.6% w/v), at room temperature (20°C), in darkness for 2 days. Seeds with >5 mm of emerged radicle were transplanted (5 seedlings per pot) to a depth of 10 mm into 260 pots of 305 mm diameter, containing standard potting mixture (25% peat moss, 25% sand and 50% mulched pine bark). Pots were maintained in the outdoor growth facility at The University of Western Australia (Perth) during their normal winter–spring growing season (June–October). All pots were watered regularly and fertilized weekly with 2 g Scotts Cal-Mg grower plus™ soluble fertilizer (N 15% [urea 11.6%, ammonium 1.4%, nitrate 2%], P 2.2%, K 12.4%, Ca 5%, Mg 1.8%, S 3.8%, Fe 120 mg kg^{-1}, Mn 60 mg kg^{-1}, Zn 15 mg kg^{-1}, Cu 15 mg kg^{-1}, B 20 mg kg^{-1}, Mo 10 mg kg^{-1}). The plants were monitored daily to observe the first signs of anthesis. At this time of first flowering, 13 plants (1%) were selected, based on the number of days from emergence to opening of the first flower (as marked by the protrusion of the corolla beyond the calyx). These selected plants were isolated to ensure cross-pollination only among the selection and to prevent ingress of foreign pollen. Once all selected plants were flowering, all earlier flowers were removed to minimize any unintended drift in selection due to differences in

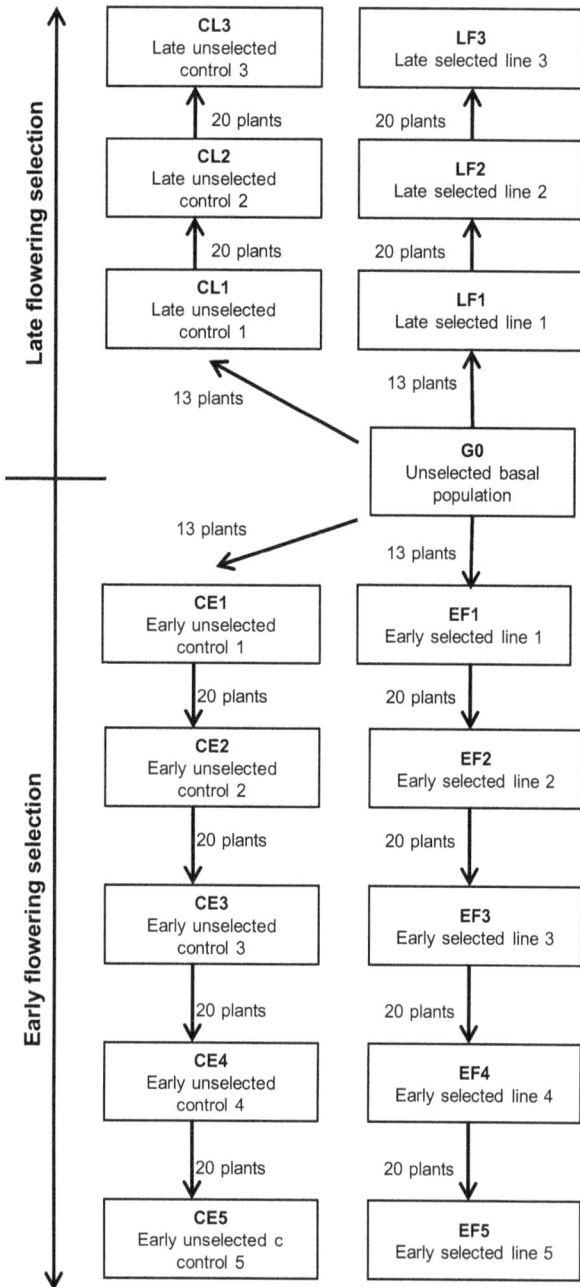

Figure 1 Hierarchy of flowering time selection applied to the commencing wild radish population (G0).

Mature siliques were then processed using a modified 'grist mill' with seed progeny representing the first selected generation (EF1; LF1) (Table 1). Concurrently, a random sample of 13 plants were selected and maintained as described above to form the first generation of the unselected control line (CE1; CL1) (Table 1).

The initial FT selection for late flowering (LF1) was made from a commencing population of 1300 plants using the previously described procedure. All plants were monitored daily to observe anthesis. Only the last 13 plants to flower among the 1300 plants (1%) were selected. These plants were isolated and cross-pollinated among themselves as previously described. At maturity, siliques were harvested and processed, with seed progeny representing the first long flowering selected generation (LF1) (Table 1).

Subsequent selections general procedure

Subsequent directional FT selections were conducted in a temperature-controlled glasshouse with natural light, where cooling was initiated above 25°C day and 15°C night. Large seed (>2.2 mm diameter) of the initial selected populations (EF1 or LF1) and the initial control populations (CE1 or CL1) were germinated on solidified water agar (0.6% w/v), at room temperature (20°C) in darkness for 2 days. After germination, 250 pregerminated seeds (>5 mm emerged radicle) were seeded into separate 220-mm-diameter pots, watered twice daily to field capacity using an automated irrigation system and fertilized weekly as previously described. The date of emergence was noted for each pot. At first flowering, 20 plants were selected from the 250 individuals based on the number of days from emergence to the opening of the first flower. These selected plants were isolated and crossed as previously described to produce early-selected generations (EF2–EF5). Using this methodology, the late-selected generations were also selected (LF2; LF3) (Fig. 1). Concurrently, 20 randomly selected seeds from each respective control line (CE1; CL1) were planted, maintained and crossed as previously described to produce unselected early-control generations (CE2–CE5) and late-control generations (CL2; CL3) (Fig. 1).

Analysis of selection and crossing lines

The rate of FT progression was evaluated by growing the G0, selected (EF1–EF5; LF1–LF3) and control (CE1–CE5; CL1–CL5) generations at the same time within temperature-regulated glasshouse conditions during a period of stable to gradually increasing day length (June onwards, 2013) (Supporting information). Large seed (>2.0 mm in diameter) from each population was pregerminated on agar (0.6% w/v)-solidified water in darkness for 2 days.

female fitness among selected individuals (Sahli and Conner 2011). Newly opened flowers were crossed using the 'Beestick method', where a bee carcass was used to cross-fertilize flowers as outlined by Williams (1980), ensuring a random pattern of cross-pollination (panmixia), where each flower was randomly crossed with a flower from a different plant. At maturity, the same number of siliques were harvested from each plant in the population and bulked.

Table 1. Flowering time advancement [days to flowering and cumulative growing degree-days (GDD)] of each wild radish accession during selection.

	Selected generations						Unselected control generations			Phenotypic advancement in first flowering between selected and control generations	
	Selected line	Selection coefficient (1-selected plants ratio)	Sowing date	Selection dates	Days to first flowering	Cumulative GDD (°C d)	Control line	Days to first flowering	Cumulative GDD (°C d)	Change in days to first flowering	Change in cumulative GDD (°C d)
Early-flowering time selection	Commencing G0	0.99	5 Sept 2011	5 Oct 2011	30	446	CE1	26	467	−5	−85
	EF1	0.95	9 Dec 2011	30 Dec 2011	21	382	CE2	30	468	−8	−99
	EF2	0.95	9 March 2012	31 March 2012	22	369	CE3	72	521	−30	−172
	EF3	0.95	23 May 2012	4 July 2012	42	349	CE4	34	506	−15	−237
	EF4	0.95	4 Dec 2012	24 Dec 2012	19	269	CE5	Final early control		—	—
	EF5			Final early selected							
Late-flowering time selection	Commencing G0	0.99	4 June 2012	4 Sept 2012	68 (92)*	447 (936)†	CL1	21	482	4	30
	LF1	0.95	8 Nov 2012	6 Jan 2013	25 (48)*	512 (1024)†	CL2	21	436	7	48
	LF2	0.95	11 Feb 2013	14 April 2013	28 (62)*	484 (2214)†	CL3	Final late control		—	—
	LF3			Final late selected							

*Bracketed data denotes days to last flowering selection (last 10% of the population).
†Bracketed data denotes cumulative degree-days until final last flowering individuals were selected (last 10% of the population).

Seventy-five seeds from each population (with >5 mm emerged radicle) were seeded 10 mm deep into individual 220-mm-diameter pots containing standard potting mixture (25% peat moss, 25% sand and 50% mulched pine bark). All selected, control and progeny generations were arranged within the glasshouse in a randomized block design (3 blocks of 25 plants per treatment) with the date of emergence noted for each pot. All pots were watered to field capacity every 2 h (during the day) using an automated irrigation system (Supporting information), with 2 g Scotts Cal-Mag grower plus™ soluble fertilizer applied weekly, as previously described. For the duration of the experiment, temperatures were maintained at temperatures of 25°C day and 15°C night, above the base temperature for wild radish growth (4.5°C) (Reeves et al. 1981). Air temperature and daylight was recorded every 15 min using an environment-controlling thermistor and light photometer (Schneider Electric; www.schneider-electric.com) located 1 m above the pots in the centre of the glasshouse. For the duration of the experiment, the date of flowering and height of the first flower were recorded daily for each individual. Above-ground biomass at the initiation of flowering was cut and dried at 65°C for 7 days before weighing.

Data analysis

To compare the FT response of recurrently selected wild radish populations, nonlinear regression analysis was performed using the DRC package in R 3.0.0 (R Development Core Team 2011; http://www.R-project.org) (Streibig et al. 1993). The observed population flowering over time was fitted to a four-parameter logistic model [1]:

$$Y = \frac{c + (d - c)}{1 + e^{b(\log x - \log e)}} \qquad (1)$$

where Y denotes cumulative flowering as a percentage of the total population, e is the FD_{50} denoting the time or accumulated temperature to flowering response is halfway between the upper limit, d (fixed to the total percentage of the population collected) and c the lower asymptotic value of Y (set to 0). The parameter b denotes the relative slope around e. FD_{50} parameter was compared between selected and unselected (G0) populations using the selectivity indices (SI) function (R 3.0.0) which determines whether the ratios between the FD_{50} values are significantly different ($P < 0.05$). A lack-of-fit test was also applied to each curve to ascertain the appropriateness of the model [1] in R3.0.0.

The experiments were conducted at different times of the year; therefore, the different selections were compared using growing degree-days (GDD) to flowering, as described by Marcellos and Single (1971) and equation [2]:

$$GDD = \frac{\sum(T_{max} + T_{min})}{2} - T_{base} \qquad (2)$$

where T_{max} is the daily maximum temperature, T_{min} is the daily minimum temperature and T_{base} is the base temperature for wild radish (4.5°C) (Reeves et al. 1981). During this study, it is assumed that T_{base} remained constant. Height of the first flower and above-ground biomass at flowering were checked for homogeneity of variance, normality and independence of residuals as described by Onofri et al. (2010). The flowering time selections were then compared for these response variables using a one-way analysis of variance (ANOVA) in Genstat version 6.1.0.200 (VSN International, www.vsni.co.uk/genstat). Aboveground biomass at flowering was $\log_{10}$-transformed prior to a two-way ANOVA. Means were estimated and separated using Tukey's protected LSD at the 5% level of significance. Biomass data were back-transformed prior to plotting. The relationship between the response variables (height of first flower and above-ground biomass) and days to flowering was plotted using SigmaPlot v.12 (Systat Software Inc., San Jose, CA, USA).

Results

Effect of recurrent early-flowering time selection

In one large final experiment, the G0 population and all successive selected (EF1–EF5; LF1–LF3) and the unselected control (CE1–CE5; CL1–CL3) populations were grown in a temperature-controlled glasshouse to evaluate the population phenotypic change of each FT selection. Analysis of all accessions showed that FT was halved at the population level (FD_{50}), following five successive generations of early-FT selection. The FD_{50} parameter is the median time for the population to initiate its first flower. Early-FT selection reduced the time from emergence to flowering from 59 days after emergence (DAE) (G0) to 29 DAE (EF5) (Fig. 2), reducing the thermal time requirement prior to flowering (GDD) from 634°C d (G0) to 344°C d (EF5) (see Supporting information). This reduction in FT was evident during each selection with thermal time requirement to flowering decreasing by 85, 99, 172 and 237°C d in the EF1 to EF4 generations, respectively, when compared to the concurrently grown but unselected controls (CE1–CE4) (Table 1). In the absence of selection, the control generations (CE1–CE5) changed FT negligibly compared with the G0 population, demonstrating that FT reductions in the selected generations (EF1–EF5) were primarily due to the effects of FT selection (Supporting information).

Flowering time reductions at the population level (FD_{50}) in the early-FT-selected generations were found to be a result of a reduction in the distribution rather than any shift towards an earlier initiation of flowering. Following

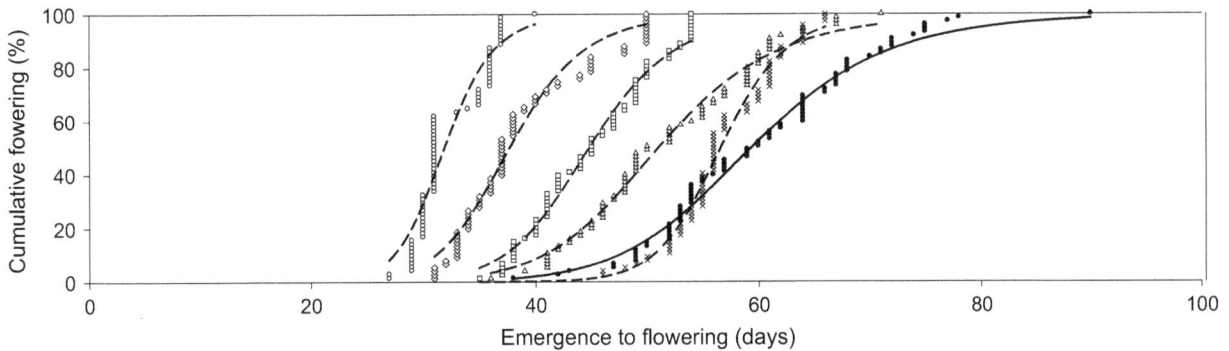

Figure 2 The observed population response to early-flowering time selection against the unselected commencing wild radish population G0 (_•_). Early-flowering time-selected generations EF1 (...Δ...), EF2 (...×...), EF3 (...□...), EF4 (...◊...) and EF5 (...O...). Each symbol represents cumulative data points of 75 replicate plants. The plotted lines are predicted cumulative flowering date curves fitted to a four-parameter logistic model [1].

five generations of early-FT selection, the initiation of flowering decreased by 11 days (EF5) (Fig. 2); however, the distribution of FT in the population decreased fourfold, from an initial range of 52 days (G0) to 13 days (EF5) (Fig. 2). This decrease in the distribution of FT resulted in 77% of the EF5 generation flowering prior to the initiation of flowering in the unselected G0 population.

As well as reductions in FT, selection also led to reduced plant height and above-ground biomass at flowering. Five generations of early-FT selection reduced mean height at the initiation of flowering 2.6-fold from 88 cm in the G0 population to 33 cm (EF5) ($P < 0.001$) (Table 2). Concurrently, mean plant biomass at the initiation of flowering decreased 5.5-fold from 22 g (G0) to 4 g (EF5) ($P < 0.001$) (Table 2). In the absence of selection (CE1–CE5), there was no significant change in population biomass or flowering height from the G0 population ($P > 0.05$) (Table 2). Both height and biomass at flowering decreased in plants that flowered earlier (Figs 3 and 4, respectively).

Effect of recurrent late-flowering time selection

Conversely, late-FT selection resulted in large stepwise increases in FT in both the first (LF1) and third (LF3) generation (Fig. 5). Following three generations of late-FT selection, the length of the vegetative stage was doubled, from 59 DAE (G0; FD_{50}) to 114 DAE (LF3; FD_{50}) (Table 1), corresponding to a 2.1-fold increase in the thermal time requirement prior to flowering (634°C d (G0) to 1314°C d (LF3) (Supporting information). The initiation of flowering was delayed by 23 days following a single generation of FT selection (LF1). Subsequent selections did not further delay the initiation of flowering (LF2; LF3) (Supporting information). Additional selections, however, progressively increased the distribution of flowering from 52 days in the G0 population to 84 days following three generations of late-FT selection (Fig. 5). In the absence of

Table 2. Mean height of first flower and aboveground biomass at flowering for the commencing (G0), early-selected (EF1–EF5) and late-selected (LF1–LF3) generations.

Selection	Selected line	Height (cm)	Biomass (g plant^{-1})
Early-flowering time selection	EF5	33 a*	4 a
	EF4	44 b	7 ab
	EF3	46 b	10 b
	EF2	72 c	17 c
	EF1	69 c	16 c
Unselected control early flowering	CE5	90 e	22 d
	CE4	83 de	22 d
	CE3	89 de	21 d
	CE2	80 d	20 d
	CE1	88 de	21 d
Unselected	Commencing G0	88 de	22 d
Unselected control late flowering	CL1	84 de	20 d
	CL2	90 e	22 d
	CL3	87 de	19 d
Late-flowering time selection	LF1	112 f	29 e
	LF2	121 f	35 f
	LF3	140 g	46 g

*Different letters indicate significant difference between means (Tukey separation) at $P \leq 0.05$.

selection, the concurrently grown generations (CL1–CL3) were not different from the G0 population ($P > 0.05$), again demonstrating that FT increases were primarily due to the effects of selection (Supporting information).

Late-FT selection progressively increased plant biomass and height of the first flower. The height of the first flower progressively increased from 88 cm in the G0 population to 112 cm, 121 cm and 141 cm in the LF1, LF2 and LF3 generations, respectively (Table 2). Late-FT selection also increased mean plant biomass at flowering from 22 g (G0) to 29 g, 35 g and 46 g per plant in the LF1, LF2 and LF3 generations, respectively (Table 2).

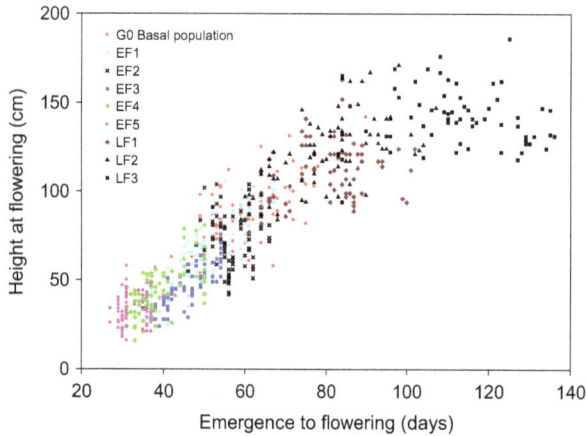

Figure 3 The relationship between the height of the first flower and days to flowering for the commencing (G0), early-selected (EF1–EF5) and late-selected (LF1–LF3) generations. Each symbol represents individual plants (each population *n* = 75; LF3 *n* = 64).

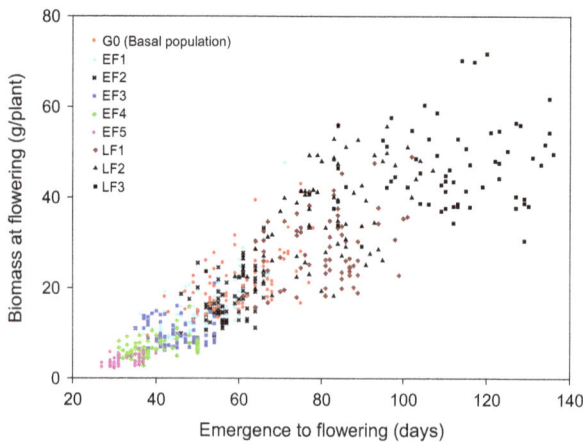

Figure 4 The relationship between biomass at flowering and days to flowering for the commencing (G0), early-selected (EF1–EF5) and late-selected (LF1–LF3) generations. Each symbol represents individual plant biomass measurements (each population *n* = 75; LF3 *n* = 64).

Discussion

Sustainable agriculture is based on achieving a balance between producing highly productive and profitable crops whilst minimizing cost and energy inputs (Gomiero et al. 2011). The evolution of herbicide resistance in weed species infesting crops poses a significant threat to crop production (Powles and Yu 2010). As a consequence, there is an increased interest in nonchemical weed management tools (Walsh and Powles 2014a,b). However, over-reliance of any single agronomic weed control practice is expected to result in rapid selection of adaptive traits, selected from the standing genetic variation within weed populations (Powles and Yu 2010).

The results of this study demonstrate that wild radish exhibits significant standing genetic variability to adapt to flowering time (FT) selection. Following five generations of early-FT selection, wild radish FT (FD$_{50}$) was halved, whilst three generations of late-FT selection doubled FT at the population level. Bidirectional selection resulted in a total FT divergence of 83 days at the population level following five early and three late generations of flowering selection. The rapid FT response in this study is consistent with previous bidirectional FT selection results in studies of Chinese daikon radish (*Raphanus sativus* L.). These studies hinted that the nature of the genetic control of FT was likely to be polygenic with incomplete dominance (Vahidy 1969). Similar flowering time shifts in response to early-FT selection has also been observed in both field and glasshouse environments in other closely related Brassica species such as wild mustard (*Brassica rapa* L.) (Franke et al. 2006; Franks et al. 2007; Franks 2011).

Adaptability of wild radish

Wild radish is a genetically diverse, highly adaptable species which has been found to consistently thrive in a diverse range of environments (Madhou et al. 2005; Snow and Campbell 2005) and production systems (Alemseged et al. 2001; Borger et al. 2012). This study shows that when selected for reduced time to initiate flowering, wild radish plants can flower at far lower thermal requirements than normally observed in field populations. Previous studies have indicated that wild radish can reach anthesis in as little as 600°C d (Reeves et al. 1981; Cheam 1986; Malik et al. 2010). However, following five generations of early-FT selection in this study, wild radish reduced its thermal requirement to 344°C d [FD$_{50}$] at the population level. At an individual level, a thermal requirement of just 281°C d was observed resulting in a wild radish plant flowering 22 days after emergence (EF5) with a biomass of just 2.4 g. Identification of these early-flowering individuals in the EF5 generation is significant as they became less sensitive to photoperiod or temperature cues, flowering under a short photoperiod of 9.5 h per day, at an average daily temperature of 15°C. Conversely, late-FT selection over three generations increased thermal time to 1314°C d [FD$_{50}$]. This study, however, understates the full adaptive response of wild radish to late-FT selection, as the analysis of the selected populations was suspended at 149 DAE (or after 1565°C d) with 14% (11 plants) from the final LF3 generation still failing to initiate flowering. A total FT divergence of 127 days from the initiation of the first flowering individual to the suspension of this study is a clear demonstration of the remarkable capacity of wild radish populations to adapt FT when selected.

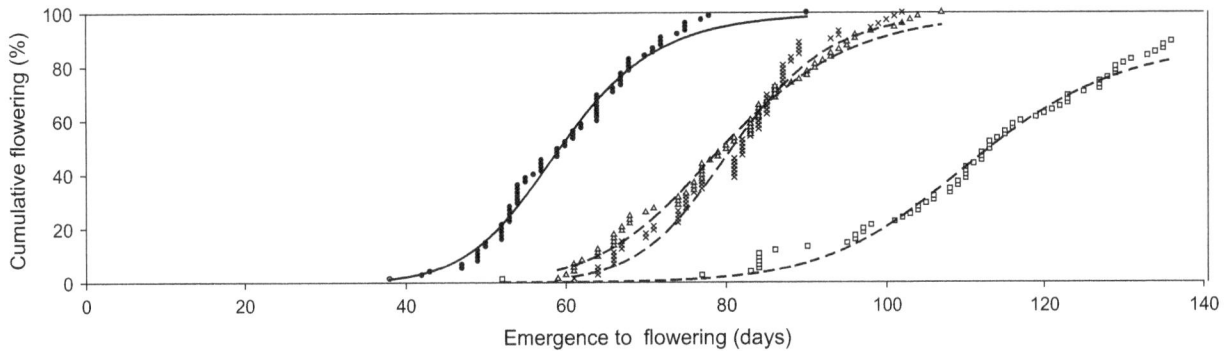

Figure 5 The observed population response to late-flowering time selection against the unselected commencing wild radish population G0 (__•__). late-selected generations LF1 (...×...), LF2 (...Δ...) and LF3 (...□...). each symbol represents cumulative data points of 75 replicate plants [except LF3 = 65 plants (max 86%)]. the plotted lines are predicted cumulative flowering date curves fitted to a four-parameter logistic model [1].

The genetic basis for FT adaptation in this study has not been determined; however, quantitative trait loci (QTL) studies in wild mustard, oilseed canola (*Brassica napus* L.) and *Arabidopsis thaliana* L. have identified that multiple loci are involved in flowering initiation in *A. thaliana* L (Osborn et al. 1997; Cai et al. 2008; Colautti and Barrett 2010; Raman et al. 2013). Over 80 different genes have been identified to affect FT initiation in response to external and endogenous cues (Simpson and Dean 2002). The progressive early-FT shifts in this study are compatible with the polygenic accumulation of minor gene traits' as other hypotheses including variation in gene editing are possible. This study is also consistent with Vahidy (1969) and Conner (1993), who observed large shifts in phenology with late-FT selection.

Ecological and evolutionary significance and implications for weed management in agro-ecosystems

The results of this study demonstrate the capacity of wild radish to adapt both phenologically and through growth traits. The selection treatments in this study mimic a strong selection force acting against plants that usually have synchronous flowering with field crops (i.e. harvest weed seed control). Recurrent selection for early flowering over five generations resulted in a halving of the flowering time (FD_{50}) as well as a fourfold reduction in the distribution of flowering, resulting in 77% of individuals flowering before the initiation of flowering in the unselected basal population (G0). Wild radish adaptation in flowering time was also accompanied by changes in plant size (biomass and height), reflecting similar physiological responses in *A. thaliana* L. (Tienderen et al. 1996). In this study, the height of the first flower and vegetative above-ground biomass were consistently reduced with early-flowering time selection. As a result of insufficient biomass accumulation prior to flowering, early-FT-selected populations grow in a more prostrate form, lacking the ability to support reproductive branches (Supporting information). Given that in annual species like wild radish plant size is a predictive value of the amount of resources to be allocated to reproductive fitness (Weiner 2004; Weiner et al. 2009), the reduced biomass observed in early-flowering plants is likely to cause reductions in the population's competitive ability for resources (Goldberg 1990) and overall fitness. Whilst not measured in this study, early-FT-selected plants are likely to have lower fecundity (Cheam 1986; Conner and Via 1993), due to lower biomass plants producing fewer and smaller flowers that produce less seeds per silique (Conner and Via 1993; Conner et al. 1996a,b; Williams and Conner 2001).

Despite this likely fitness cost, it is anticipated that early-flowering plants will also have a fitness advantage whilst harvest weed seed control selection of retained seeds at crop maturity is occurring (Vila-Aiub et al. 2009). The rate of fruit abscission was not determined in this study; however, as a consequence of early flowering time, the number of individuals in the population carrying well-matured pods at the time of crop maturity would rapidly increase as flowering time is reduced. This increase in the proportion of well-matured pods at harvest is expected to increase the probability of silique abscission prior to harvest, especially during periods of water deficit, high temperature or wind (Taghizadeh et al. 2012). However, seed retention amongst early-flowering time-selected populations is speculated to vary according to climatic conditions, which is likely to vary across differing agro-ecosystems (Taghizadeh et al. 2012). Similarly, the reduction in wild radish flowering height and the resultant prostrate growth habit (associated with low biomass accumulation) is likely to reduce seed interception at the time of crop harvest as a greater proportion of siliques are likely to be located on unsupported stems, below the required height for harvest interception (Supporting information).

Within an agricultural context where weed species are selected with harvest weed seed control, the evolution of early-maturing ecotypes can be seen as an optimal survival strategy. However, the presence of a steep fitness gradient favouring late-flowering ecotypes implies that early-flowering time-selected populations are likely to rapidly moderate flowering time back to an ecological optimum when selection is relaxed, as later-flowering ecotypes are likely to have a greater reproductive capacity (Baker 1974; Conner et al. 1996b). From a weed management perspective, in order to maintain long-term effectiveness of harvest weed seed control techniques, it may be prudent to periodically stop the use of this selective tool once weed seed banks have been reduced to manageable levels. Any relaxation in selection is expected to allow for the recovery of the standing genetic variation in flowering time traits within wild radish populations, therefore restoring phenological traits that are important for the interception of weed seeds at harvest.

Our results also demonstrate the evolutionary capacity of wild radish populations to rapidly adapt when late-flowering individuals are favoured. Within agro-ecosystems, other human-mediated selective tools that target early-maturing phenotypes are used. Seed set reduction techniques such as herbicidal flower sterilization (termed as crop-topping) (Madafiglio et al. 2006) target early-flowering individuals before the maturity of the crop, effectively favouring the proliferation of late-maturing phenotypes. Three generations of late-FT selection resulted in a doubling of FT along with a 2.2-fold increase in biomass and a 1.6-fold increase in plant height at anthesis. However, FT adaptation favouring lateness is likely to be moderated by moisture stress during reproduction in Australia, reducing the fitness of these extremely late-flowering ecotypes. Therefore, it is again speculated that the evolutionary effect of late-FT selection would vary according to the climatic conditions to be more likely in agro-ecosystems of higher rainfall and a longer growing season (Conner et al. 1996a).

The significant reductions and delays in wild radish flowering time as a result of FT selection further highlight the genetic potential of wild radish populations for rapid adaptive evolution in response to selection agents that act both prior and at the time of crop harvest.

Acknowledgements

This research was funded by the Grains Research Development Corporation of Australia (GRDC) and the Australian Government through an Australian Post Graduate Award to M Ashworth. Thank you to AHRI and plant growth facilities staff at The University of Western Australia who provided technical assistance.

Literature cited

Alemseged, Y., R. E. Jones, and R. W. Medd 2001. A farmers survey of weed management and herbicide problems on winter crops in Australia. Plant Protection Quarterly **16**:21–25.

Ashworth, M. B., M. J. Walsh, K. C. Flower, and S. B. Powles 2014. Identification of the first glyphosate-resistant wild radish (*Raphanus raphanistrum L.*) populations. Pest Management Science **70**:1432–1436.

Baker, H. G. 1974. The evolution of weeds. Annual Review of Ecology and Systematics **5**:1–24.

Blackshaw, R. E., D. Lemerle, and K. R. Young 2002. Influence of wild radish on yield and quality of canola. Weed Science **50**:344–349.

Borger, C. P. D., P. J. Michael, R. Mandel, A. Hashem, D. Bowran, and M. Renton 2012. Linking field and farmer surveys to determine the most important changes to weed incidence. Weed Research **52**:564–574.

Cai, C. C., J. X. Tu, T. D. Fu, and B. Y. Chen 2008. The genetic basis of flowering time and photoperiod sensitivity in rapeseed *Brassica napus* L. Russian Journal of Genetics **44**:326–333.

Cheam, A. H. 1986. Seed production and seed dormancy in wild radish (*Raphanus raphanistrum*) and some possibilities for improving control. Weed Research **26**:405–413.

Code, G. R., and T. W. Donaldson 1996. Effect of cultivation, sowing methods and herbicides on wild radish populations in wheat crops. Australian Journal of Experimental Agriculture **36**:437–442.

Colautti, R. I., and S. C. H. Barrett 2010. Natural selection and genetic constraints on flowering phenology in an invasive plant. International Journal of Plant Sciences **171**:960–971.

Conner, J. 1993. Tests for major genes affecting quantitative traits in wild radish (*Raphanus raphanistrum L.*). Genetica **90**:41–45.

Conner, J., and S. Via 1993. Patterns of phenotypic and genetic correlations among morphological and life history traits in wild radish, *Raphanus raphanistrum*. Evolution **47**:704–711.

Conner, J. K., S. Rush, and P. Jennetten 1996a. Measurements of natural selection on floral traits in Wild Radish (*Raphanus raphanistrum*). I. Selection through lifetime female fitness. Evolution **50**:1127–1136.

Conner, J. K., S. Rush, S. Kercher, and P. Jennetten 1996b. Measurements of natural selection on floral traits in Wild Radish (*Raphanus raphanistrum*). II. Selection through lifetime male and total fitness. Evolution **50**:1137–1146.

Conner, J. K., R. Franks, and C. Stewart 2003. Expression of additive genetic variances and covariances for Wild Radish floral traits: comparison between field and greenhouse environments. Evolution **57**:487–495.

Duke, S. O. 2012. Why have no new herbicide modes of action appeared in recent years? Pest Management Science **68**:505–512.

Fernandez-Quintanilla, C. 1988. Studying the population dynamics of weeds. Weed Research **28**:443–447.

Ferris, D. G. 2007. Evolutionary Differentiation in *Lolium L.* (Ryegrass) in Response to the Mediterranean Type Climate and Changing Farming Systems of Western Australia. School of Plant Biology, Faculty of Natural and Agricultural Sciences, University of Western Australia, Perth.

Franke, D., A. T. Ellis, M. Dharjwa, M. Freshwater, M. Fujikawa, A. Padron, and A. Weis 2006. A steep decline in flowering time for *Brassica rapa* in Southern California: population level variation in the field and the greenhouse. International Journal of Plant Sciences **167**:83–92.

Franks, S. J. 2011. Plasticity and evolution in drought avoidance and escape in the annual plant *Brassica rapa*. New Phytologist **190**:249–257.

Franks, S. J., S. Sim, and A. E. Weis 2007. Rapid evolution of flowering time by an annual plant in response to a climate fluctuation. Proceedings of the National Academy of Sciences of the USA **104**:1278–1282.

Goldberg, D. E. 1990. Components of resource competition in plant communities. In: J. B. Grace, and D. Tilman, eds. In Perspectives in Plant Competition. Academic Press, San Diego.

Gomiero, T., D. Pimentel, and M. G. Paoletti 2011. Is there a need for a more sustainable agriculture? Critical Reviews in Plant Sciences **30**:6–23.

Grime, J. P. 1977. Evidence for the existence of three primary strategies in plants and its relevance to ecological and evolutionary theory. American Naturalist **111**:1169–1194.

Harper, J. L. 1977. Population Biology of Plants. Academic Press, London.

Hashem, A., H. S. Dhammu, S. B. Powles, D. G. Bowran, T. J. Piper, and A. H. Cheam 2001. Triazine resistance in a biotype of wild radish (*Raphanus raphanistrum*) in Australia. Weed Technology **15**:636–641.

Heap, I. 2015. *The international survey of herbicide resistant weeds*. Available from www.weedscience.com (accessed on 29 June 2015).

Jordan, N. R., and J. L. Jannink 1997. Assessing the practical importance of weed evolution: a research agenda. Weed Research **37**:237–246.

Kercher, S., and J. K. Conner 1996. Patterns of genetic variability within and among populations of wild radish, *Raphanus raphanistrum* (Brassicaceae). American Journal of Botany **83**:1416–1421.

Kleemann, S. G. L., and G. S. Gill. 2013. Seed dormancy and seedling emergence in ripgut brome (*Bromus diandrus*) populations in Southern Australia. Weed Science **61**:222–229.

Madafiglio, G. P., R. W. Medd, P. S. Cornish, and R. V. De Ven 2006. Seed production of *Raphanus raphanistrum* following herbicide application during reproduction and effects on wheat yield. Weed Research **46**:50–60.

Madhou, P., A. Wells, E. Pang, and T. Stevenson 2005. Genetic variation in populations of Western Australian wild radish. Australian Journal of Agricultural Research **56**:1079–1087.

Malik, M. S., J. K. Norsworthy, M. B. Riley, and W. Bridges 2010. Temperature and Light Requirements for Wild Radish (Raphanus raphanistrum) Germination over a 12-Month Period following Maturation. Weed Science **58**:136–140.

Marcellos, H., and W. V. Single 1971. Quantitative responses of wheat to photoperiod and temperature in the field. Australian Journal of Agricultural Research **22**:343–357.

Mortimer, A. M. 1997. Phenological adaptation in weeds—an evolutionary response to the use of herbicides? Pesticide Science **51**:299–304.

Murphy, C., D. Lemerle, and R. Medd. 1998. The ecology of wild radish: the key to sustainable management. In *Proceedings of the 9th Australian Agronomy Conference*. Wagga Wagga, Australia.

Onofri, A., E. A. Carbonell, H. P. Piepho, A. M. Mortimer, and R. D. Cousens 2010. Current statistical issues in Weed Research. Weed Research **50**:5–24.

Osborn, T. C., C. Kole, I. A. P. Parkin, A. G. Sharpe, M. Kuiper, D. J. Lydiate, and M. Trick 1997. Comparison of flowering time genes in *Brassica rapa*, *B. napus* and *Arabidopsis thaliana*. Genetics **146**:1123–1129.

Owen, M. J., D. E. Goggin, and S. B. Powles. 2014. Intensive cropping systems select for greater seed dormancy and increased herbicide resistance levels in Lolium rigidum (annual ryegrass). Pest Management Science **71**:966–971.

Panetta, F. D., D. J. Gilbey, and M. F. Dantuono 1988. Survival and fecundity of wild radish (*Raphanus raphanistrum* L) plants in relation to cropping, time of emergence and chemical control. Australian Journal of Agricultural Research **39**:385–397.

Powles, S., and Q. Yu 2010. Evolution in action: plants resistant to herbicides. Annual Review of Plant Biology **61**:317–347.

R Core Team. 2011. R: A Language and Environment for Statistical Computing. R Foundation for Statistical Computing, Vienna, Austria. ISBN: 3-900051-07-0, URL: http://www.Rproject.org/

Raman, H., R. Raman, P. Eckermann, N. Coombes, S. Manoli, X. Zou, D. Edwards et al. 2013. Genetic and physical mapping of flowering time loci in canola (*Brassica napus* L.). Theoretical and Applied Genetics **126**:119–132.

Reeves, T. G., G. R. Code, and C. M. Piggin 1981. Seed production and longevity, seasonal emergence and phenology of Wild Radish (*Rahanus raphanistrum*). Australian Journal of Experimental Agriculture **21**:524–530.

Sahli, H. F., and J. K. Conner 2011. Testing for conflicting and nonadditive selection: floral adaptation to multiple pollinators through male and female fitness. Evolution **65**:1457–1473.

Simpson, G., and C. Dean 2002. *Arabidopsis*, the rosetta stone of flowering time? Science **296**:285–289.

Snow, A., and L. Campbell. 2005. Can feral radishes become weeds. In J. Gressel, ed. Crop Ferality and Volunteerism, pp. 193–208. CRC Press, Boca Raton, FL.

Streibig, J. C., M. Rudemo, and J. E. Jensen 1993. Dose-response curves and statistical models. In: P. Kudsk, and J. C. Streibig, eds. Herbicide Bioassays, pp. 29–55. CRC Press, Boca Raton.

Taghizadeh, M. S., M. E. Nicolas, and R. D. Cousens 2012. Effects of relative emergence time and water deficit on the timing of fruit dispersal in Raphanus raphanistrum L. Crop and Pasture Science **63**:1018–1025.

Tienderen, P. H. V., I. Hammad, and F. C. Zwaal 1996. Pleiotropic effects of flowering time genes in the annual crucifer *Arabidopsis thaliana* (Brassicaceae). American Journal of Botany **83**:169–174.

Tremblay, R., and J. Colasanti. 2007. Floral Induction. In Charles Ainsworth, ed. Annual Plant Reviews Volume 20: Flowering and its Manipulation, pp. 28–62. Blackwell Publishing Ltd, Oxford.

Vahidy, A. A. 1969. The Genetics of Flowering time in *Raphanus sativus* L. (Chinese Radish). PhD thesis, University of Hawaii.

Vila-Aiub, M. M., P. Neve, and S. B. Powles 2009. Fitness costs associated with evolved herbicide resistance alleles in plants. New Phytologist **184**:751–767.

Walsh, M. J., and S. B. Powles 2007. Management strategies for herbicide-resistant weed populations in Australian dryland crop production systems. Weed Technology **21**:332–338.

Walsh, M. J., and S. Powles. 2014a. High seed retention at maturity of annual weeds infesting crop fields highlights the potential for harvest weed seed control. Weed Technology **28**:486–493.

Walsh, M. J., and S. B. Powles 2014b. Management of herbicide resistance in wheat cropping systems: learning from the Australian experience. Pest Management Science **70**:1324–1328.

Walsh, M. J., S. B. Powles, B. R. Beard, and S. A. Porter 2004a. Multiple-herbicide resistance across four modes of action in wild radish (*Raphanus raphanistrum*). Weed Science **52**:8–13.

Walsh, M., B. Westphal, and S. Powles. 2004b. Multiple herbicide resistance in Wild Radish (*Raphanus raphanistrum*) populations in Western Australia. In *4th International Weed Society Congress*. Durban, South Africa.

Walsh, M. J., R. B. Harrington, and S. B. Powles 2012. Harrington seed destructor: a new non-chemical weed control tool for global grain crops. Crop Science **52**:1343–1347.

Walsh, M., P. Newman, and S. Powles 2013. Targeting weed seeds in-crop: a new weed control paradigm for global agriculture. Weed Technology **27**:431–436.

Weiner, J. 2004. Allocation, plasticity and allometry in plants. Perspectives in Plant Ecology, Evolution and Systematics **6**:207–215.

Weiner, J., L. G. Campbell, J. Pino, and L. Echarte 2009. The allometry of reproduction within plant populations. Journal of Ecology **97**:1220–1233.

Williams, P. H. 1980. Bee-sticks, an aid in pollinating *Cruciferae*. HortScience **15**:802–803.

Williams, J., and J. Conner 2001. Sources of phenotypic variation in floral traits in Wild Radish (*Raphanus raphanistrum*). American Journal of Botany **88**:1577–1581.

PERMISSIONS

The contributors of this book come from diverse backgrounds, making this book a truly international effort. This book will bring forth new frontiers with its revolutionizing research information and detailed analysis of the nascent developments around the world.

We would like to thank all the contributing authors for lending their expertise to make the book truly unique. They have played a crucial role in the development of this book. Without their invaluable contributions this book wouldn't have been possible. They have made vital efforts to compile up to date information on the varied aspects of this subject to make this book a valuable addition to the collection of many professionals and students.

This book was conceptualized with the vision of imparting up-to-date information and advanced data in this field. To ensure the same, a matchless editorial board was set up. Every individual on the board went through rigorous rounds of assessment to prove their worth. After which they invested a large part of their time researching and compiling the most relevant data for our readers.

The editorial board has been involved in producing this book since its inception. They have spent rigorous hours researching and exploring the diverse topics which have resulted in the successful publishing of this book. They have passed on their knowledge of decades through this book. To expedite this challenging task, the publisher supported the team at every step. A small team of assistant editors was also appointed to further simplify the editing procedure and attain best results for the readers.

Apart from the editorial board, the designing team has also invested a significant amount of their time in understanding the subject and creating the most relevant covers. They scrutinized every image to scout for the most suitable representation of the subject and create an appropriate cover for the book.

The publishing team has been an ardent support to the editorial, designing and production team. Their endless efforts to recruit the best for this project, has resulted in the accomplishment of this book. They are a veteran in the field of academics and their pool of knowledge is as vast as their experience in printing. Their expertise and guidance has proved useful at every step. Their uncompromising quality standards have made this book an exceptional effort. Their encouragement from time to time has been an inspiration for everyone.

The publisher and the editorial board hope that this book will prove to be a valuable piece of knowledge for researchers, students, practitioners and scholars across the globe.

LIST OF CONTRIBUTORS

Marco Archetti
School of Biological Sciences, University of East Anglia, Norwich, UK

Lesley G. Campbell
Department of Chemistryand Biology, Ryerson University, Toronto, ON, Canada

Zachary Teitel
Department of Chemistryand Biology, Ryerson University, Toronto, ON, Canada
Department of Integrative Biology, University of Guelph, Guelph, ON Canada

Maria N. Miriti
Department of Evolution, Ecology and Organismal Biology, The Ohio State University, Columbus, OH, USA

Rickey D. Cothran and Rick A. Relyea
Department of Biological Sciences and Pymatuning Laboratory of Ecology, University of Pittsburgh, Pittsburgh, PA, USA

Jenise M. Brown
Department of Biological Sciences and Pymatuning Laboratory of Ecology, University of Pittsburgh, Pittsburgh, PA, USA
Department of Integrative Biology, University of South Florida, Tampa, FL, 33620, USA

Frédéric Fabre
UMR 1065 Unité Santé et Agroécologie du Vignoble, INRA, Villenave d'Ornon Cedex, France

Elsa Rousseau
Biocore Team, INRIA, Sophia Antipolis, France
UMR 1355 Institut Sophia Agrobiotech, INRA, Sophia Antipolis, France
UMR 7254 Institut Sophia Agrobiotech, Université Nice Sophia Antipolis, Sophia Antipolis, France
UMR 7254 Institut Sophia Agrobiotech, CNRS, Sophia Antipolis, France
UR 407 Pathologie Végétale, INRA, Montfavet, France

Ludovic Mailleret
Biocore Team, INRIA, Sophia Antipolis, France
UMR 1355 Institut Sophia Agrobiotech, INRA, Sophia Antipolis, France

UMR 7254 Institut Sophia Agrobiotech, Université Nice Sophia Antipolis, Sophia Antipolis, France
UMR 7254 Institut Sophia Agrobiotech, CNRS, Sophia Antipolis, France

Benoît Moury
UR 407 Pathologie Végétale, INRA, Montfavet, France

Zachariah Gompert
Department of Biology, Utah State University, Logan, UT, USA

C. Alex Buerkle
Department of Botany, University of Wyoming, Laramie, WY, USA

Mato Lagator, Tom Vogwill and Paul Neve
School of Life Sciences, University of Warwick Coventry, UK

Nick Colegrave
School of Biological Sciences, Institute of Evolutionary Biology, University of Edinburgh Edinburgh, UK

Ben Libberton
Department of Integrative Biology, University of Liverpool, Liverpool, UK
Karolinska Institute, SE-171 77 Stockholm, Sweden

Malcolm J. Horsburgh
Department of Integrative Biology, University of Liverpool, Liverpool, UK

Michael A. Brockhurst
Department of Biology, University of York, York, UK

Luisa Listmann, Maxime LeRoch, Lothar Schlüter and Thorsten B. H. Reusch
Evolutionary Ecology of Marine Fishes, GEOMAR Helmholtz-Centre for Ocean Research Kiel, Kiel, Germany

Mridul K. Thomas
Department of Aquatic Ecology, Eawag, Swiss Federal Institute of Aquatic Science and Technology, Dübendorf, Switzerland

Ville Ojala and Jarkko Laitalainen
Centre of Excellence in Biological Interactions, Department of Biological and Environmental Science and Nanoscience Center University of Jyväskylä, Jyväskylä, Finland

Matti Jalasvuori
Centre of Excellence in Biological Interactions, Department of Biological and Environmental Science and Nanoscience Center University of Jyväskylä, Jyväskylä, Finland
Division of Ecology, Evolution and Genetics, Research School of Biology, Australian National University, Canberra, ACT, Australia

Julien Papaïx
UMR 1290 BIOGER, INRA, Thiverval-Grignon, France
UR 341 MIA, INRA, Jouy-en-Josas, France
UR 546 BioSP, INRA, Avignon, France
CSIRO Agriculture Flagship, Canberra, ACT, Australia

Jeremy J. Burdon and Peter H. Thrall
CSIRO Agriculture Flagship, Canberra, ACT, Australia

Jiasui Zhan
Fujian Key Lab of Plant Virology, Institute of Plant Virology, Fujian Agriculture and Forestry University, Fuzhou, China

Victoria L. Sork
Department of Ecology and Evolutionary Biology, University of California, Los Angeles, CA, USA
Institute of Environment and Sustainability, University of California, Los Angeles, CA, USA

Nevra Özer and Türkan Haliloğlu
Polymer Research Center and Chemical Engineering Department, Bogazici University, Bebek, Istanbul, Turkey

Ayşegül Özen and Celia A. Schiffer
Department of Biochemistry and Molecular Pharmacology, University of Massachusetts Medical School, Worcester, MA, USA

Pascal Campagne
Laboratoire Évolution, Génome et Spéciation, CNRS UPR9034, Unité de Recherche IRD 072, Gif-sur-Yvette, France
Université Paris-Sud 11, Orsay, France

Department of Ecology, Evolutionand Natural Resources, School of Environmentaland Biological Sciences, Rutgers University, New Brunswick, NJ, USA
Noctuid Stem Borers Biodiversity in Africa Project, Environmental Health Division, International Centre for Insect Physiologyand Ecology, Nairobi, Kenya
Institute of Integrative Biology, University of Liverpool, Liverpool, UK

Peter E. Smouse
Department of Ecology, Evolutionand Natural Resources, School of Environmentaland Biological Sciences, Rutgers University, New Brunswick, NJ, USA

Rémy Pasquet and Bruno Le Ru
Laboratoire Évolution, Génome et Spéciation, CNRS UPR9034, Unité de Recherche IRD 072, Gif-sur-Yvette, France
Université Paris-Sud 11, Orsay, France
Noctuid Stem Borers Biodiversity in Africa Project, Environmental Health Division, International Centre for Insect Physiologyand Ecology, Nairobi, Kenya

Jean-François Silvain
Laboratoire Évolution, Génome et Spéciation, CNRS UPR9034, Unité de Recherche Université Paris-Sud 11, Orsay, France

Johnnie Van den Berg
School of Biological Sciences - Zoology, North-West University, Potchefstroom, South Africa

Piero Calosi
Département de Biologie Chimie et Géographie, Université du Québec à Rimouski, Rimouski, QC, Canada

Pierre De Wit
Department of Marine Sciences, University of Gothenburg, Strömstad, Sweden

Peter Thor
Norwegian Polar Institute, Fram Centre, Tromsø, Norway

Sam Dupont
Department of Biological and Environmental Sciences, University of Gothenburg, Fiskebäckskil, Sweden

Mónica Betancourt
Centro de Biotecnología y Genómica de Plantas UPM-INIA and E.T.S.I. Agrónomos, Universidad Politécnica de Madrid, Campus de Montegancedo Madrid, Spain
Facultad de Ciencias Agrícolas, Universidad Santa Rosa de Cabal, UNISARC, Campus el Jazmín Santa Rosa, Colombia

Fernando Escriu
Centro de Investigación y Tecnología Agroalimentaria de Aragón, Unidad de Sanidad Vegetal Zaragoza, Spain

Aurora Fraile and Fernando García-Arenal
Centro de Biotecnología y Genómica de Plantas UPM-INIA and E.T.S.I. Agrónomos, Universidad Politécnica de Madrid, Campus de Montegancedo Madrid, Spain

Michael B. Ashworth
Australian Herbicide Resistance Initiative, School of Plant Biology, The University of Western Australia, Crawley, WA, Australia

Department of Agriculture and Environment, School of Science, Curtin University, Bentley, WA, Australia

Michael J. Walsh and Stephen B. Powles
Australian Herbicide Resistance Initiative, School of Plant Biology, The University of Western Australia, Crawley, WA, Australia
School of Plant Biology, The University of Western Australia, Crawley, WA, Australia

Ken C. Flower
School of Plant Biology, The University of Western Australia, Crawley, WA, Australia

Martin M. Vila-Aiub
Australian Herbicide Resistance Initiative, School of Plant Biology, The University of Western Australia, Crawley, WA, Australia
IFEVA-CONICET, Facultad de Agronomia, Universidad de Buenos Aires, Buenos Aires, Argentina

Index

www.ingramcontent.com/pod-product-compliance
Lightning Source LLC
Chambersburg PA
CBHW082030190326
41458CB00010B/3325